Engineering Applications of Fluid Dynamics

Engineering Applications of Fluid Dynamics

Edited by
Henley Fisher

Willford Press
www.willfordpress.com

Published by Willford Press,
118-35 Queens Blvd., Suite 400,
Forest Hills, NY 11375, USA

ISBN: 978-1-64728-533-3

Cataloging-in-Publication Data

Engineering applications of fluid dynamics / edited by Henley Fisher.
p. cm.
Includes bibliographical references and index.
ISBN 978-1-64728-533-3
1. Fluid dynamics. 2. Fluid mechanics. 3. Mechanics, Applied. I. Fisher, Henley.
TA357 .E54 2023
620.106--dc23

For information on all Willford Press publications
visit our website at www.willfordpress.com

Willford Press

Contents

Preface

This book has been an outcome of determined endeavour from a group of educationists in the field. The primary objective was to involve a broad spectrum of professionals from diverse cultural background involved in the field for developing new researches. The book not only targets students but also scholars pursuing higher research for further enhancement of the theoretical and practical applications of the subject.

Fluid dynamics is a branch of fluid mechanics, which describes the flow of fluids including gases and liquids. It focuses on the study of movements of fluids, including the interactions that occur when two fluids are exposed to one another. There are many subfields of fluid dynamics such as hydrodynamics and aerodynamics. Fluid dynamics plays an important role in the study of ocean tides, plate tectonics, planet evolution, weather patterns and blood circulation. Wind turbines, air conditioning systems, rocket engines and oil pipelines are some of the significant technological uses of fluid dynamics. It also has several engineering applications such as refrigerators and air conditioners, hydraulic equipment, automobiles, hydroelectric power plants, nuclear power plants, and thermal power plants. This book aims to shed light on some of the unexplored engineering applications of fluid dynamics and the recent researches in this field. It elucidates new techniques and their applications in a multidisciplinary manner. This book includes contributions of experts and scientists, which will provide innovative insights into this field.

It was an honour to edit such a profound book and also a challenging task to compile and examine all the relevant data for accuracy and originality. I wish to acknowledge the efforts of the contributors for submitting such brilliant and diverse chapters in the field and for endlessly working for the completion of the book. Last, but not the least; I thank my family for being a constant source of support in all my research endeavours.

Editor

1

The use of CFD for the Design and Development of Innovative Configurations in Regenerative Glass Production Furnaces

Carlo Cravero * and Davide De Domenico

Department of Mechanical Engineering, DIME, Università di Genova, 16145 Genova, Italy; davide.dedomenico@edu.unige.it

* Correspondence: cravero@unige.it

Abstract: The limitation of nitrogen oxides emissions is nowadays a challenge in several engineering fields. Recent European regulations have reduced the maximum NO_x emissions and therefore forced the glass production sector to develop emission reduction strategies. Two different systems have been developed within the framework of the European LIFE project and are currently applied to glass regenerative furnaces: the Waste Gas Recirculation (WGR) and the Hybrid Air Staging (HyAS). The above systems are primary NO_x reduction strategies because they both operate to control the combustion evolution. Both WGR and HyAS systems have been conceived with the extensive use of Computational Fluid Dynamics (CFD) models: design strategies for both systems have been developed based on the use of CFD and are currently under use by glass furnace designers. In the present work, the CFD procedures routinely used for the design of the above systems are described. The systems effectiveness, due to the harsh conditions in the industrial installation, can be tested with oxygen concentration measurements inside the regenerators. The oxygen concentration is correlated to the flame evolution and therefore to the nitrogen oxides formation. For the above reason, the models have been validated with experimental data from pilot furnaces using measured values of O_2 mole fraction. The CFD procedures are described in the paper together with their application to different configurations.

Keywords: CFD; NO_x reduction; glass production plants; thermal regenerator

---

1. Introduction

The glass industry has grown in recent years, as glass is an increasingly used material due to its recyclability. However, due to the high temperature inside the furnace (commonly obtained with combustion of natural gas) the energy consumption is very high and the environmental impact from greenhouse gases and pollutants is significant. The efforts of nitrogen oxides reduction in energy systems are very high in various engineering fields from internal combustion engines to power plants to industrial plants. Zeldovich [1] firstly postulated the formation mechanism of thermal NO_x, which refers to high temperature reactions. However high temperatures are not sufficient to trigger NO formation: a crucial role is played by the residence time of the reactive species at high temperature and by the local air fuel ratio (i.e., the O_2 concentration).

In order to decrease NO_x emissions, several technologies and techniques have been developed [2]. They can be divided into primary and secondary reduction methods. The primary methods act on the combustion setup to limit nitrogen oxides formation. The most common techniques employed for the primary reduction are: exhaust gas recirculation systems (EGR) and staged combustion. Secondary methods tend to decrease the concentration of the pollutant gas in the exhausts after the combustion process is complete. The most widely used technologies consider urea injection into exhaust gases. Recently, new European limits for emissions [3] have forced many industrial sectors, including glass production, to adopt NO_x limitation measures. Within the Life PRIMEGLASS project [4,5] two different primary methods for NO_x reductions have been studied and developed: Waste Gas Recirculation system (WGR) and Hybrid Air Staging (HyAS).

Both are applied to regenerative glass production plants. They consist of a Heat Recovery System (HRS) based on regenerative chambers. Both WGR and HyAS systems have been conceived and developed with the extensive use of Computational Fluid Dynamics (CFD) models; design strategies for both systems have been developed based on the use of CFD and are currently under use by glass furnace designers. Several authors in different fields [6–8] have tackled the problem of modeling the HRS. The first application of CFD to the study of regenerative chambers in glass production plants was reported by Reboussin et al. [9]; they proposed an open model, validated with experimental data, and introduced an original method to take into account the effect of temperature vertical gradients by using a fictitious expansion coefficient. The research team from the University of Genoa, partner of the PRIMEGLASS project, has gained relevant experience in the simulation of regenerative glass production plants. Basso et al. [10] set up a CFD approach for the regenerative chamber where the solid phase is modeled by porous medium. They simulated the flow field and the thermal behavior both for the air and the hot gases that alternatively feed the regenerator. From the analysis of the results, they confirmed the reliability of the model in predicting thermodynamic quantities and its effectiveness in new geometry design phase. A further development of the model [11] considered the introduction of the non-equilibrium thermal model. It allowed simulating the regenerator without the need of temperature experimental data through the modelling of heat transfer coefficient distributions in the cold and hot phases. A model of gas emissivity has been developed [12] and introduced into the CFD procedure to take into account the significant radiative heat transfer from the exhaust gases (that contain CO_2 and H_2O molecules). In the present paper, the CFD procedures for the development of WGR and HyAS are discussed. The former is applied and tested with three configurations and validated with experimental data of O_2 fraction measured in few accessible sections in the system. The procedure is based on the detection of O_2 concentrations because NO_x cannot be measured in those sections; the measurements of oxygen can give an indirect evaluation of the zones with the highest probability of pollutants formation. The overall effectiveness of the systems has been demonstrated in the pilot furnaces with the direct measure of nitrogen oxides at the stack [5]. The second procedure is applied to the identification of the proper injector flow rates for the HyAS system. The scope is to demonstrate the use of the CFD simulation technology to optimize the WGR system and to support the design from its systematic use.

2. End-Port Glass Production Plant Layout and Regenerative Chambers

One of the most frequent regenerative furnaces is the End-Port type, as seen in Figure 1, which is composed of two main structures: the combustion chamber (CC) and the heat recovery system (HRS).

The CC is a refractory oven where the raw materials introduced are melted by a combustion process. The high temperatures required for the glass production (the glass bath reach 1700–1800 K) lead to the use of natural gas as fuel. The temperature of the exhaust gas leaving the CC is very high (1400–1500 K). In order to improve the overall efficiency a heat recovery from waste gas is mandatory. In glass production, regenerative chambers are used. The system stems from the Martin-Siemens open-hearth furnace. It is composed of two or more regenerative towers, usually situated at the end of the CC (End-Port scheme), which are alternatively fed with air and exhaust gas (cold phase and hot phase), by cycles of 20 min. The regenerative chambers can be divided into three main zones, as depicted in Figure 2a: top chamber, checkers zone and bottom chamber.

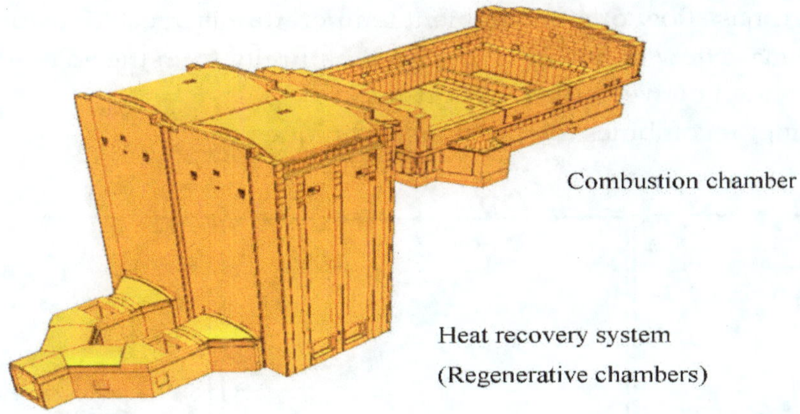

Figure 1. Regenerative End-Port glass production furnace.

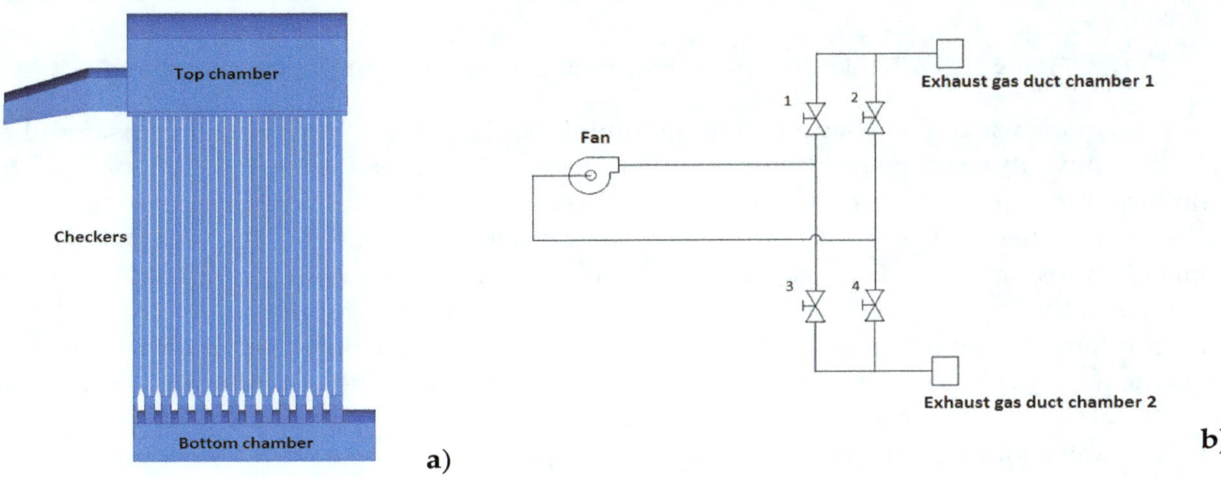

Figure 2. (**a**) Regenerative tower and (**b**) Waste Gas Recirculation (WGR) system scheme.

The checker zone is the crucial part of the regenerator for the heat recovery [10]: it consists of the refractory material bricks, that absorb the heat of the waste gas during the hot phase and release it to the incoming air during the cold phase. Bricks are assembled in a modular way and present different shapes in order to promote the heat exchange and to ease the maintenance operations. The WGR system is based on the recirculation of a waste gas fraction extracted from the regenerative chamber in the hot phase and injected in the bottom chamber of the tower fed with air. However, due to the adverse pressure gradient, the gas cannot naturally reach the other regenerator and the flow should be inverted according to the heat recovery cycles. For these reasons, a specific ducting system supported by a fan is needed, as reported in the scheme of Figure 2b. The recirculated waste gases allow to partially dilute the incoming air thus reducing the NO_x formation during combustion. The recirculated gases enter the chamber at the bottom side and flow with the combustion air up to the upper exit where they meet the natural gas and the resulting flame. The recirculated gas injection phase (flow rate and position) is of utmost importance in order to get a required gas stratification over the flame to control its temperature and therefore the NO_x formation.

An alternative method to decrease the NO_x emissions is the air staging. The technique consists of injecting air inside the top chamber when it is fed with the burned gas. In order to decrease the nitrogen oxides, the combustion is performed in a reducing environment, which leads to the formation of a large amount of CO. Hence, the combustion process is completed in a second stage outside the CC [13] and inside the top chamber at a lower temperature (NO_x limitation). Three different types of air staging can be considered [4] (Figure 3):

- Cold Air Staging: mass flow of air at ambient temperature is injected inside the chamber;
- Hot Air Staging: mass flow of hot air is conveyed naturally from the adjacent chamber through a U-shaped duct located between the two port-neck;
- Hybrid Air Staging: it combines the previous techniques.

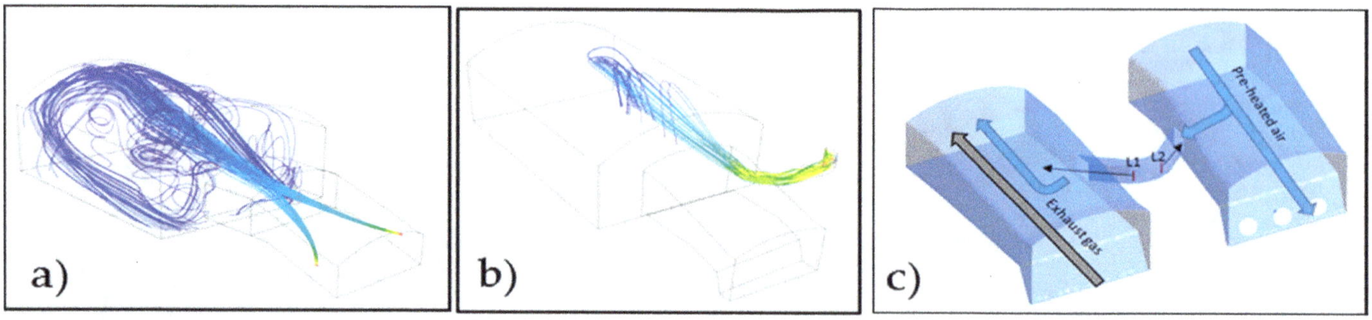

Figure 3. Schematic view of air stage solutions: (**a**) cold, (**b**) hot, (**c**) hybrid.

The systematic use of CFD has allowed the understanding of the flow mechanisms and thermal effects of the above three strategies [4]. It was demonstrated that the first one is the most effective in reducing emissions; however a considerable amount of air at low temperature in the top chamber affects the effectiveness of the HRS. The second solution overcomes the problem by the injection of a percentage of air coming from the adjacent top chamber; the by-pass is provided by a duct connecting the two ports of the top chambers. In these conditions, the flow is naturally driven by the favorable pressure gradient but the mixing of combustion gases and air is uncontrolled and insufficient. Best results are achieved with the third solution. The Hybrid Air Staging differs from the Hot Air Staging for the installation of two injectors inside the connecting duct. The injectors are fed by compressed air and allow the control of the mass flow rate to be bypassed. Moreover, the jet of compressed air enhances the mixing of the gases bypassed into the regenerative chamber. They are alternatively called pros and cons injectors with respect to the air/gas cycle: the pros provides air in the same direction of the by-passed flow, the cons in the opposite (Figure 3c).

3. Reference Geometries and CFD Models

3.1. WGR Waste Gas Recyrculation Systems

Three different existing regenerative chamber geometries are considered with main geometric features, reported in Table 1. The configuration differs, as seen in Figure 4, also for the lower part geometry, (RC1 and RC3 have two separated ducts for inlet/outlet sections connected to the checkers by the arches, built with a particular shape, RC2 has one bottom chamber with single section). The checkers zone has different constructive solutions: in RC2 and RC3 bricks have a cruciform shape, whereas, in RC1 octagonal shaped bricks are present.

Table 1. Geometrical parameters of the tested regenerative chambers.

Geometrical Parameters	RC1	RC2	RC3
Aexhaust/Aair	0.1	0.05	0.08
Aexhaust/Aexit	0.16	0.1	0.17
Bottom chamber aspect ratio	2.2	3.9	4
Top chamber aspect ratio	1.6	1.5	1.6

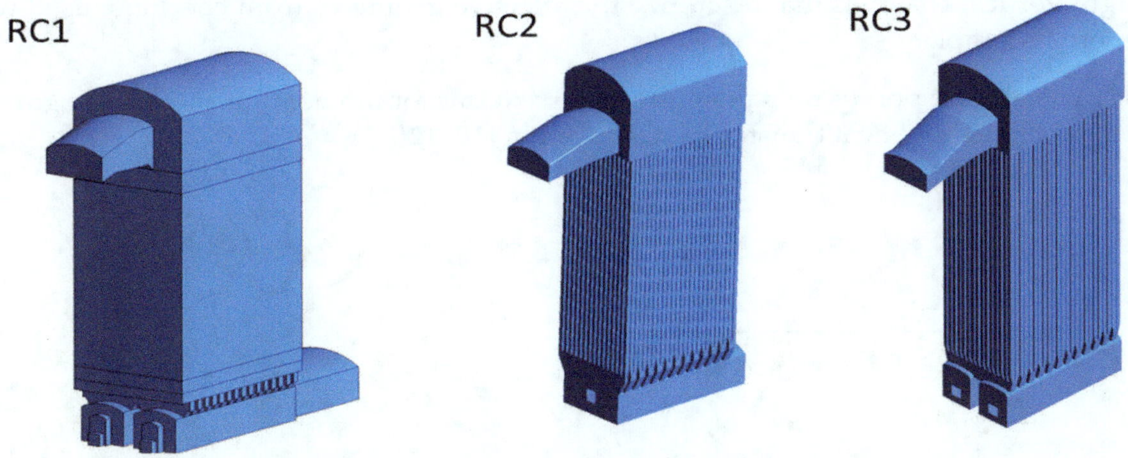

Figure 4. Tested regenerative chambers.

The regenerative domain is divided into four sub-domains where different grids are generated, as summarized in Table 2: bottom chamber, arches (i.e., the structure that links the bottom chamber to the checkers), checkers, and top chamber. Each sub-domain is meshed with a block structured approach using Ansys ICEM® software. The reference dimension of the single mesh element for the sub-domains is approximately 10 × 10 × 10 mm. In Figure 5, a typical mesh for the regenerative chamber at both lower and upper sections is shown.

Table 2. Subdomains and mesh characteristics.

Subdomains		Mesh Type	n° of Elements
Bottom chamber	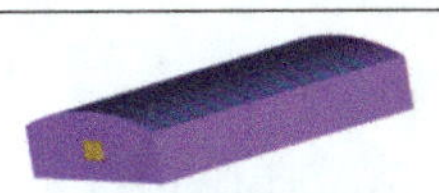	Struct.	2M–4M
Arches		Struct.	500 k
Checkers zone		Struct.	1M–2M
Top chamber		Struct.	1M–2M

The Ansys Fluent v17.2 (ANSYS Inc., Canonsburg, PA, USA) flow solver has been used for the RANS CFD simulations with SIMPLE scheme and second order spatial discretization for both governing equations and turbulence model. The SIMPLE scheme has been preferred, over more sophisticated numerical schemes, for its robustness in simulating complex flow structures with incompressible flow. The waste gasses and the air have been modelled as gas mixtures composed by N_2, O_2, CO_2 and water vapor. The species transport model has been activated and the proper composition of the gasses has been provided. The standard k-ε turbulence model with scalable wall functions has been used. At inlet sections, the mass flow rate (air or waste gases) is fixed as boundary conditions together with temperature, gas composition and turbulence intensity of 5%. The outflow condition (zero gradient) is fixed at the outlet sections. The steady flow option with automatic flow domain initialization has been used. Depending on the target, the checker zone domain can be modelled with the following options:

- As a porous domain, used when the geometry of the stackers is too complicated and

- Straight vertical channels that discretize the effective geometry from checkers, used when the geometry is simpler.

In the first case, the porous domain has the source terms for momentum and energy governing to model flow resistance and heat transfer, as described in [10–12].

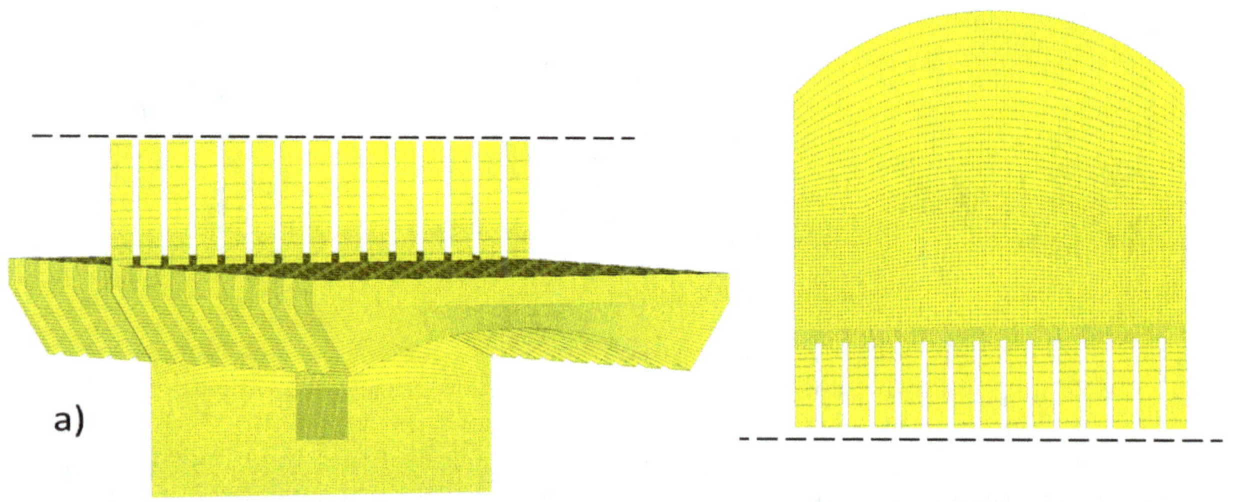

Figure 5. Mesh section of the RC3 regenerative chamber—(**a**) bottom, (**b**) top.

According to the checkers configurations, RC1 has been modeled with a porous domain while RC2 and RC3, have been modelled with a series of vertical channels.

3.2. HyAS—Hybrid Air Staging Configurations

The Air Staging configuration considered is obtained from case RC2. Both chambers have been considered connected by a U-shaped duct for the air bypass. As previously remarked, in the HyAS, two additional injectors are positioned inside the duct; they have a dual purpose: to provide the cold air flow rate and to regulate the amount of air recirculated and control the air-gases mixing.

Due to the presence of the injectors, an unstructured grid has been used to discretize the U-shaped duct. The same settings for the Fluent solver previously described are fixed. Mass flow rate, temperature and gas composition have been fixed at the regenerative chamber inlet, while outlet static pressure has been fixed at the other chamber exit. The mass flow rate and temperature of the injection air are fixed at the injection orifices.

4. Results and Discussion

4.1. WGR—Waste Gas Recyrculation Systems

4.1.1. Simulation Campaign and Experimental Data

The CFD model and the data post-processing have been organized in order to be routinely used for design and development of the WGR systems. In the following sections, the applications to the configuration considered are described together with a validation process with a direct comparison with experimental data. The gas recirculation strategies and their effects are monitored using the O_2 distributions inside some strategic sections of the domain where monitor points are considered. The experimental campaign has been provided for configuration RC1 and the O_2 concentration has been measured on the sections reported in Figure 6. Section nr.1, left, is at the port outlet and section nr.2, right, is at the top of the checkers domain. Section nr.2 gives the distribution of the gasses inside the checkers and it is strategic for the stratification of the exhausts at the outlet port (section nr.1) of the top chamber, which is fundamental for the effective reduction of the NO_x emissions. The above sections have been divided into zones where the measuring point has been centered. The O_2 concentrations (detection of waste gas distribution) from the experimental campaign are compared to those obtained by the CFD analysis for three different exhaust gas mass flow rate percentage ratios, with respect to incoming air, 10%, 20%, 30%. The above percentages set the exhaust gas recirculation (EGR) rate of the operating point.

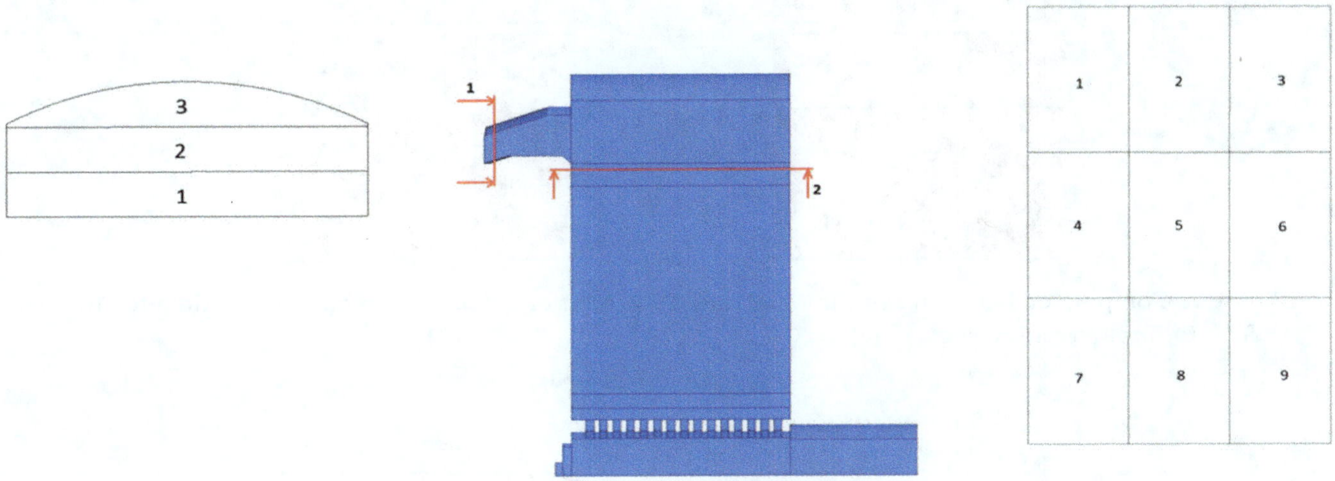

Figure 6. Control sections and experimental grid.

For each monitor section, the percentage error between O_2 concentration from experimental data and CFD results are computed:

$$percentage_{error} = \frac{[O_2]_{exp} - [O_2]_{CFD}}{[O_2]_{exp}} \tag{1}$$

The O_2 concentration has been measured experimentally at one probe position at the centroid of the control section numbered in Figure 6, while the CFD data are average values of O_2 concentration in the above areas.

The probe consists of a water refrigerated duct inserted at different depths inside the regenerator; the aspirated gas is conveyed to the oxygen concentration meter. An accuracy of 2% and a precision of 0.2% are guaranteed for the O_2 concentration in the measured data.

This is clearly an approximation and must be taken into account for the data comparison. Figure 7 shows the comparison for a waste gas recirculation percentage of 10%.; the numbers in the zones are the O_2 mole fractions. As seen, the trends and data are in good agreement apart from zone 1 where the percentage error (Equation (1)) is about 10%. The area with highest discrepancy is close to the wall and measurements could have been distorted by the presence of air leakage from outside (pressure inside the checkers is always lower than atmospheric pressure). The other two cases considered show also a good agreement, especially in terms of flow distribution, between data from the numerical simulation and measurements, as reported in Figures 8 and 9 for 20% and 30% of gas recirculation, respectively. It can be observed that experimental values of mole fraction can sometimes differ significantly in some sections. However, it had to be considered that probes provide a punctual measurement of the oxygen concentration, while the CFD value is the area averaged datum over the discretized area. Nevertheless, it is interesting to note that the same flow structures, O_2 distribution patterns, are obtained with CFD and experiments.

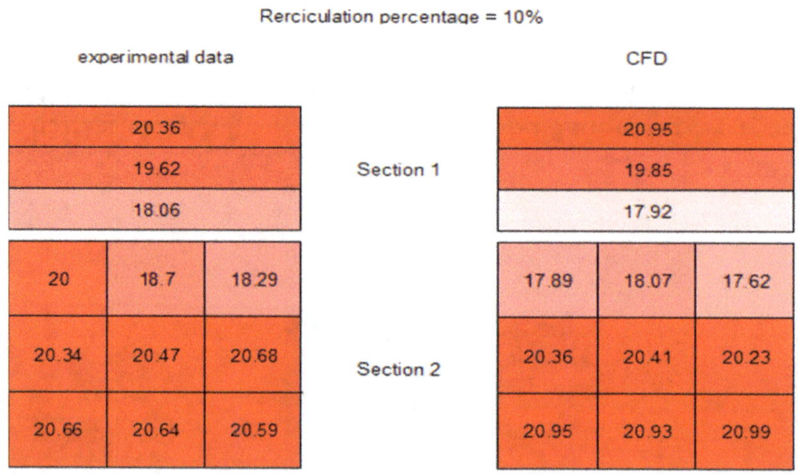

Figure 7. Comparison between experimental and Computational Fluid Dynamics (CFD) data for 10% WGR—O_2 mole fraction.

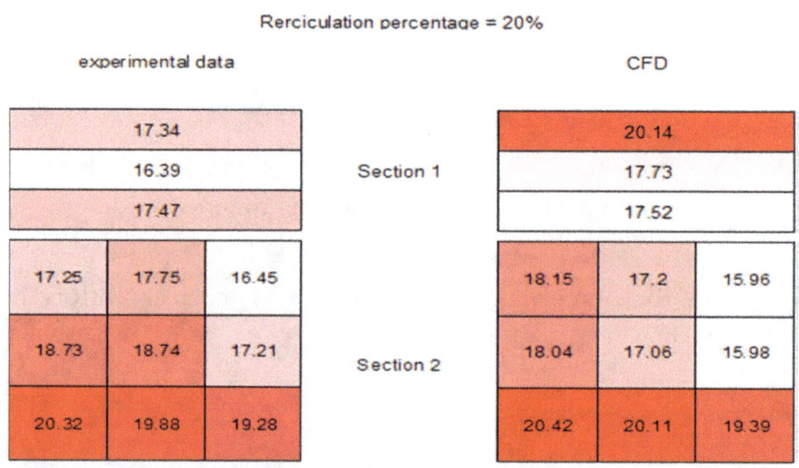

Figure 8. Comparison between experimental and CFD data for 20% WGR—O_2 mole fraction.

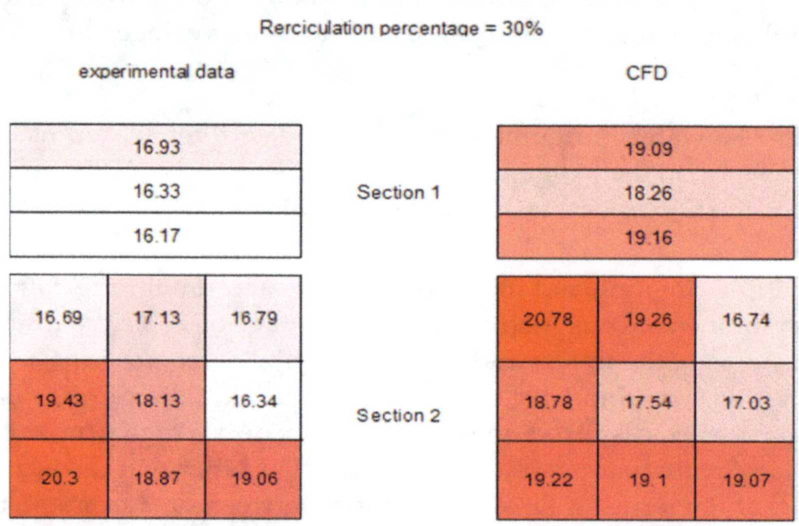

Figure 9. Comparison between experimental and CFD data for 30% WGR—O_2 mole fraction.

Taking into account the harsh conditions for the experimental data acquisition, and the approximations discussed it can be nevertheless concluded that the CFD approach is adequate to predict the flow structure inside the chambers in order to develop the EGR strategies. In the following, the results obtained for the three configurations using the CFD models are presented with the post-processing approach that has been specifically conceived to understand the EGR performance in order to systematically use the CFD technology for the design and the development of these systems. The monitor sections nr.1 and nr.2 are discretized with 8 and 12 control surfaces respectively where the O_2 mass fraction and the specific mass flow are averaged. These quantities are useful monitor to understand the EGR system performance. To reduce the NO_x formation, a low fraction of O_2 in the lowest part of section nr.1 is needed and a proper O_2 distribution is therefore required on section nr.2. To guarantee for a good thermal efficiency of the regenerative chamber a uniform distribution of the specific mass flow rate on Section 2 is a requirement. All the configurations considered (RC1-RC2-RC3) have been simulated in a range of different WGR percentages. In the following paragraphs, a selection of relevant results for each configuration is described.

4.1.2. Configuration RC1

The 20% recirculation case is considered. In Figure 10, the O_2 mass fractions on the control surfaces are reported together with the contours on the same sections.

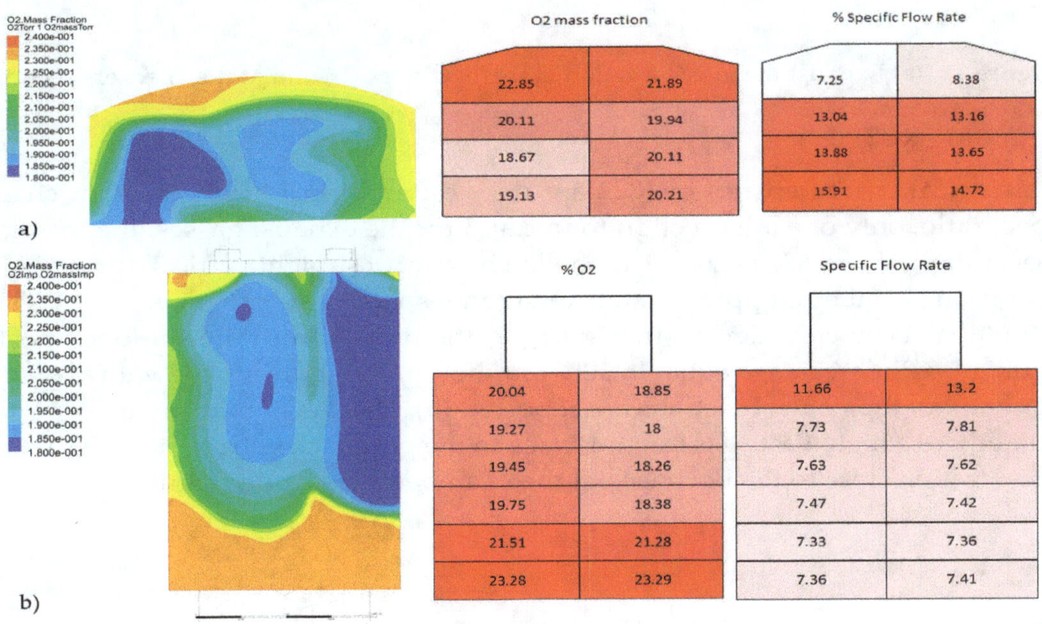

Figure 10. O_2 mass fraction and specific flow rate for 20% WGR—(**a**) port neck section (**b**) checkers section.

A peculiar flow characteristics inside the regenerative chamber is observed. If recirculated mass flow increases the gases tend to reach the deepest part of the bottom chamber and, as a consequence, they flow through a larger area of the checkers. This results in a more uniform distribution at the port outlet section. However, the aim of the strategic exhaust gas recirculation is to get a gas stratification in the lower part of the port: in fact, the burners are located in this zone inside the CC, hence the presence of gases in that area is preferable since they reduce the oxygen content during the combustion. There is a best compromise among the above effects at a given mass recirculation percentage.

In order to quantify the gas distribution in a systematic way, to compare different solutions, the following O_2 fraction is averaged in each reference control sections previously identified [14]:

$$f = \frac{Y_{O_2 air} - Y_{O_2}}{Y_{O_2 air} - Y_{O_2 gas}} * 100 \tag{2}$$

The above index gives the local recirculated gas fraction (difference between fixed O_2 air value to local value referred to the fixed O_2 air value minus the O_2 fraction in the incoming mass flow of WGR). The gas composition of the recirculated exhaust gases has been kept constant and therefore the above index f is related to the local amount of gases. Figure 11 reports the averaged values of f in the control areas for both port neck (a) and checkers top section (b). It can be observed that with 20% of WGR the gases stratify in the bottom part of the port neck outlet section with higher values than with 10% WGR. Increasing the percentage (30%) a large amount of gases stratifies to the top. The distributions of f at the chamber top section (Figure 11b) confirm that a larger fraction of gases tends to feed the sections opposite to the port neck at a higher WGR percentage; this is an undesired feature.

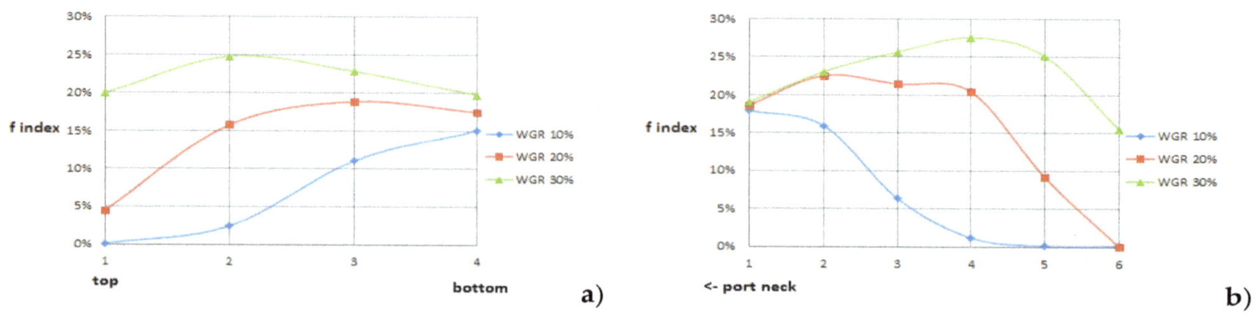

Figure 11. (**a**) Exhaust mean gas fraction at the port neck (**b**) and at the checkers section.

4.1.3. Configuration RC2

The geometrical characteristics of RC2 are very different from RC1 (bottom chamber with a different aspect ratio, only one inlet section for air and for the exhaust gasses, the top chamber has a different aspect ratio). The checkers zone is modelled by a set of straight vertical ducts. The air flow rate entering the bottom chamber is approximately a half in respect to the previous case. WGR percentage from 5% to 20% has been considered. In Figure 12, the results from the standard post-processing procedure with a WGR percentage equal to 10% are shown. In Figure 13, the values of f for both port neck (a) and chamber top control sections are reported (b). The WGR percentage of 10% is optimum because it guarantees a high waste gas fraction close to the port neck bottom side (Figure 13a) and also a higher concentration close to the port neck at the chamber top section (Figure 13b).

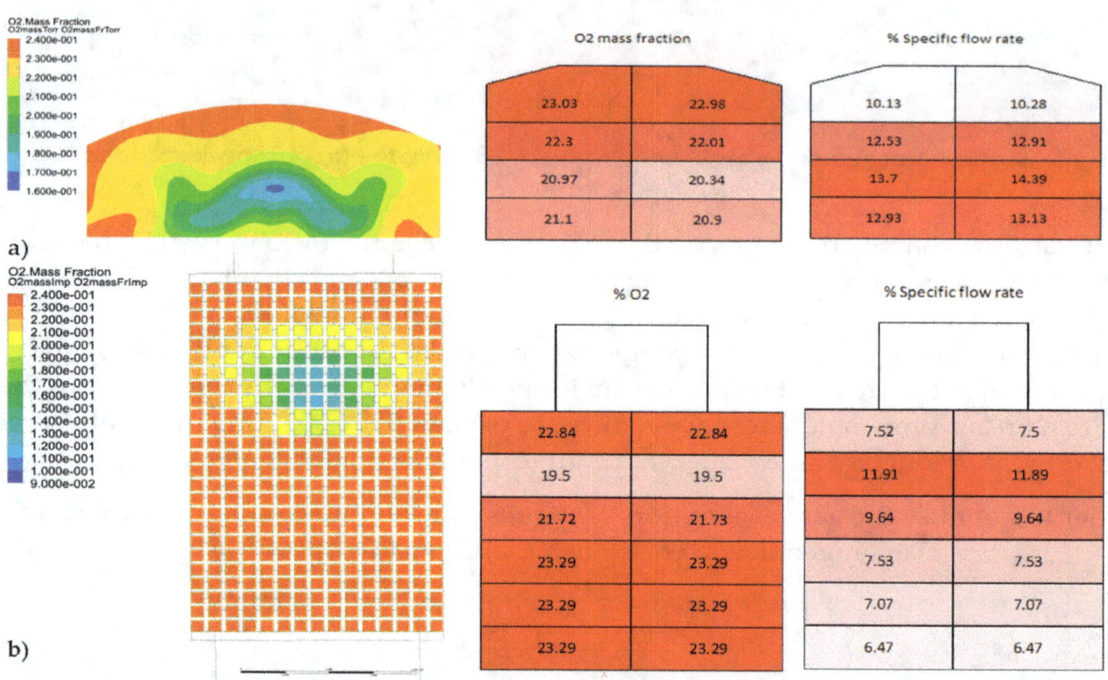

Figure 12. O_2 mass fraction and specific flow rate for 10% WGR—(**a**) port neck section (**b**) checkers section.

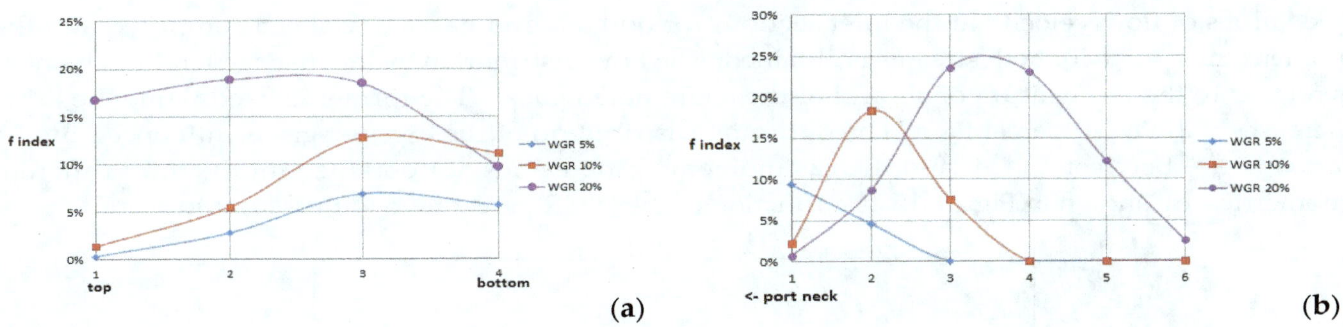

Figure 13. (**a**) Exhaust mean gas fraction at the port neck and (**b**) at the checkers section.

The appropriate distribution of recirculated gases is also confirmed by the contours of Figure 12, where low values of O_2 fraction are identified at the port neck bottom and close to the port neck in the chamber top section.

4.1.4. Configuration RC3

The configuration RC3 has two air/exhaust gas entries and the checkers are modelled as vertical ducts. Results WGR percentage of 20% are discussed. In this case, the required flow structure at both port neck exit and top chamber have been partially obtained as shown in Figure 14.

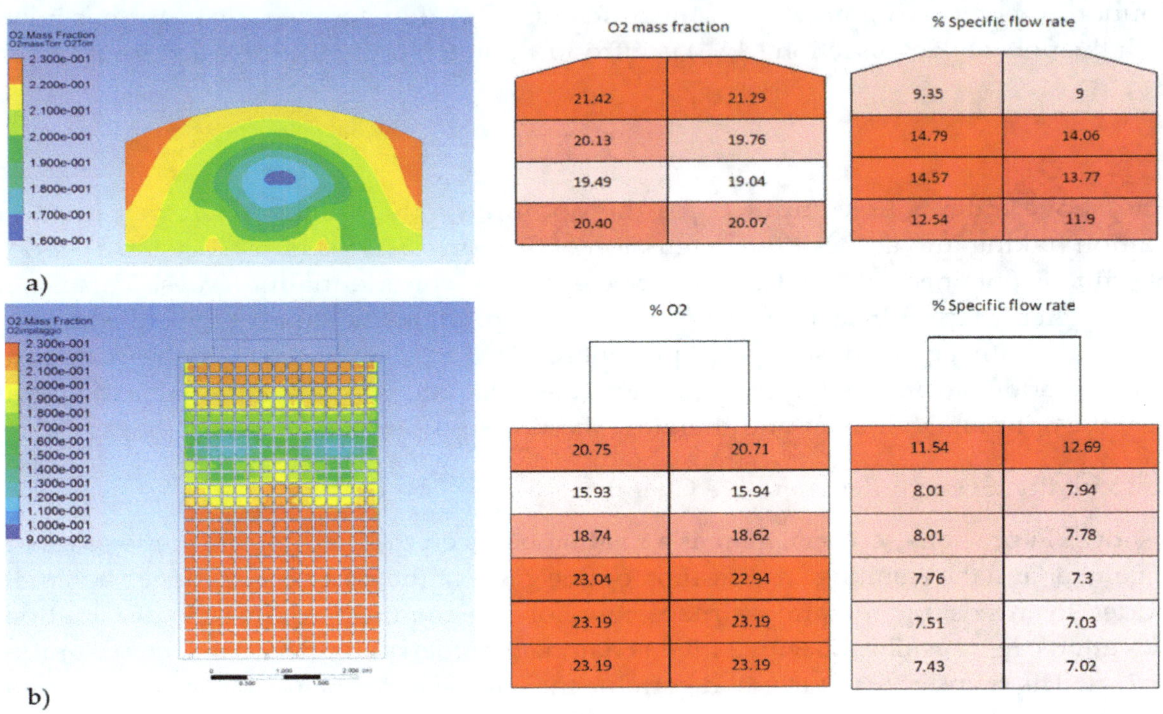

Figure 14. O_2 mass fraction and specific flow rate for 20% WGR—(**a**) port neck section (**b**) checkers section.

The above applications have confirmed that the proposed CFD approach with its post-processing procedure are useful to identify the optimal WGR percentage for a given configuration. It is also evident that each configuration needs a tailored WGR system.

4.1.5. Momentum Index

In order to extract from a 3D CFD simulation as much synthetic and quantitative information as possible (to drive the design of the systems) different indices and correlations have been defined.

The values of flow velocity at the inlet sections for both air and gases have been computed and the momentum $q = \dot{m}v$ for both streams is obtained. The flow distribution inside the regenerative chamber is related to the momentum of air and of the recirculated gases. The former is fixed while the latter increases with WGR percentage. The optimum distribution of the exhaust gas is influenced by the ratio $r = \frac{q_{air}}{q_{gas}}$ between the air and the gas momenta. The values of r obtained for the three different geometries are shown in Figure 15 as a function of the WGR percentage with respect to air.

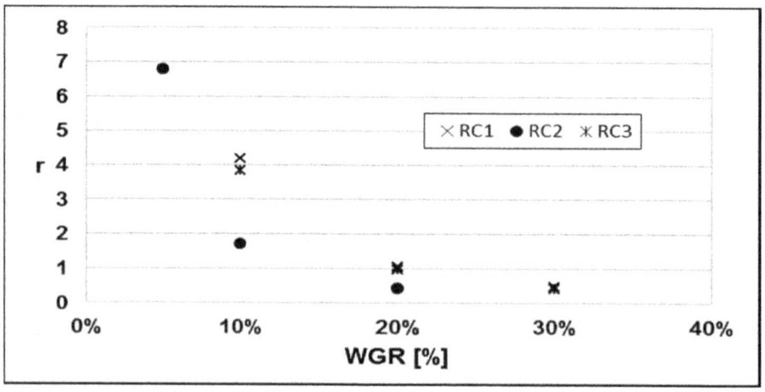

Figure 15. Momentum ratio between air and recirculated gases at inlet sections.

The above figure shows a good correlation of r for the different configurations. In order to take into account the geometric parameters of the different RC examined, the ratio of the gas inlet section (A_{gas}) with the port neck exit section (A_{exit}) is introduced in a modified expression for r:

$$r' = r\frac{A_{gas}}{A_{exit}} \tag{3}$$

If the above index r' is computed for the optimum value of the WGR percentage of each configuration (taking into account the discussion of Section 4.1.2, Section 4.1.3, and Section 4.1.4) a constant value is obtained. The value cannot be given for confidentiality issues. This is a relevant aspect for design purposes. In fact, to design a new configuration the value of the index can be used to set the area ratio with a given design WGR percentage (that in turns fixes r). On the other hand, if the WGR system is added to an existing configuration (the area ratio is known), the value of optimum r' is used to set the optimum value of r and therefore the WGR percentage.

4.1.6. Recirculation System: Regenerative Chamber Inlet Section Details

As shown in previous sections, there is a relation between the distribution of the gases inside the bottom chamber and the resulting stratification of flue gases at the port neck section. The WGR system can be added to an existing operating furnace; therefore the required exhaust gas distribution during WGR activation can be achieved (with a given chamber geometry) with a proper configuration of the gas inlet duct. The shape of the duct will determine the gas penetration inside the chamber and the gas throughflow up to the port neck outlet section. CFD can be effectively used to optimize the inlet duct layout. In the following the application of the CFD model to configuration RC3 with WGR percentage of 20% is discussed for three different duct shapes. The first configuration is a simple rectangular duct that is not introduced into the chamber. The same duct which penetrates into the chamber section for a certain amount defines the second case. A duct with a divergent layout (diffuser) introduced into the chamber defines the third case. The iso-surface of O_2 fraction equal to 16% has been used to detect and show the flow structure of the gases inside the regenerative chamber. In Figure 16, the above iso-surface is depicted for the three duct configurations; the picture shows the details at the chamber bottom part and at the chamber top in the port neck zone. This post processing approach, used to systematically compare different solutions, is very effective. The flow structure from configuration RC2, with the straight duct inserted into the chamber, due to the high flow velocity at the duct exit and the squared corners of the duct, develops a vertical structure. Moreover, the high speed at the duct exit give a gas stratification toward the top surface of the port neck, which is not desired. The first and third configurations are preferable. The former is the easiest to consider for the application of the WGR system to an existing operating furnace, the latter is optimized for a new furnace design including WGR system.

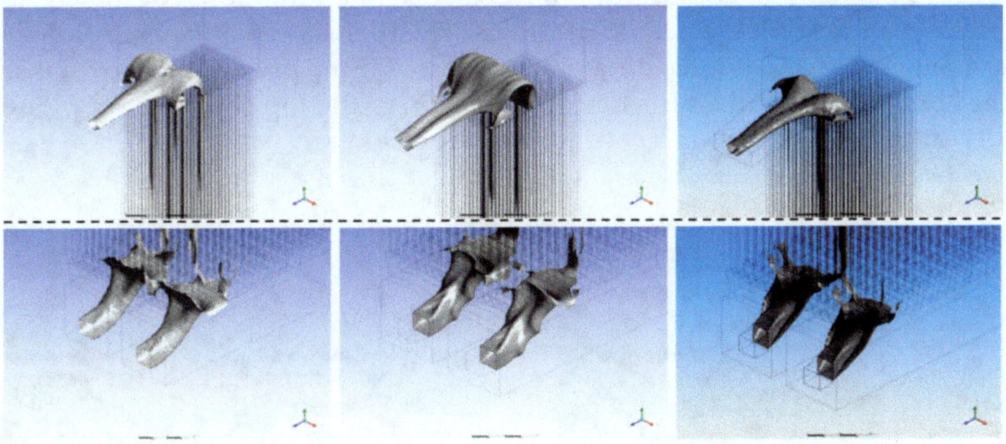

a) b) c)

Figure 16. Iso-surface 16% O_2 mass fraction top and bottom—(**a**) simple duct (**b**) long duct (**c**) divergent duct.

In order to have a quantitative comparison between the different solutions, the O_2 mass fraction contours at the usual control sections (port neck exit and checkers top surface) are investigated. In Figure 17 the above contours for all configurations are shown.

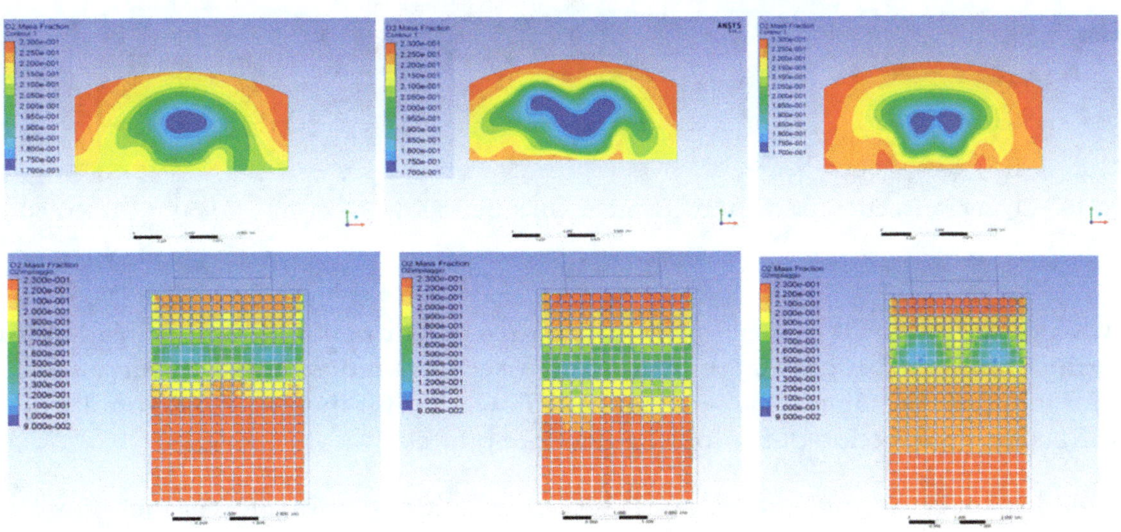

a) b) c)

Figure 17. O_2 mass fraction contours—(**a**) first case (**b**) second case (**c**) third case.

As it can be seen both the first and the third configurations are quite effective to get the required exhaust gas distribution at the port neck exit; the case with the divergent duct inserted into the chamber gives a very interesting symmetric flow structure at the port neck exit with a core of gases toward the base surface.

4.2. HyAS—Hybrid Air Staging Solution

The hybrid air staging is a gas recirculation strategy operated at the port neck level. In order to control both the recirculated gas mass flow and their mixing into the chamber, a set of two injectors, as previously described, is introduced. For a given geometrical layout, the ratio of the mass flow rates of injected air in the pros and cons injectors needs to be optimized. This can be done with the CFD models developed. For chamber configuration, RC3 series of simulations for the HyAS system with different compressed air ratios have been performed. The first set has a fix mass flow rate of 30 m^3/h for the cons injector and a series of three different flow rates for the pros (50 m^3/h, 60 m^3/h and 80 m^3/h). In Figure 18, the contours of O_2 mass fraction for a section of the top chamber are plotted.

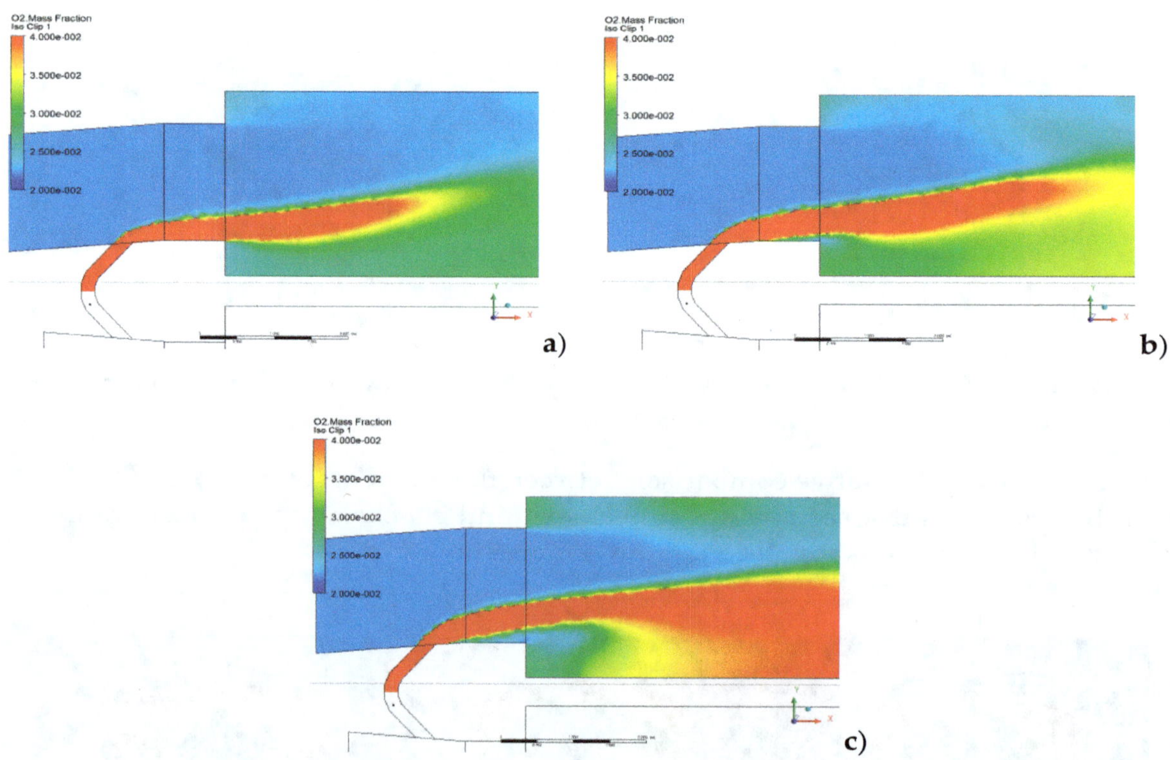

Figure 18. O_2 mass fraction at the top chamber cons massflow 30 m^3/h—(**a**) 50 m^3/h (**b**) 60 m^3/h (**c**) 80 m^3/h.

It can be noticed that as the pros flow rate increases the air penetrates more inside the chamber, and the mixing of the exhaust gasses is enhanced. The second set of simulations has a fixed flow rate of 16 m^3/h for the cons injector and the same range of flow rates for the pros. In Figure 19, the contours of O_2 mass fraction for a section of the top chamber are plotted.

The beneficial effect of cons mass flow reduction is evident; with a lower mass flow rate of the pros injector a significant mixing in the chamber is obtained.

In Figure 20, a detail of the injectors flow structure is depicted: the cons injector (on the right) is activated because of the cooling purpose, while the pros injector (on the left) drives the recirculating flow and entrains it into the chamber. The CFD model is effective and routinely used for the design and development of the above systems.

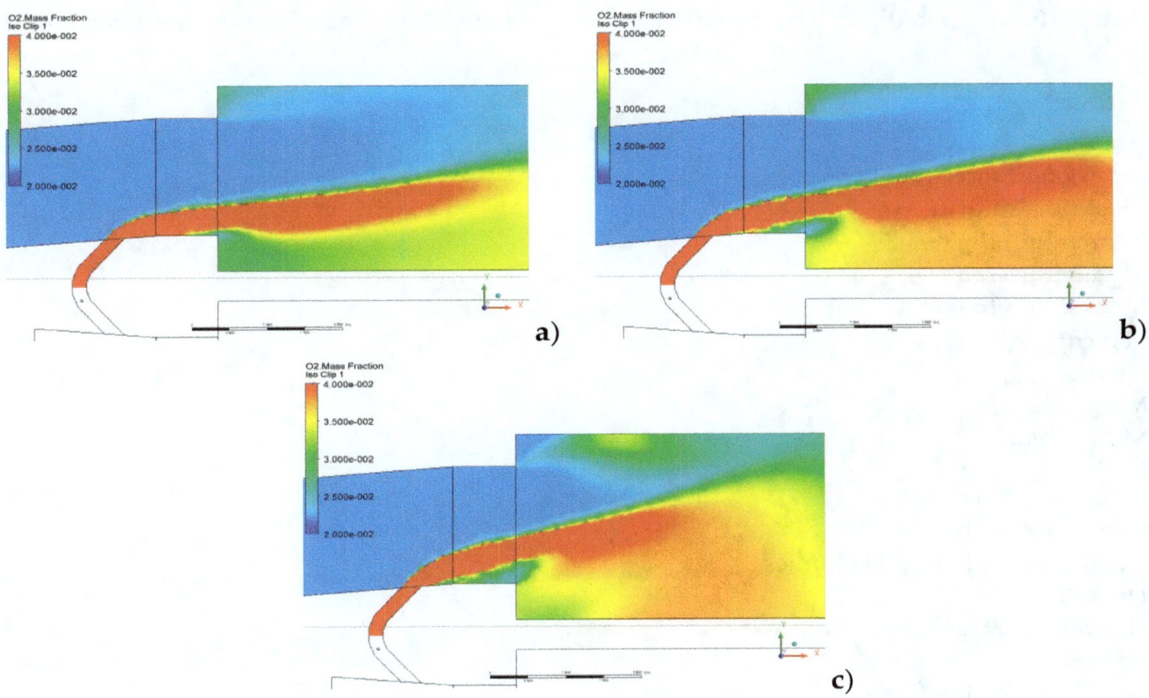

Figure 19. O_2 mass fraction at the top chamber cons massflow 16 m^3/h—(**a**) 50 m^3/h (**b**) 60 m^3/h (**c**) 80 m^3/h.

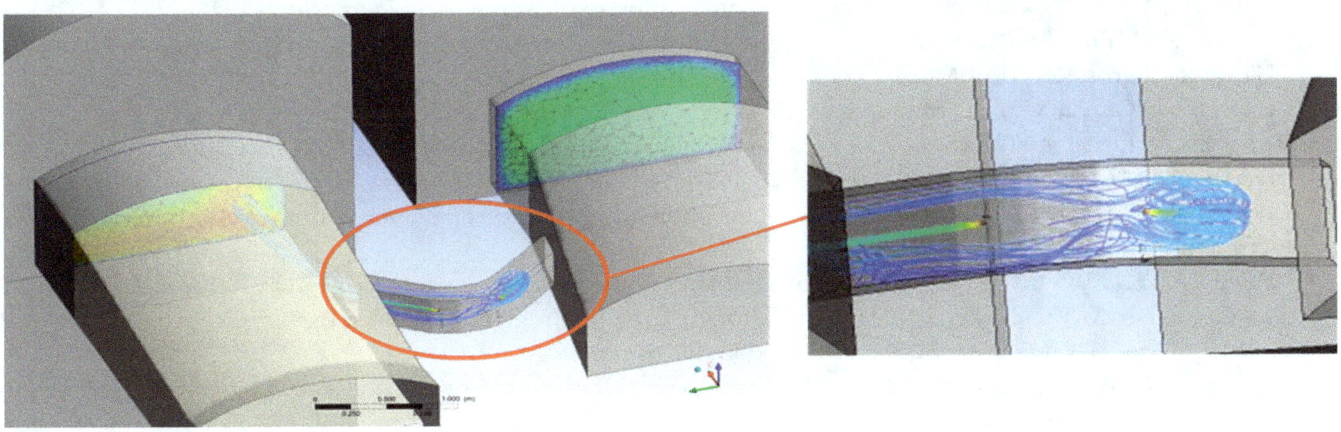

Figure 20. Detail of the flow structure—pros and cons injector flow streamlines.

5. Conclusions

CFD applied to the glass production field reveals to be an effective approach useful for the design and development of innovative solutions and techniques for NO_x reduction based on gas recirculation systems. The authors developed several CFD procedures for the simulation of the regenerative chambers and the exhaust gas recirculation systems. Special attention has been paid on the development of systematic approach for post-processing of the CFD results in order to quantitatively compare different possible solutions. With this approach, the CFD technology becomes an effective tool for design and development. In the case of WGR systems, the recirculated mass flow, for a given geometrical layout, can be determined using CFD. A synthetic index to drive the WGR design has been identified from the systematic use of the simulation approach. For the HyAs, CFD is useful to set the amount of compressed air necessary for a proper mixing and post-combustion of the exhaust gases. Also, for these configurations, the systematic use of CFD supports the design process and is effective in understanding the management strategies of the HyAS system.

Author Contributions: All authors have equally contributed to the development of this research paper.

Nomenclature

A	Area
f	Index of local recirculated gas
m	Mass flow rate [kg/s]
q	Momentum [kg m/s^2]
r	Momentum ratio
r′	Momentum index
v	Velocity [m/s]
ρ	Density [kg/m^3]
[X]	Molar fraction
Y_X	Mass fraction

Acronyms

CC	Combustion Chamber
CFD	Computational Fluid Dynamics
HRS	Heat Recovery System
HyAS	Hybrid Air Staging
RC	Regenerative Chamber
WGR	Waste Gas Recirculation

References

1. Zeldovich, Y.B. The oxidation of nitrogen in combustion explosions. *Acta Physiochim.* **1946**, *21*, 577–628.
2. Skalska, K.; Miller, J.S.; Ledakowicz, S. Trends in NOx abatement: A review. *Sci. Total Environ.* **2010**, *408*, 3976–3989. [CrossRef] [PubMed]
3. Scalet, B.M.; Garcia Muñoz, M.; Sissa, A.Q.; Roudier, S.; Delgado Sancho, L. *"Best Available Tecniques (BAT) Reference Document for The Manufacture of Glass", Industrial Emission Directive 2010/75/EU. Integrated Pollution Prevention and Control*; Joint Research Centre (JRC): Seville, Spain, 2013.
4. LIFE Project "LIFE 12 ENV/IT/001020". Available online: http://ec.europa.eu/environment/life/project/Projects (accessed on 18 June 2019).
5. PRIMEGLASS LIFE Project. Available online: http://www.primeglass.it/ (accessed on 18 June 2019).
6. Zarrinehkafsh, M.T.; Sadrameli, S.M. Simulation of fixed bed regenerative heat exchangers for flue gas heat recovery. *Appl. Therm. Energy* **2004**, *24*, 373–382. [CrossRef]
7. Yakinthos, K.; Missirlis, D.; Sideridis, A.; Vlahostergios, Z.; Seite, O.; Goulas, A. Modelling operation of system of recuperative heat exchangers for aero engine with combined use of porosity model and thermo-mechanical model. In *Engineering Applications of Computational Fluid Mechanics*; Taylor & Francis: Boca Raton, FL, USA, 2012; Volume 6, pp. 608–621.
8. Sardeshpande, V.; Anthony, R.; Gaitonde, U.N.; Banerjee, R. Performance analysis for glass furnace regenerator. *Appl. Energy* **2011**, *88*, 4451–4458. [CrossRef]
9. Reboussin, Y.; Fourmigué, J.F.; Marty, P.; Citti, O. A numerical approach for the study of glass furnace regenerators. *Appl. Therm. Energy* **2005**, *25*, 2299–2320. [CrossRef]
10. Basso, D.; Cravero, C.; Reverberi, A.P.; Fabiano, B. CFD Analysis of regenerative chambers for energy efficiency improvement in glass production plants. *Energies* **2015**, *8*, 8945–8961. [CrossRef]
11. Cravero, C.; Marsano, D. Numerical simulation of regenerative chambers for glass production plants with a non-equilibrium heat transfer model. *WSEAS Trans. Heat Mass Transf.* **2017**, *12*, 21–29.
12. Cogliandro, S.; Cravero, C.; Marini, M.; Spoladore, A. Simulation strategies for regenerative chambers in glass production plants with strategic exhaust gas recirculation system. *IIETA Int. J. Heat Technol.* **2017**, *35*, S449–S455. [CrossRef]
13. Spliethoff, H.; Greul, U.; Rodiger, H.; Hein, K.R.G. Basic effects on NOx emissions in air staging and reburning at a bench-scale test facility. *Fuel* **1996**, *75*, 560–564. [CrossRef]
14. De Domenico, D. Development of CFD Procedure for Exhaust Gas Recirculation Strategies in Glass Production Plants. Master's Thesis, University of Genova, Genova, Italy, 2018.

2

CFD Modeling and Experimental Validation of an Alkaline Water Electrolysis Cell for Hydrogen Production

Jesús Rodríguez and Ernesto Amores *

Centro Nacional del Hidrógeno (CNH2), Prolongación Fernando El Santo s/n, 13500 Puertollano, Ciudad Real, Spain; jesus.rodriguez@cnh2.es

* Correspondence: ernesto.amores@cnh2.es

Abstract: Although alkaline water electrolysis (AWE) is the most widespread technology for hydrogen production by electrolysis, its electrochemical and fluid dynamic optimization has rarely been addressed simultaneously using Computational Fluid Dynamics (CFD) simulation. In this regard, a two-dimensional (2D) CFD model of an AWE cell has been developed using COMSOL® software and then experimentally validated. The model involves transport equations for both liquid and gas phases as well as equations for the electric current conservation. This multiphysics approach allows the model to simultaneously analyze the fluid dynamic and electrochemical phenomena involved in an electrolysis cell. The electrical response was evaluated in terms of polarization curve (voltage vs. current density) at different operating conditions: temperature, electrolyte conductivity, and electrode-diaphragm distance. For all cases, the model fits very well with the experimental data with an error of less than 1% for the polarization curves. Moreover, the model successfully simulates the changes on gas profiles along the cell, according to current density, electrolyte flow rate, and electrode-diaphragm distance. The combination of electrochemical and fluid dynamics studies provides comprehensive information and makes the model a promising tool for electrolysis cell design.

Keywords: energy storage; green hydrogen; alkaline water electrolysis; polarization curve; void fraction; CFD analysis; numerical simulation; renewable energy

1. Introduction

The current fossil fuel-based energy system has to face the most important challenge of our time: climate change. The energy transition to a zero-carbon model requires an ambitious growth plan for renewable energies [1]. However, renewable energy sources (mainly wind and solar power) involve a critical issue that needs to be addressed [2]; their dependence on weather conditions makes them very intermittent, causing a mismatch between supply and demand. Furthermore, in the medium-long term, it is expected that these discrepancies can produce failures in the electricity grid. So, to prevent these problems, adequate energy storage systems are required.

Regarding long-term storage of renewable energies, only chemical energy carriers can be considered nowadays as real solution. Among them, hydrogen has been identified as the best candidate because it can be more efficiently produced from renewable energy surplus in combination with water electrolysis [3]. Furthermore, hydrogen can be transported by its own hydrogen distribution grid as admixture in the natural gas grid, or it can be also stored in appropriate facilities. When required, hydrogen can be efficiently transformed into electricity by fuel cells and returned to the electricity grid or used as fuel for the transport sector or even for new industrial processes [4,5]. In this way, electrolysis enables large scale production of green hydrogen and electrical energy storage.

With respect to water electrolysis, different technologies are commercially available: alkaline, proton exchange membrane (PEM) and solid oxide (SO) electrolysis cells, according to the electrolyte used. Among them, alkaline water electrolysis (AWE) stands out as being a highly developed technology in the industry and the main way to obtain "green hydrogen". Compared to other technologies, the key advantages of alkaline electrolysis are its proven durability, maturity, and low specific costs. Nowadays, electrolyzers are commercially available in the MW range, offering high hydrogen production capacities. The cost of alkaline electrolysis systems is currently in the range of 600–1000 €/kW at the MW level. Regarding the disadvantages, these are mainly the low current densities and the problems derived from working in dynamic conditions. In practice, hydrogen production is limited to an operating range of 20% to 100% of the nominal power in order to prevent the formation of flammable mixtures due to gas diffusion (crossover) through the diaphragm at very low current densities (<0.1 A/cm^2) [2,6–8].

For these reasons, the research and development efforts in this technology are mainly focused on increasing the current density and solving the problems derived from working in dynamic conditions. The fluctuations in power supplies to the electrolyzer could cause problems such as the generation of explosive mixtures, corrosion of materials, lower efficiency, pressure drops, and temperature changes, etc. Therefore, the design of alkaline electrolyzers powered by renewable energy is a critical issue [9]. In this sense, the optimization of the cell design and the space between electrodes is a fundamental step to reduce ohmic overpotentials and increase the nominal current density of the electrolyzer [6,10,11].

Therefore, in the last few decades, a significant effort has been made to model, characterize, and analyze the operation of alkaline electrolyzers [2]. In this sense, Olivier et al. [12] conducted an exhaustive and comprehensive analysis of the existing modelling works concerning low temperature electrolysis systems. Therefore, numerous empirical and semi-empirical models have been proposed. One of the most widely used models was proposed by Ulleberg [13], which provides a mathematical description of the polarization curve with parameters fitted empirically. However, these models usually do not include geometric aspects, nor are they capable of modelling those effects that depend on the complete design of the cell.

In this context, Computational Fluid Dynamics (CFD) simulation is a critical and powerful design tool that can be used to improve flow distribution and minimize energy consumption, allowing efficient electrolyte inlet and gas removal from the cell as soon as possible [9]. CFD models allow for the description of thermal, electrochemical, and fluid dynamic phenomena that occur at the same time, from a multiphysics perspective, according to the considered cell geometry. In this way, the voltage required to carry out electrolysis, the generation of H_2/O_2 bubbles at the electrodes, the flow distribution within each electrolysis cell, and the heat generation according to current density are some of the processes that happen simultaneously in an electrolyzer during its operation.

However, few analytical models based on CFD simulation that describe the behavior of alkaline electrolysis cells have been reported and validated. With this in mind, Aldas [14] applied a two-phase mathematical model to the numerical investigation of gas evolution in a vertical electrochemical cell using the PHOENICS computer code. In other work, Mat et al. [15] investigated the hydrogen evolution, flow field, and current density distribution in an electrochemical cell using a two-phase flow model. The predicted results satisfactorily agreed with data available in the literature. Later, Hawkes at al. [16] used a three-dimensional (3D) CFD electrochemical model to study a high-temperature electrolysis stack using the commercial code FLUENT. Alternatively, Hreiz et al. [17] simulated bubble hydrodynamics on vertical plane electrodes using an Euler–Lagrange CFD approach, and the results were compared with gas velocity fields obtained by a Particles Image Velocimetry (PIV) algorithm, which were in good agreement. Later, El-Askary et al. [18] modeled the hydrodynamics characteristics of the hydrogen evolution process to predict the flow characteristics, gas release rate, and void fraction distribution in electrolysis cells; subsequently, an experimental study was also carried out to verify the calculations made. Zarghami et al. [19] used an Euler-Euler model in FLUENT to simulate the multiphase flow in an alkaline water electrolyzer and compared the results to existing experimental data. Recently, Le Bideau et al. [20] developed a two-phase hydrodynamics model and validated the results with the experimental velocity profiles measured using the Laser Doppler Velocimetry (LDV) method.

Considering the increasing industrial development of water electrolysis that is expected in the coming years [21], it is crucial to consider the effect of renewable energies on the performance of electrolyzers. For these reasons, in the present paper, a two-dimensional (2D) CFD model of an alkaline electrolysis cell implemented in COMSOL is reported based on a work previously published by the authors [9]. The aim of the work is to show the capacity of the model to predict the behavior of the electrolysis cell under different operational conditions and also show how it can influence its performance. Thus, a novel CFD model that simultaneously simulates the electrochemical and fluid dynamic behavior is described, taking a real alkaline water electrolysis cell as a reference. In this way, the response of an AWE cell is simulated under different temperatures and electrolyte concentrations. Polarization curve and gas-liquid fraction can be analyzed at the same time using the presented model. Furthermore, the model shows a wide versatility regarding geometrical parameters, being able to consider the effect of modifying the electrode-diaphragm distance. In order to evaluate the accuracy of the model, simulated results were validated against the real response of the AWE cell at different operating conditions, reporting a very good agreement with a mean relative absolute difference error lower than 1%. In addition, fluid dynamics phenomena such as gas-liquid distribution, turbulence, or formed gas profile were predicted. Therefore, the capacity to simulate both phenomena (electrochemistry and fluid dynamics) in the same model make it a complete design tool.

2. Methodology: Modeling and Experimental

2.1. Alkaline Water Electrolysis

Water electrolysis deals with decomposition of water into hydrogen and oxygen by passing an electric current (DC) between two electrodes separated by an aqueous electrolyte with high ionic conductivity [22]. The overall reaction for water splitting is:

$$H_2O \rightarrow H_2 + \frac{1}{2}\, O_2 \tag{1}$$

Figure 1 shows a scheme of an alkaline electrolysis cell. The two electrodes (anode and cathode) are immersed in a KOH aqueous solution (electrolyte), and they are separated by a porous diaphragm that allows the ionic transport (OH^-) but is impermeable to gases. In a typical operation, electrolyte enters into the anodic and cathodic compartments by the bottom inlets. At the anode, oxygen bubbles are produced and, at the cathode, hydrogen is generated. These bubbles grow until they detach from the electrode surface. The mixture of generated gases and electrolyte then leaves each compartment through the upper outlets [9].

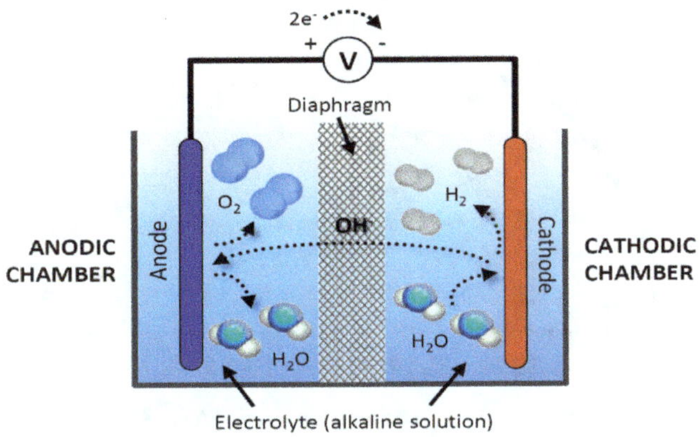

Figure 1. General scheme and operation of an alkaline electrolysis cell [6].

In this way, in an alkaline electrolysis cell, when the required potential between the electrodes is applied, the following semi-reactions take place simultaneously in the cathode (hydrogen evolution reaction (HER)) and anode (oxygen evolution reaction (OER)) [6]:

$$\text{CATHODE (HER): } 2\,H_2O + 2\,e^- \rightarrow H_2 + 2\,OH^-, \qquad E^{0,HER} = -0.829 \text{ V vs. SHE} \tag{2}$$

$$\text{ANODE (OER): } 2\,OH^- \rightarrow \frac{1}{2}\,O_2 + H_2O + 2\,e^-, \qquad E^{0,OER} = +0.401 \text{ V vs. SHE} \tag{3}$$

Thus, for alkaline water electrolysis to occur, a minimum voltage is required (reversible voltage, U_{rev}), which, according to the semi-reactions potential (E^0) from Equations (2) and (3), is equal to 1.23 V ($E^{0,OER}$–$E^{0,HER}$) at standard conditions (1 bar and 25 °C). This value, in agreement with the second law of thermodynamics, corresponds to $\Delta g^\circ = 273.2$ kJ/mol.

However, the real cell voltage (U) is always higher than the latter because of irreversibilities or overpotentials. Therefore, the real cell voltage can be defined as the sum of reversible voltage (U_{rev}) and the overpotentials (η), as shown in Equation (4):

$$U = U_{rev} + \sum \eta \tag{4}$$

The term $\sum \eta$ is the sum of activation, ohmic, and concentration overpotentials. These overpotentials are defined as follows [23]:

(1) Activation overpotentials: related to activation energies of hydrogen and oxygen formation reactions on the surface of electrodes;

(2) Ohmic overpotentials: sum of the electrical resistance of several components such as electrodes, current collectors, etc., and the transport resistance related to gas bubbles, ionic transfer in the electrolyte, and resistivity of the diaphragm;

(3) Concentration overpotentials: due to mass-transport limitations occurring on the surface of the electrodes at high currents.

The total contribution of these overpotentials to the cell voltage (U) can be analyzed through the polarization curve of an electrolysis cell, as represented in Figure 2.

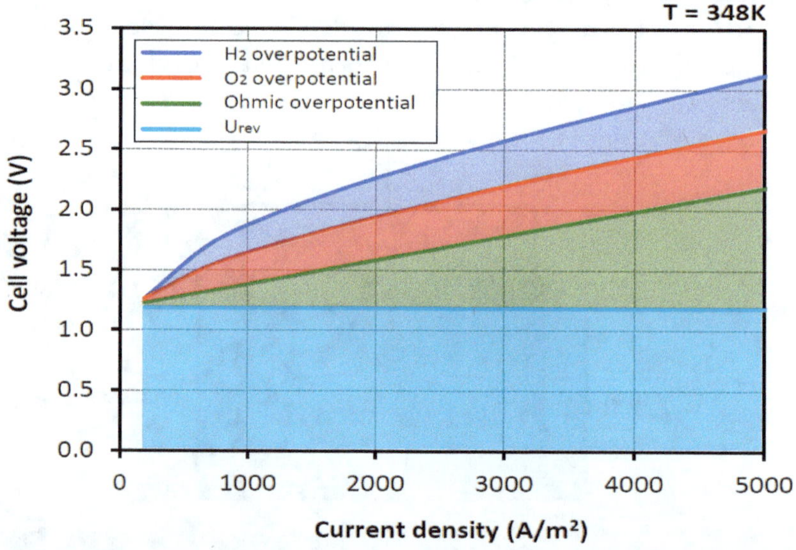

Figure 2. Graphical representation of a polarization curve with the contribution of each overpotential.

2.2. *CFD Modeling of an Alkaline Electrolysis Cell*

2.2.1. Model Geometry and Mesh

As mentioned previously in Figure 1, a conventional alkaline water electrolysis cell is divided into two compartments separated by a diaphragm. In the cathodic chamber, the reduction of water to produce H_2 takes places, and the oxygen evolution reaction occurs at the anode. The function of the diaphragm is to avoid the mixture of the two gases while maintaining a low resistivity.

Taking as reference a rectangular laboratory electrolysis cell (Figure 3a), the geometry of the model was built following fluid dynamic requirements. In this way, simplifications were made in order to reduce the model complexity (Figure 3b). As result, a good approximation can be made just by 2D geometry, which allows an optimal study of the main involved variables (Figure 3c). The electrode surfaces (33 mm height) were considered by two simple boundary conditions (1 and 5 in Figure 3c) on both sides of the cell. These domains work as current collectors where potential was applied. Regarding the separator (3 in Figure 3c), it was introduced in the geometry as a thin rectangle (0.5 mm) that separates both compartments. In this case, only ionic conductivity is allowed in this component. Concerning the anodic and cathodic compartments (2 and 4 in Figure 3c), they were defined as two thick rectangles (1.5, 4, or 10 mm depending on the model considered) between the separator and the corresponding electrode with the electrolyte inlet at the bottom and the biphasic mixture (electrolyte and gases) outlet at the top.

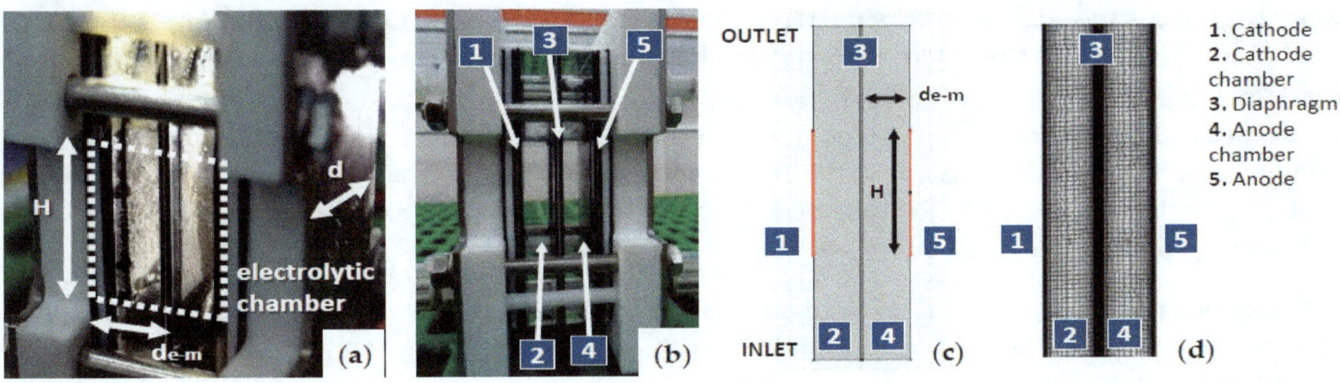

Figure 3. Alkaline electrolysis cell used in the model: (**a**) Detail image of the electrolysis cell; (**b**) Main components; (**c**) Simplified geometry and control domain; (**d**) Mesh applied in the model.

Regarding the meshing of the control domain, a rectangular mesh was generated, taking advantage of the regular geometry of the cell, with a total of 7200 elements (when the electrode-diaphragm distance was 10 mm). Previously, a mesh independence study was carried out in order to determine the number of nodes and elements that should be used. In the anode and cathode chamber, a maximum element size of 0.8 mm was established, reducing this value to 0.1 mm in the diaphragm domain (these values in the mesh were scaled when the electrode-diaphragm distance decreased from 10 mm to 1.5 mm). Furthermore, at the boundary of the electrodes and on the surfaces between the diaphragm and the electrolytic chambers, inflation layers were incorporated to correctly model the effects that take place in these contours (Figure 3d).

2.2.2. Mathematical Procedure and Governing Equations

As previously mentioned, the proposed model is based on previous work carried out by the authors [9] using the CFD simulation program COMSOL Multiphysics v4.3a. In this way, the model has been built using the "electric currents" module to calculate both the current distribution and the polarization curve, and the "two-phase turbulent bubbly flow" module was used to study the generation and distribution of gas, one for each electrolytic chamber. An implicit backward differentiation formula

(BDF) method was used for the numerical integration of the differential equations. The tolerance was fixed at 0.001 in the solver to control the absolute error.

The coupling between the electrochemical and fluid dynamic results was linked by the current density. The intensity supplied to the electrolysis cell allows the electric field to be resolved and at the same time determines the generation of hydrogen and oxygen. The gas formation in the cell is studied by means of the conservation equations for the biphasic mixture (hydrogen/electrolyte in cathode and oxygen/electrolyte in anode) that solves, among other variables, the fraction of gas (Φ_g) in each electrolytic chamber. This fraction of gas is used in each instance to determine how the conductivity of the electrolyte varies and again to resolve the electric field inside the cell.

Electrical Current Conservation

The electric charge transport was studied using the "electric currents" module of COMSOL, which solves the current conservation problem (J in A/m^2) for the scalar electric potential (U in V). For time dependent studies, the equation takes the following form [24]:

$$\nabla \cdot \vec{J} = Q_j \Rightarrow -\nabla \cdot \left[\sigma(\nabla U) - \frac{\partial \vec{D}}{\partial t} - \vec{J}_e \right] \cdot d = Q_j \cdot d \tag{5}$$

where σ (S/m) is the equivalent electrical conductivity, Q (A/m^3 in a 3D approximation) is the current source and J_e (A/m^2) is the externally generated current density. Regarding the electric displacement ($\partial \vec{D} / \partial t$), it is a constitutive relation that describes the macroscopic properties of the medium. Otherwise, in planar 2D, it is assumed that the model has symmetry and the electric potential varies only in the "x" and "y" directions and is constant in the "z" direction. This implies that the electric field is tangential to the "xy" plane. To solve it, the thickness in the "z" direction (electrode depth) is introduced in the equation by the parameter d (m) [24], as shown in Figure 3a.

On the other hand, to define the equivalent electrical conductivity (σ), the following three equations were incorporated to define the conductivity of the medium (electrolyte) and the diaphragm (see parameters in Table 1):

Table 1. Constants, initial, and boundary conditions for the two-dimensional (2D) model [25–28].

Symbol	Value	Unit	Description
d_{e-m}	1.5 to 10	mm	Electrode-diaphragm distance
F	96485	C	Faraday constant
i	250 to 4000	$A \cdot m^{-2}$	Current density (polarization curve)
$i_{0,c}$	21.1 to 93.5	$A \cdot m^{-2}$	Exchange current density (cathode) [25]
$i_{0,a}$	1.1 to 9.3	$A \cdot m^{-2}$	Exchange current density (anode) [26]
M_{H2}	2	$g \cdot moL^{-1}$	Hydrogen molecular weight
M_{O2}	32	$g \cdot moL^{-1}$	Oxygen molecular weight
p	1	bar	Pressure operation
R	8.314	$J \cdot K^{-1} \cdot moL^{-1}$	Ideal gas constant
T	303 to 343	K	Temperature operation
v_0	700	$ml \cdot min^{-1}$	Electrolyte flow rate (inlet velocity)
w	5 to 10	M	Electrolyte concentration (22–42 wt% KOH)
α_a	0.60	-	Charge transfer coefficient (anode)
α_c	0.77		Charge transfer coefficient (cathode)
ε_m	0.55	-	Diaphragm porosity [27]
τ_m	1.89	-	Diaphragm tortuosity [28]

1. An empirical relationship for specific conductivity of electrolyte (σ_0 in S/m) with respect to temperature (T) and KOH concentration (w) was used [29]:

$$\sigma_0 = -204.1 \cdot w - 0.28 \cdot w^2 + 0.5332 \cdot (w \cdot T) + 20720 \cdot \frac{w}{T} + 0.1043 \cdot w^3 - 0.00003 \cdot \left(w^2 \cdot T^2\right) \tag{6}$$

2. The conductivity of the diaphragm (σ_m in S/m) was defined as a function of the conductivity of electrolyte (σ_0) and geometric parameters such as porosity (ε_m) and tortuosity (τ_m), according to [28]:

$$\sigma_m = \sigma_0 \cdot \frac{\varepsilon_m}{\tau_m} \tag{7}$$

3. The Bruggeman equation (σ_e in S/m) relates the variation of conductivity of electrolyte (σ_0) with the volume fraction of gas (Φ_g) inside the cell [30]. The gas fraction for both electrolytic chambers at each current density value was calculated according to the equations described in the "liquid-gas flow distribution" section:

$$\sigma_e = \sigma_0 \cdot \left(1 - \Phi_g\right)^{1.5} \tag{8}$$

Finally, according to Equation (5), the electric transport charges on the electrodes were not considered (only ionic transport was studied inside the electrolysis cell). In order to incorporate the activation overpotentials and the reversible voltage at different temperatures in the polarization curve, the following two auxiliary equations, whose parameters are defined in Table 1, were introduced into the model (as boundary conditions in the electrode):

1. The reversible potential (U_{rev} in V) was defined according to the LeRoy et al. [31] equation as a function of temperature (T):

$$U_{rev} = 1.5184 - 1.5421 \cdot 10^{-3} \cdot T + 9.523 \cdot 10^{-5} \cdot T \cdot ln(T) + 9.84 \cdot 10^{-8} \cdot T^2 \tag{9}$$

2. Activation overpotentials (η_{act} in V) were defined for the cathode and the anode by the Butler–Volmer equation (Tafel equation form) for each current density (i), according to:

$$\eta_{act} = 2.3 \cdot \frac{R \cdot T}{\alpha \cdot F} \cdot log\left(\frac{i}{i_0}\right) \tag{10}$$

Liquid-Gas Flow Distribution

The gas and liquid flow distribution was modeled, applying the "two-phase turbulent bubbly flow" module of COMSOL. This mathematical approach describes the two-phase flow using an Euler-Euler model. The module solves the volume fraction occupied by each of the two phases, without defining each bubble in detail. This mathematical approach is suitable for modeling the macroscopic behavior of many gas bubbles rising through a liquid. It treats the two phases as interpenetrating media, tracking the averaged concentration of the phases. One velocity field is associated with each phase, and the dynamics of each of the phases are described by a momentum balance equation and a continuity equation [9,32].

Based on these assumptions, the sum of the momentum equations for the two phases gives a momentum equation for the liquid velocity, a continuity equation, and a transport equation for the volume fraction of the gas phase. The momentum equation is [32]:

$$\begin{aligned} &\Phi_l \cdot \rho_l \cdot \frac{\partial \vec{u}_l}{\partial t} + \Phi_l \cdot \rho_l \cdot \left(\vec{u}_l \cdot \nabla\right)\vec{u}_l \\ &= -\nabla p + \nabla \cdot \left[\Phi_l \cdot (\psi_l + \psi_T) \cdot \left(\nabla \vec{u}_l + \nabla \vec{u}_l^{\,T} - \tfrac{2}{3} \cdot \left(\nabla \vec{u}_l\right) \cdot \vec{I}\right)\right] + \Phi_l \cdot \rho_l \cdot \vec{g} + \vec{F} \end{aligned} \tag{11}$$

where u is velocity (m/s), p is the pressure (Pa), Φ is the phase volume fraction, ρ is the density (kg/m^3), g is the gravity vector (m/s^2), F is any additional volume force (N/m^3), ψ_l is the dynamic viscosity of the liquid (Pa·s), and ψ_T is the turbulent viscosity (Pa·s). The subscripts "l" and "g" denote quantities related to the liquid phase and the gas phase, respectively [32].

The continuity equation is:

$$\frac{\partial}{\partial t}\left(\Phi_l \cdot \rho_l + \Phi_g \cdot \rho_g\right) + \nabla \cdot \left(\Phi_l \cdot \rho_l \cdot \vec{u}_l + \Phi_g \cdot \rho_g \cdot \vec{u}_g\right) = 0 \tag{12}$$

And the transport of the volume fraction of gas is given by Equation (13):

$$\frac{\partial \Phi_g \cdot \rho_g}{\partial t} + \nabla \cdot \left(\Phi_g \cdot \rho_g \cdot \vec{u}_g\right) = -\dot{m}_{gl} \tag{13}$$

where $\dot{m}_{gl}$ is the mass transfer rate from gas to liquid (g·m^{-3}·s^{-1}). In the simplified 2D geometry considered in this study, the gas flows generated (g·m^{-2}·s^{-1}) on the active surfaces of the electrodes (H_2 and O_2) were defined by the Faraday equation for each current density (i):

$$\dot{m}_{H2} = \frac{M_{H2}}{2 \cdot F} \cdot i \tag{14}$$

$$\dot{m}_{O2} = \frac{M_{O2}}{4 \cdot F} \cdot i \tag{15}$$

Table 1 summarizes the parameters corresponding to Equation (14) for hydrogen production (boundary condition at cathode) and to Equation (15) for oxygen production (boundary condition at the anode).

Regarding the turbulence, a $k - \varepsilon$ model was used. This model solves two extra transport equations for two additional variables: the turbulent kinetic energy, k (m^2/s^2), and the dissipation rate of turbulent energy, ε (m/s^3). The turbulent viscosity (ψ_T), the transport equations for the turbulent kinetic energy (k), and the evolution of the turbulent energy's dissipation (ε) are defined as [32]:

$$\psi_T = \rho_l \cdot C_\mu \cdot \frac{k^2}{\varepsilon} \tag{16}$$

$$\rho_l \cdot \frac{\partial k}{\partial t} - \nabla \cdot \left[\left(\psi + \frac{\psi_T}{\Phi_k}\right)\nabla k\right] + \rho_l \cdot \vec{u}_l \cdot \nabla k = \frac{1}{2}\psi_T \cdot \left(\nabla \vec{u}_l + \left(\nabla \vec{u}_l\right)^T\right)^2 - \rho_l \cdot \varepsilon + S_k \tag{17}$$

$$\begin{aligned} &\rho_l \cdot \tfrac{\partial \varepsilon}{\partial t} - \nabla \cdot \left[\left(\psi + \tfrac{\psi_T}{\Phi_\varepsilon}\right)\nabla \varepsilon\right] + \rho_l \cdot \vec{u}_l \cdot \nabla \varepsilon \\ &= \tfrac{1}{2} C_{\varepsilon 1} \cdot \tfrac{\varepsilon}{k} \cdot \psi_T \cdot \left(\nabla \vec{u}_l + \left(\nabla \vec{u}_l\right)^T\right)^2 - \rho_l \cdot C_{\varepsilon 2} \cdot \tfrac{\varepsilon^2}{k} + \tfrac{\varepsilon}{k} C_\varepsilon \cdot S_k \end{aligned} \tag{18}$$

where C_ε, C_μ, Φ_k, and Φ_ε are constants of the turbulent model. The term S_k is related to the bubble-induced turbulence. Finally, the gas velocity ($\vec{u}_g$ in m/s) is calculated according to:

$$\vec{u}_g = \vec{u}_l + \vec{u}_{slip} + \vec{u}_{drift} \tag{19}$$

where $\vec{u}_{slip}$ is the relative velocity between the phases and $\vec{u}_{drift}$ is an additional contribution from the turbulence model used [32].

2.2.3. Initial and Boundary Conditions

To solve the different equations described above, the following simplifications were considered in order to improve convergence and reduce the computational cost of the model [5,9,32]:

(1) The gas density is negligible compared to the liquid density;
(2) The movement of gas bubbles in relation to the liquid phase is determined by a balance between viscous drag and pressure forces;
(3) The two phases share the same pressure field;
(4) Hydrogen and oxygen crossover through the diaphragm is negligible;
(5) The electrolysis cell works with a high enough flow rate of electrolyte to avoid the accumulation of gas bubbles in the cell;
(6) The bubbles have a diameter of less than 1 mm, so the Hadamard–Rybczynski drag law for spherical gas bubbles in liquid is used for the gas velocity;
(7) The electrical resistance of the electrodes (Ni) is negligible with respect to the rest of the elements of the electrolysis cell.

Figure 4 shows the boundary conditions implemented in this study. Regarding the different materials used in the model, the COMSOL material library was used to define hydrogen, oxygen, and water. In the latter case, the density and dynamic viscosity were also corrected to take into account the KOH dissolution (electrolyte). Table 1 summarizes, the different parameters and constants considered in the model.

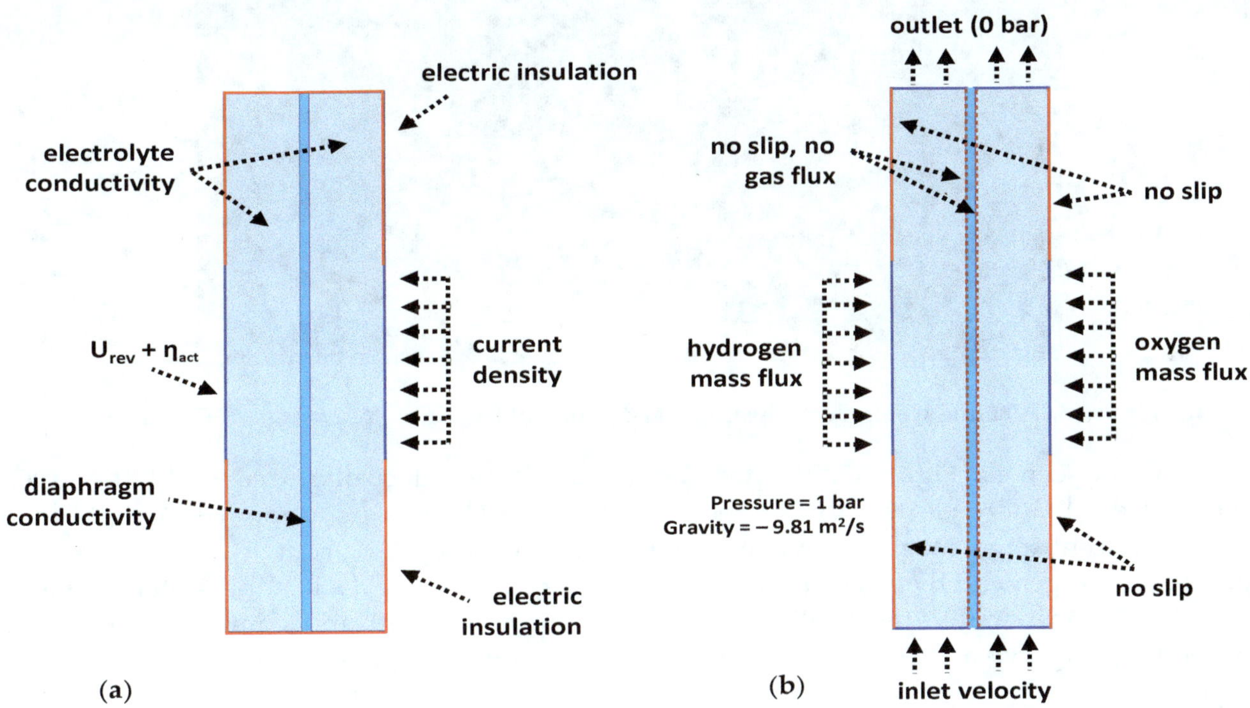

Figure 4. Initial and boundary conditions for the 2D Computational Fluid Dynamics (CFD) model of the alkaline electrolysis cell: (**a**) Electrochemical model; (**b**) Fluid dynamic model.

Regarding the charge transfer coefficient (α), it is worth noting that 0.5 is the value usually reported in most of the models in which electrochemical kinetics are considered. According to the literature [33], this value is usually used in the absence of actual measurements. However, α, in most systems, turns out to be between 0.3 and 0.7 [33]. In this work, as can be seen from Table 1, 0.60 and 0.77 were used for anodic and cathodic reactions, respectively. These values were chosen because they reported the best fit with the experimental values. In the case of the anode, α_a is in the common experimental range. On the other hand, the cathodic charge transfer coefficient ($\alpha_c = 0.77$) is slightly higher than the upper limit of the usually experimental range. However, this value well agrees with experimental values reported by authors working on alkaline water electrolysis [34]. In fact, some authors [35] working with real electrolysis stacks have noted that α cannot be considered as a constant, because it varies with temperature, electrode materials, etc., and, thus, it is expected to change even with the experimental system. Furthermore, differences are expected when α is determined in an electrochemical laboratory cell than when it is done in a real water electrolysis cell or even a commercial stack.

2.3. Experimental Test Facility and Methodology

2.3.1. Alkaline Water Electrolysis Test-Bench

Figure 5 shows the alkaline water electrolysis test bench used in this study, located in the *Centro Nacional del Hidrógeno* (www.cnh2.es). The system is constituted by all the elements of a typical electrolyzer, but on a laboratory scale, which allows for the study of a wide range of parameters. Therefore, the system is suitable for analyzing electrochemical and fluid dynamic processes.

Figure 5. Alkaline water electrolysis cell test bench of the *Centro Nacional del Hidrógeno*.

As can be seen in Figure 5, the test bench has three electrolysis cells (ELECTROCELL, Micro Flow Cell) that are electrically connected in series, a DC power supply (Elektro-Automatik, EA-PSI 6000), a centrifugal magnetic drive pump (IWAKI, MD-30RVM-220N), two gas-liquid separators to separate the oxygen and hydrogen produced from the electrolyte and a heating system (Hillesheim, H300 DN12). Additionally, the facility has different measurement and control devices that are monitored by a Supervisory Control and Data Acquisition (SCADA) system [10].

2.3.2. Experimental Protocol

The polarization curves were obtained according to a previously established experimental procedure [5,10]. These curves describe the electrochemical behavior of an electrolysis cell and allow determination of the values of voltage and current in which it works. For this purpose, several temperatures (from 30 to 70 °C) were tested. The electrolyte used in the test bench was a concentrated KOH solution in water (from 22 to 42 wt% KOH), and the overall flow rate was fixed at 1.4 l/min per cell. Regarding the electrodes, Ni > 99% was used for the anodes and cathodes. In all cases, the active area was 10 cm^2. The diaphragms used were Zirfon® Perl 500 UTP (AGFA) with a pore size of 0.15 μm and a thickness of 500 μm [10]. The electrode-diaphragm distance was fixed from 1.5 to 10 mm according to the case considered (see Figure 3b).

For each of the tests, the following protocol was used: Once the cell was assembled, both the electrolyte pump and the heating system were turned on. Once the flow rate and temperature were stable, the power supply was switched on and a gradual current sweep from 50 to 400 mA/cm^2 was performed. For each current intensity value, different operating parameters, such as the cell voltage to obtain the different polarization curves, were monitored and recorded [5,10].

3. Results and Discussion

3.1. Polarization Curve

The electrochemical response of the model was validated by comparing simulated polarization curves (U_{cell}^{model}) against those experimentally obtained for the electrolysis cell (U_{cell}^{test}). The accuracy of the CFD model was also studied by varying some of the most critical operating parameters in an alkaline water electrolysis system. Additionally, the mean relative absolute difference error ($MRAD_{error}$) was calculated according to Equation (20) in order to ensure the accuracy and validity of the proposed model [5]:

$$MRAD_{error} = \frac{1}{N} \cdot \sum_{j=1}^{N} \left(\frac{U_{cell}^{model} - U_{cell}^{test}}{U_{cell}^{test}} \right) \tag{20}$$

The error calculated was lower than 0.51%, which indicates an excellent correlation between the experimental and modeled results. Figure 6 shows a parity chart between the model and the experimental voltage results. From this figure, it is confirmed that the main discrepancy between the model and real data occurs at low current densities (corresponding to the lowest voltages in the polarization curve). In this zone, activation overpotentials represent the major contribution to the energy required for the electrolysis process. In other words, at these potentials the cathodic and anodic reactions are the limiting steps and, thus, it can be considered that the differences are related to electro-kinetics. In the model, the main electrochemical kinetics parameters involved are i_0 and α. It is well known [35] that both concepts strongly depend on operating conditions (temperature, pressure, reaction, electrode material, etc.). While the influence of i_0 has been widely studied on electrolysis [25,26,35], the charge transfer coefficient has not been treated in much detail. In fact, as described above, in the literature, typically a constant value of 0.5 is given to α for both reactions. In the present work, different values of i_0, depending on temperature and reaction, have been used (see Table 1). In the case of α, although in the model different values were used for oxidation and reduction, these chosen values were constant for all the operating conditions. The use of these approximations, instead of the real value for each temperature, can be the cause of the discussed deviations at low current densities. Nevertheless, it is worth noting that, despite these differences, the error reported value is very low and the model works satisfactory, as shown in the following sections.

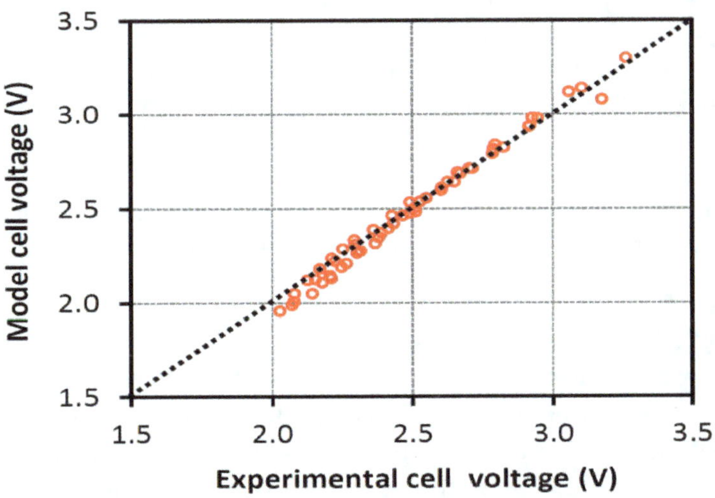

Figure 6. Parity chart between the model and experimental voltage results for all the conditions of temperature, electrolyte conductivity, and electrode-diaphragm distance analyzed.

3.1.1. Influence of Temperature

Polarization curves simulated by the model were compared with experimental ones in Figure 7, at different temperatures. As expected, the cell potential at a certain current density value decreases as temperature increases within the analyzed range (Figure 7a). This trend is due to the fact that an increase in temperature favors the reaction kinetics, decreasing the reversible voltage and therefore the required energy. This behavior was previously reported by other authors for AWE systems [2,5,10,13,22,23]. From Figure 7b, it is evidenced that the best correlation between experimental and calculated results can be seen at 30 °C, 50 °C, and 70 °C.

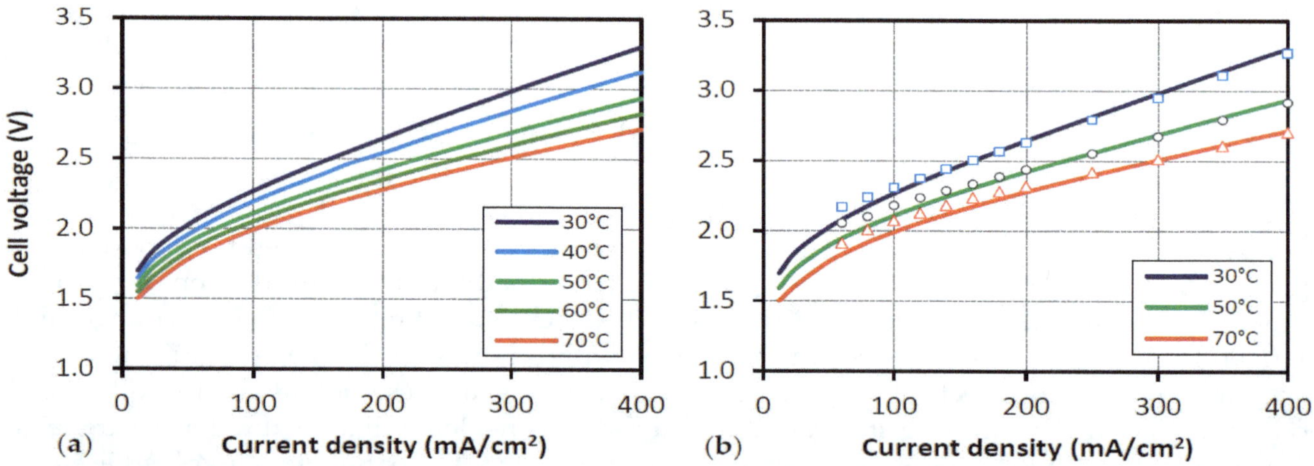

Figure 7. Polarization curve at different temperatures (30–70 °C) at 32 wt% KOH, 10 mm, and 1.4 l/min: (**a**) model; (**b**) model vs. experimental electrolysis cell voltage. Experimental data are indicated with dots.

3.1.2. Influence of Electrolyte Conductivity

Electrolyte conductivity is a parameter that has a significant effect on the ohmic overpotential. As described from Equation (6) to Equation (8), it is a critical aspect to determine not only the electrolyte ohmic losses, but also those related to diaphragm and generated gas bubbles in both cathodic and anodic compartments. In the case of the diaphragm, it must be taken into account that it is just a physical barrier, without ionic or electronic conductivity, so its conductivity is determined by the conductivity of the electrolyte inside the diaphragm channels. Regarding the gas bubbles, when H_2 and O_2 are generated, a biphasic mixture gas-electrolyte is formed, in which the gas fraction is a non-conductor. The greater the number of gas bubbles in this biphasic mixture, the greater the fraction of gas (Φ_g). As a result, the electrolyte conductivity will be lower according to Equation (8), especially in the close electrode region.

In Figure 8, the polarization curves for different electrolyte concentration values calculated with the model are compared with those experimentally obtained. From this figure, the following points are discussed:

(1) As previously reported [10], increasing the electrolyte conductivity implies a better electrolysis performance due to a reduction of ohmic losses, which turns out in a lower required energy.
(2) An optimum value of electrolyte concentration is identified at 32 wt% KOH, which corresponds to 94.54 S/m at 50 °C [29]. Above this value, mass transfer limitations can occur, but below it, the ohmic losses are too high because the electrolyte conductivity is very low [29].
(3) The model shows an accurate agreement with the experimental polarization curves for the different electrolyte conductivity values considered (Figure 8b).

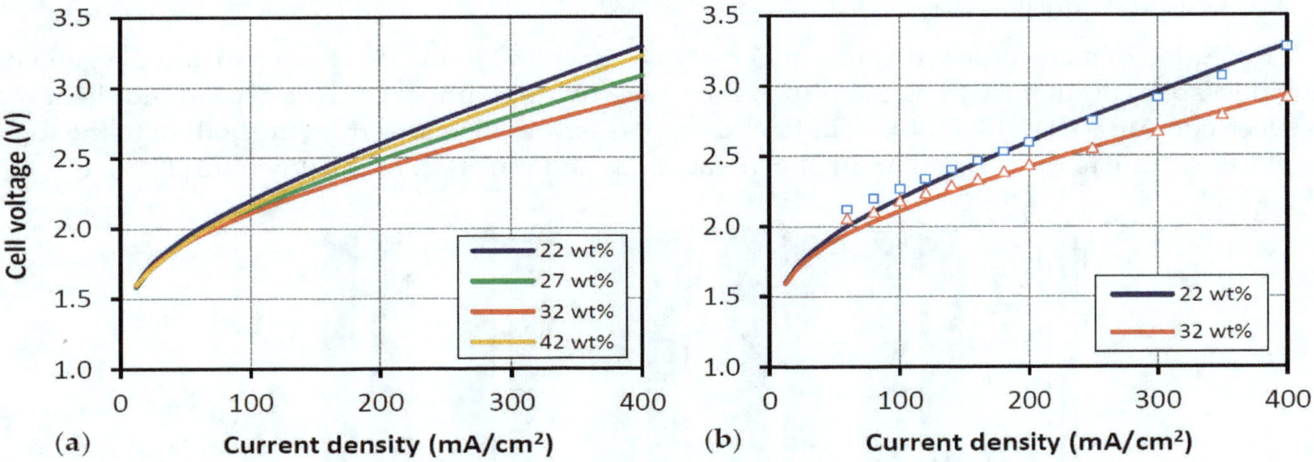

Figure 8. Polarization curve at different electrolyte concentrations (22–42 wt% KOH) at 50 °C, 10 mm, and 1.4 l/min: (**a**) model; (**b**) model vs. experimental electrolysis cell voltage. Experimental data are indicated with dots.

3.1.3. Influence of Electrode-Diaphragm Distance

In AWE cells, the ohmic overpotential is strongly affected by the proximity of the electrode to the diaphragm. The reason for this is that a reduction of this distance means a lower quantity of electrolyte between both components, and thus the ohmic contribution of this compartment decreases. However, very narrow cells can favor high gas fraction during the electrolysis, which can result in a significant increasing of ohmic overpotential. Hence, it seems reasonable that there is an optimal electrode-diaphragm distance to achieve the best AWE cell performance. In fact, several works have calculated this optimum distance [11,36].

Figure 9 shows the simulated and experimental polarization curves when different electrode-diaphragm distances were used, for an electrolyte flow rate of 1.4 l/min. The tendency of the modeled curves well coincide with the real ones; the decrease of the distance between electrode and diaphragm leads to a lower electrolysis potential.

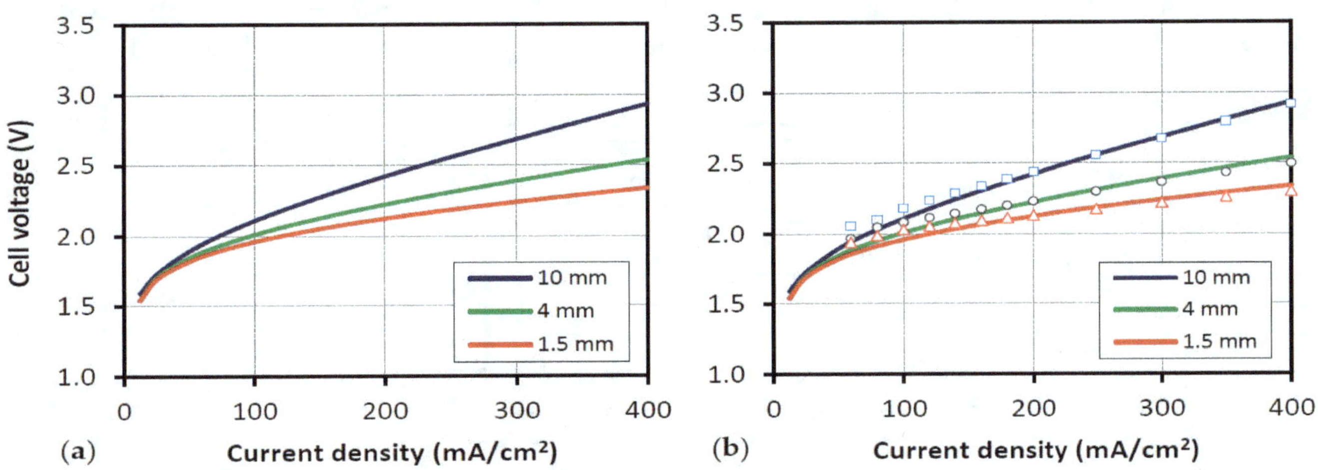

Figure 9. Polarization curve at different electrode-diaphragm distances (1.5–10 mm) at 50 °C, 32 wt% KOH, and 1.4 l/min: (**a**) model; (**b**) model vs. experimental electrolysis cell voltage. Experimental data are indicated with dots.

3.2. Gas Generation Profile

Evolution of gas profile with current density is shown in Figure 10. Hydrogen and oxygen void fractions (space occupied by gas bubbles in each electrolytic chamber) draw a typical profile next to the electrode surface [17,18,37,38]: void fraction progressively increases from the bottom to the top of the electrode due to the accumulation of generated gas and the effect of the flow rate [15].

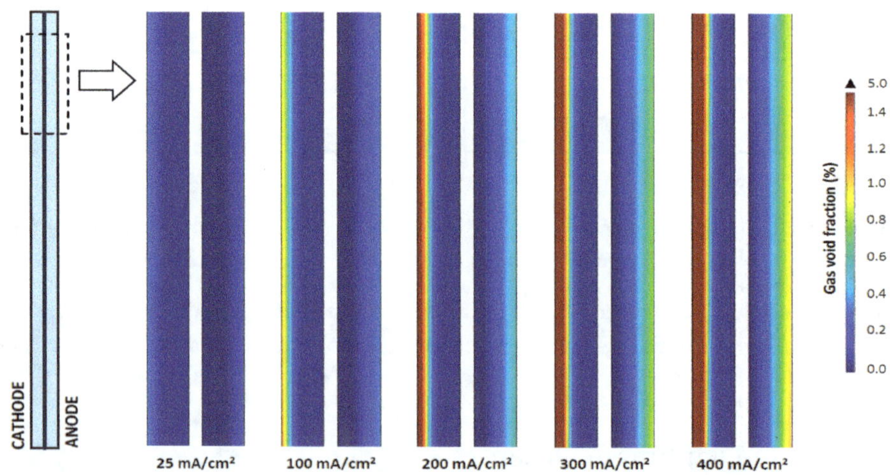

Figure 10. Void fraction distribution in cathodic and anodic compartments at different current density values (model conditions: 50 °C, 32 wt% KOH, 1.5 mm, and 1.4 l/min).

Hydrodynamic characteristics of the biphasic flow (generated gas/electrolyte) in the AWE cell strongly influence the performance of the system. In these systems the generated gas bubbles have a multiple and critical role. On one hand, when they are produced, bubbles form a curtain of increasing thickness along the surface of the electrode in a vertical direction, causing a turbulence phenomenon that contributes to the mixing and distribution of present species and so too to mass transfer. On the other hand, when bubbles are produced, they can adhere to some sections of the electrode surface, making them inactive for electrochemical reaction. This causes an increase in the ohmic overpotential, a higher required energy and, as a result, a worse cell performance [19].

According to the previous discussion, the model results show that gas fraction reaches the highest value next to the electrode surface (x = 0 cm) and it decreases towards the diaphragm (x = 0.15 cm) for all studied current densities (Figure 11). In addition to the expected increase in the fraction of gas with the current, from this graph it can also be deduced that the higher the current density, the greater the penetration of gas in the lateral direction, due to the increase in the lateral gas velocity [14].

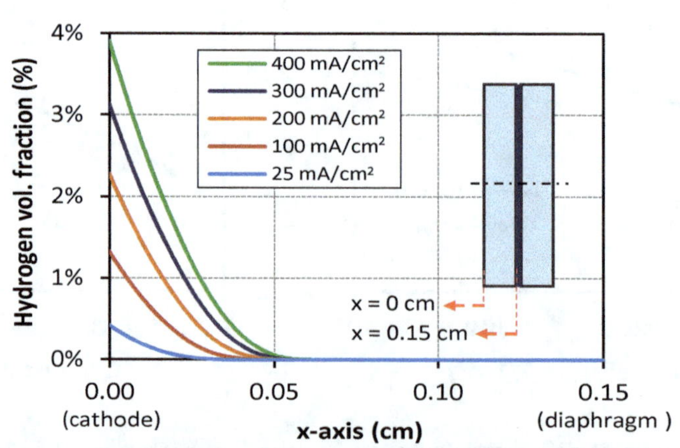

Figure 11. Hydrogen fraction distribution at different current densities in the x-axis at the middle of the electrode height (model conditions: 50 °C, 32 wt% KOH, 1.5 mm, and 1.4 l/min).

4. Conclusions and Future Work

In the present work, the development and experimental validation of an alkaline water electrolysis cell 2D CFD model was reported. The strong points of the model are its versatility to simulate different operating conditions as well as the simultaneous analysis of fluid dynamic and electrochemical phenomena. These aspects make it a powerful and cheap design tool, particularly suitable for the study of new cell configurations.

Regarding the variation of electrolysis parameters, the response of the model was evaluated in terms of polarization curves at different values of temperature, electrode-diaphragm distance, and electrolyte concentration. Performance of the electrolysis cell was improved for high temperatures, small electrode-diaphragm distances, and high electrolyte conductivity, as observed from both experimental and simulated data. In this regard, the mean relative absolute difference error was calculated in order to ensure the accuracy and validity of the proposed model. The error calculated was lower than 0.51% for the polarization curve, which indicates an excellent correlation between the experimental and modeled results. Therefore, the proposed model is suitable for predicting the behavior of the AWE cell under a wide range of operating conditions.

In addition to electrochemical studies, the proposed model also offers a fluid dynamic analysis. From simulation results, it was concluded that produced gases generate a "curtain profile" in the electrode, increasing the void fraction in a vertical direction due to the accumulation of gas. Furthermore, the model simulates the gas profile along the x-axis of the electrolytic chamber (from the electrode to the diaphragm), proving that the higher the current, the greater the gas distribution in the electrolytic chamber space. These results provide very useful information for cell design purposes.

On the other hand, although the results confirm that the model well fits the experimental results, various points still need to be studied:

(1) Charge transfer coefficient (α): the influence of this term over the performance of real electrolysis systems has hardly been studied. Even though it is possible to find different values in the literature, some authors have claimed significant differences when working with a laboratory electrochemical cell and with an electrolysis stack. Experimental values up to 1.035 [35] have been reported, which contrast with that which is considered to be the common range (0.3–0.7). In fact, it greatly depends on operating conditions: electrode materials, electrolyte, etc. For these reasons, it would be very interesting to experimentally determine this parameter in the test bench shown in Figure 5 in order to improve the accuracy of the model, especially in the activation overpotentials zone where the model fits worse with real data. This is extensible to exchange current density, which is another critical kinetic parameter.

(2) Zero-gap cell design: this is one of the most extended alkaline water electrolysis cell designs in commercial systems. This configuration allows us to "ideally" eliminate bubbles between the electrodes because cathode and anode are placed directly over the diaphragm. For this to happen, the electrodes have to be porous. Therefore, in order to improve the model applicability to real systems, this geometry should be implemented in the model, making it a more useful design tool.

Author Contributions: Conceptualization, J.R. and E.A.; methodology, E.A. and J.R.; software, J.R. and E.A.; Validation, E.A.; formal analysis, J.R.; data curation, E.A. and J.R.; writing—original draft preparation, J.R. and E.A.; writing—review and editing, E.A. and J.R. All authors have read and agreed to the published version of the manuscript.

References

1. Armaroli, N.; Balzani, V. Solar electricity and solar fuels: Status and perspectives in the context of the energy transition. *Chem. Eur. J.* **2016**, *22*, 32–57. [CrossRef]

2. Sánchez, M.; Amores, E.; Rodríguez, L.; Clemente-Jul, C. Semi-empirical model and experimental validation for the performance evaluation of a 15 kW alkaline water electrolyzer. *Int. J. Hydrog. Energy* **2018**, *43*, 20332–20345. [CrossRef]
3. Mais, L.; Palmas, S.; Mascia, M.; Sechi, E.; Casula, M.F.; Rodriguez, J.; Vacca, A. Porous Ni photocathodes obtained by selective corrosion of Ni-Cu films: Synthesis and photoelectrochemical characterization. *Catalysts* **2019**, *9*, 453. [CrossRef]
4. Rodríguez, J.; Rojas, N.; Sánchez-Molina, M.; Rodriguez, L.G.; Campana, R.; Rodríguez, L. Hybrid membranes based in Nafion-metallic oxides: Performance evaluations. *Chem. Eng. Trans.* **2016**, *47*, 415–420. [CrossRef]
5. Amores, E.; Rodríguez, J.; Oviedo, J.; de Lucas-Consuegra, A. Development of an operation strategy for hydrogen production using solar PV energy based on fluid dynamic aspects. *Open Eng.* **2017**, *7*, 141–152. [CrossRef]
6. Amores, E.; Sánchez, M.; Rojas, N.; Sánchez-Molina, M. Renewable hydrogen production by water electrolysis. In *Sustainable Fuel Technologies Handbook*, 1st ed.; Dutta, S., Hussain, C.M., Eds.; Academic Press: London, UK, 2021; pp. 271–313. [CrossRef]
7. Lehner, M.; Tichler, R.; Steinmüller, H.; Koppe, M. *Power-To-Gas: Technology and Business Models*, 1st ed.; Springer: Cham, Switzerland, 2014. [CrossRef]
8. Bertuccioli, L.; Chan, A.; Hart, D.; Lehner, F.; Madden, B.; Standen, E. *Development of Water Electrolysis in the European Union*. FCH-JU Funded Studies 2014: Element Energy, E4tech Sarl. Available online: https://www.fch.europa.eu/sites/default/files/FCHJUElectrolysisStudy_FullReport%20(ID%20199214).pdf (accessed on 10 November 2020).
9. Amores, E.; Rodríguez, J.; Merino, C.; García, P. Study of an alkaline electrolyzer powered by renewable energy. In Proceedings of the COMSOL Conference, Stuttgart, Germany, 26–28 October 2011; COMSOL AB: Burlington, VT, USA, 2011.
10. Amores, E.; Rodríguez, J.; Carreras, C. Influence of operation parameters in the modeling of alkaline water electrolyzers for hydrogen production. *Int. J. Hydrog. Energy* **2014**, *39*, 13063–13078. [CrossRef]
11. Nagai, N.; Takeuchi, M.; Nakao, M. Influences of bubbles between electrodes onto efficiency of alkaline water electrolysis. In Proceedings of the PSFVIP-4, Chamonix, France, 3–5 June 2003.
12. Olivier, P.; Bourasseau, C.; Bouamama, B. Low-temperature electrolysis system modelling: A review. *Renew. Sustain. Energy Rev.* **2017**, *78*, 280–300. [CrossRef]
13. Ulleberg, Ø. Modeling of advanced alkaline electrolyzers: A system simulation approach. *Int. J. Hydrog. Energy* **2003**, *28*, 21–33. [CrossRef]
14. Aldas, K. Application of a two-phase flow model for hydrogen evolution in an electrochemical cell. *Appl. Math. Comput.* **2004**, *154*, 507–519. [CrossRef]
15. Mat, M.D.; Aldas, K.; Ilegbusi, O.J. A two-phase flow model for hydrogen evolution in an electrochemical cell. *Int. J. Hydrog. Energy* **2004**, *29*, 1015–1023. [CrossRef]
16. Hawkes, G.; O'Brien, J.; Stoots, C.; Hawkes, B. 3D CFD model of a multi-cell high-temperature electrolysis stack. *Int. J. Hydrog. Energy* **2009**, *34*, 4189–4197. [CrossRef]
17. Hreiz, R.; Abdelouahed, L.; Fünfschilling, D.; Lapicque, F. Electrogenerated bubbles induced convection in narrow vertical cells: PIV measurements and Euler–Lagrange CFD simulation. *Chem. Eng. Sci.* **2015**, *134*, 138–152. [CrossRef]
18. El-Askary, W.A.; Sakr, I.M.; Ibrahim, K.A.; Balabel, A. Hydrodynamics characteristics of hydrogen evolution process through electrolysis: Numerical and experimental studies. *Energy* **2015**, *90*, 722–737. [CrossRef]
19. Zarghami, A.; Deen, N.G.; Vreman, A.W. CFD modeling of multiphase flow in an alkaline water electrolyzer. *Chem. Eng. Sci.* **2020**, *227*, 115926–115935. [CrossRef]
20. Le Bideau, D.; Mandin, P.; Benbouzid, M.; Kim, M.; Sellier, M.; Ganci, F.; Inguanta, R. Eulerian two-fluid model of alkaline water electrolysis for hydrogen production. *Energies* **2020**, *13*, 3394. [CrossRef]
21. Hydrogen Roadmap Europe, FCH-JU 2019. Available online: https://www.fch.europa.eu/sites/default/files/Hydrogen%20Roadmap%20Europe_Report.pdf (accessed on 10 November 2020).
22. Divisek, J. Water electrolysis in a low- and medium-temperature regime. In *Electrochemical Hydrogen Technologies: Electrochemical Production and Combustion of Hydrogen*; Wendt, H., Ed.; Elsevier: Oxford, UK, 1990; pp. 137–212.
23. Zeng, K.; Zhang, D. Recent progress in alkaline water electrolysis for hydrogen production and applications. *Prog. Energy Combust. Sci.* **2010**, *36*, 307–326. [CrossRef]

24. *AC/DC Module User's Guide v4.3. Part n° CM020101*; COMSOL AB: Stockholm, Sweden, 2012.
25. Kibria, M.F.; Mridha, M.S.; Khan, A.H. Electrochemical studies of a nickel electrode for the hydrogen evolution reaction. *Int. J. Hydrog. Energy* **1995**, *20*, 435–440. [CrossRef]
26. Kibria, M.F.; Mridha, M.S. Electrochemical studies of the nickel electrode for the oxygen evolution reaction. *Int. J. Hydrog. Energy* **1996**, *21*, 179–182. [CrossRef]
27. AGFA: Zirfon H2 Advanced. Available online: https://www.agfa.com/specialty-products/solutions/membranes/zirfon/ (accessed on 30 October 2020).
28. Rodríguez, J.; Palmas, S.; Sánchez-Molina, M.; Amores, E.; Mais, L.; Campana, R. Simple and precise approach for determination of ohmic contribution of diaphragms in alkaline water electrolysis. *Membranes* **2019**, *9*, 129. [CrossRef]
29. Gilliam, R.J.; Graydon, J.W.; Kirk, D.W.; Thorpe, S.J. A review of specific conductivities of potassium hydroxide solutions for various concentrations and temperatures. *Int. J. Hydrog. Energy* **2007**, *32*, 359–364. [CrossRef]
30. Weijs, M.P.M.G.; Janssen, L.J.J.; Visser, G.J. Ohmic resistance of solution in a vertical gas-evolving cell. *J. Appl. Electrochem.* **1997**, *27*, 371–378. [CrossRef]
31. LeRoy, R.L.; Bowen, C.T.; LeRoy, D.J. The thermodynamics of aqueous water electrolysis. *J. Electrochem. Soc.* **1980**, *127*, 1954–1962. [CrossRef]
32. *Chemical Engineering Module User's Guide v3.5a. Part n° CM020501*; COMSOLAB: Stockholm, Sweden, 2008.
33. Bard, A.J.; Faulkner, L.R. *Electrochemical Methods: Fundamentals and Applications*, 2nd ed.; John Wiley & Sons: New York, NY, USA, 2001.
34. Santos, D.M.F.; Sequeira, C.A.C.; Macciò, D.; Saccone, A.; Figuereido, J.L. Platinum–rare earth electrodes for hydrogen evolution in alkaline water electrolysis. *Int. J. Hydrog. Energy* **2013**, *38*, 3137–3145. [CrossRef]
35. Tijani, A.S.; Kamarudin, N.A.B.; Mazlan, F.A.B. Investigation of the effect of charge transfer coefficient (CTC) on the operating voltage of polymer electrolyte membrane (PEM) electrolyzer. *Int. J. Hydrog. Energy* **2018**, *43*, 9119–9132. [CrossRef]
36. Nagai, N.; Takeuchi, M.; Oka, T. Existence of optimum space between electrodes on hydrogen production by water electrolysis. *Int. J. Hydrog. Energy* **2003**, *28*, 35–41. [CrossRef]
37. Sasaki, T.; Nagai, N.; Murai, Y.; Yamamoto, F. Particle image velocimetry measurement of bubbly flow induced by alkaline water electrolysis. In Proceedings of the PSFVIP-4, Chamonix, France, 3–5 June 2003.
38. Aldas, K.; Pehlivanoglu, N.; Mat, M.D. Numerical and experimental investigation of two-phase flow in an electrochemical cell. *Int. J. Hydrog. Energy* **2008**, *33*, 3668–3675. [CrossRef]

3

Design and Optimization of Multiple Circumferential Casing Grooves Distribution Considering Sweep and Lean Variations on the Blade Tip

Weimin Song, Yufei Zhang and Haixin Chen *

School of Aerospace Engineering, Tsinghua University, Beijing 100084, China; songweimin1989@163.com (W.S.); zhangyufei@tsinghua.edu.cn (Y.Z.)

* Correspondence: chenhaixin@tsinghua.edu.cn.

Abstract: This paper focuses on the design and optimization of the axial distribution of the circumferential groove casing treatment (CGCT). Effects of the axial location of multiple casing grooves on the flow structures are numerically studied. Sweep and lean variations are then introduced to the blade tip, and their influences on the grooves are discussed. The results show that the ability of the CGCT to relieve the blockage varies with the distribution of grooves, and the three-dimensional blading affects the performance of both the blade and the CGCT. Accordingly, a multi-objective optimization combining the CGCT design with the sweep and lean design is conducted. Objectives, including the total pressure ratio and the adiabatic efficiency, are set at the design point; meanwhile, the choking mass flow and the near-stall performance are constrained. The coupling between the CGCT and the blade is improved, which contributes to an optimal design point performance and a sufficient stall margin. The sweep and lean in the tip redistribute the spanwise and chordwise loading, which enhances the ability of the CGCT to improve the blade's performance. This work shows that the present CGCT-blade integrated optimization is a practical engineering strategy to develop the working capacity and efficiency of a compressor blade while achieving the stall margin extension.

Keywords: circumferential groove casing treatment; sweep and lean; CGCT-blade integrated optimization

---

1. Introduction

The compressor of modern aero-engines has a high stage loading to fulfill the requirement of a high thrust–weight ratio. A high stage loading increases the potential risks of tip stall in some rotors [1]. A casing treatment is usually adopted to enhance the rotor's stability when the stall margin of the tip-critical rotor is insufficient. Two types of casing treatment, namely, axial slots and circumferential grooves (see Figure 1), are commonly used. Although the circumferential groove casing treatment (CGCT) typically generates less stall margin improvement, it has a smaller efficiency penalty and greater mechanical integrity than axial slots [2,3]. Consequently, a CGCT is more practical than axial slots if the efficiency cannot be sacrificed.

The mechanism of a CGCT for extending the stall margin is relevant to the alteration of the flow structures near the blade tip. Rabe and Hah [4] conducted experimental and numerical investigations on a rotor with three different CGCT configurations. They found that the CGCT could reduce the incidence angle and suppress the flow separation. Müller et al. [5] simulated four CGCT configurations under both design and off-design conditions, and they noticed that the grooves segmented the tip leakage vortex (TLV) and alleviated the blockage, which delayed the spillage of the low-energy fluid from the leading edge of the adjacent blade. Sakuma et al. [6] studied the effects of a single circumferential groove with different axial locations on NASA Rotor 37. They concluded that the CGCT reduced both the tip loading and the momentum of the tip leakage flow.

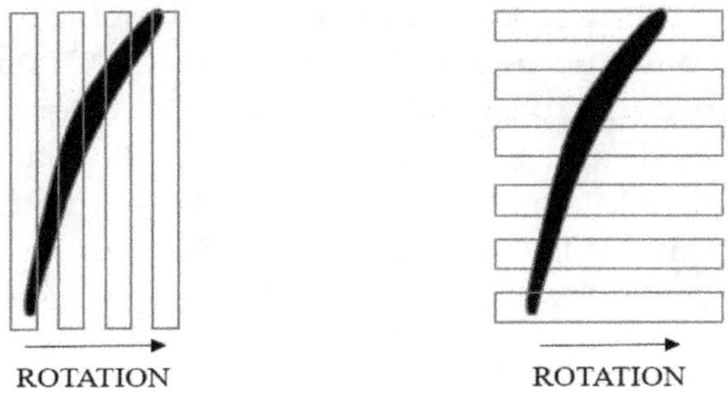

(a) (b)

Figure 1. Illustration of axial slots and circumferential grooves in a top view: (**a**) axial slots; (**b**) circumferential grooves.

The axial location of the circumferential grooves significantly affects the rotor's performance. However, the design guidelines for the CGCT's axial distribution have not been unified. Houghton and Day [7] examined the effects of a single circumferential groove at different axial locations. They reported that whereas the groove located at the 8% and 50% points of the axial tip chord length ($C_{ax,\ tip}$) achieved the maximum stall margin improvement, the groove near the leading edge, near the trailing edge, and at the 18% point of C_{ax} were inefficient. The research carried out by Du et al. [8] showed that the groove located between the 50% and 60% points of C_{ax} was the most effective for the stall margin extension; however, the groove located at approximately 20% point of C_{ax} was actually harmful to the rotor's stability. Choi [9] tested a single casing groove at different positions. The groove installed near the leading edge was found to be the most effective for expanding the stable operating range with a small efficiency loss. The numerical study of Mao et al. [10] shows that a single groove located within the front 40% point of C_{ax} had obvious benefits to the efficiency and stall margin. Although the axial location of the CGCT plays a significant role in affecting the rotor's performance and stability, there is not an acknowledged conclusion on this issue. Nevertheless, most of the previous studies used a single circumferential groove to simplify the analysis of the effects of the CGCT's axial location on a stall. Those researchers [4,11,12] who used multiple grooves to seek a larger stall margin improvement and higher peak efficiency usually chose to adjust the depth and width of each groove while fixing their axial locations. Because the effects of multiple grooves are by no means a simple accumulation of the individual grooves' effects [13], it is necessary to fill the gap between the single groove study and the practical application of multiple grooves. In this paper, efforts are made to design and optimize the axial distribution of multiple grooves, which also contributes to the understanding of the CGCT's influences on the stall.

The benefits of a CGCT are obtained by altering the tip flow structures. The flow structures near the blade tip are also affected by the 3D shape of the blade. Researchers [14–18] have found that sweeps and leans can suppress the shockwave-boundary layer interaction, redistribute the spanwise loading, reduce the accumulation of the low energy fluids near the casing and impact the strength of the TLV and the secondary flow. Thus, the three-dimensional (3D) blade design is supposed to interact with the CGCT design. Houghton and Day [7] proposed the idea combining the casing treatment design with the blade design to maximize the effectiveness of the casing treatment. Recently, Hah [19] studied the inner workings of axial slots and suggested that casing treatments could be designed by optimizing the casing groove shape and the blade loading near the tip. However, the topic of the CGCT-blade integrated design or optimization has been rarely studied in the open literature. It is noteworthy that only Kim et al. [11] and Goinis and Nickle [20] have done some pioneer work in this field. Moreover, with the increasing use of 3D blading techniques in modern compressor rotors and fans, the CGCT design should consider the interactions between the grooves and a swept or leaned blade. With this motivation, effects of the sweep and lean on the CGCT design are studied in this paper. The CGCT-blade integrated optimization is further performed to achieve an ideal CGCT-blade combination.

This paper is organized as follows. First, the numerical methods are described and validated. Second, the flow fields of the baseline blade with different CGCTs are simulated and compared. Third, the sweep and lean are introduced to the blade tip. The performance of the modified blade with the CGCT is predicted, and the influences of the 3D blading on the CGCT design are discussed. Finally, the CGCT-blade integrated optimization is conducted. By this process, an optimal CGCT-blade combination is obtained. The mechanism by which the sweep and lean variations affect the effectiveness of CGCT is revealed.

2. Parameterization of the Blade and CGCT

The compressor studied in this paper is the Notre Dame Transonic Axial Compressor (ND-TAC) that has been experimentally tested at the University of Notre Dame. The details of the experimental setup can be found in the papers of Cameron et al. [21] and Kelly et al. [22]. The parameters of the original rotor blade are listed in Table 1.

Table 1. Design parameters of the Notre Dame Transonic Axial Compressor (ND-TAC) rotor.

Symbol	Value	Symbol	Value
N_c (rpm)	14,686	Axial Chord (mm)	35.56
U_{tip} (m/s)	352	R_{casing} (mm)	228.6
Solidity	1.21	R_{hub} (mm)	171.45
N_b	20	τ (mm)	0.762

At the full rotational speed, the corrected choking mass flow of the rotor is approximately 10.1 kg/s and the design mass flow rate is 9.67 kg/s. In addition, the actual mean tip clearance used in the computational fluid dynamics (CFD) simulation is 0.6858 mm. The ND-TAC rotor is tip-critical and it exhibits a spike-type stall behavior [23]. The application of a CGCT to enhance the stability of the ND-TAC [24,25] has demonstrated the effectiveness of CGCTs in this rotor. However, the variation of the design point performance resulting from CGCT was not discussed, or of concern. This paper focuses on the design and optimization of the axial distribution of a CGCT with consideration of the 3D blading design, which aims at developing a novel CGCT-blade integrated optimization strategy to improve the design point performance while extending the stall margin.

In this paper, the sweep and lean are introduced to the blade tip to investigate the effects of 3D blading on the CGCT design. The sweep is defined as the shift of the airfoil section along the local chord line. The upstream shift is called the forward sweep, and the downstream shift is the backward sweep. The movement of the section that is perpendicular to the local chord line is defined as the lean. If the direction of the lean is towards the pressure side, the lean is defined as a positive lean; otherwise, it is a negative lean.

The blade is generated by a third-order B-spline interpolation based on fourteen airfoils stacked from the hub to the casing. The sweep and lean are manipulated by two third-order Bezier curves. The values of the sweep and lean are defined by the percentage of the baseline-blade mean chord length L and the baseline-blade mean leading-trailing edge deviation $\Delta(r{\cdot}\theta)$, respectively. Initially, the leading edge (LE) and corresponding trailing edge (TE) for each baseline 2D airfoil are expressed in cylindrical coordinates $(r,\ \theta, z)$. Then, L and $\Delta(r{\cdot}\theta)$ are calculated by Equations (1) and (2), respectively:

$$L = \sqrt{(z_{TE} - z_{LE})^2 + [(r{\cdot}\theta)_{TE}^2 - (r{\cdot}\theta)_{LE}^2]} \tag{1}$$

$$\Delta(r{\cdot}\theta) = |(r{\cdot}\theta)_{TE} - (r{\cdot}\theta)_{LE}| \tag{2}$$

Next, the values of L and $\Delta(r\cdot\theta)$ of all of the 2D airfoils are averaged to obtain the mean value for the baseline blade. The above definitions of the sweep and lean adopted in this paper are essentially consistent with the conventional method that defines the sweep and lean by the absolute axial displacement and the circumferential angle. Still, the geometry changes defined by the relative variation can be more intuitive. Figure 2a,b illustrate the distributions of the sweep and lean along the blade span. The abscissa is the percentage of the mean chord length and the mean deviation. Both the sweep and the lean are regulated by five equally distributed control points from root to tip. In this study, only the control point at the blade tip is variable, which causes the 3D shape of the last 35% of the blade span to be affected by the sweep and lean variations.

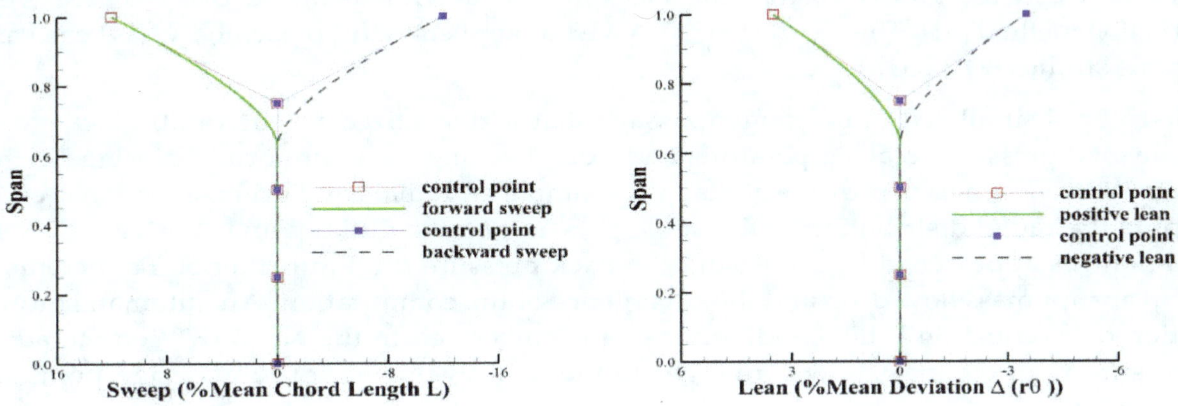

(a) (b)

Figure 2. Illustration of the parameterization of the sweep and lean: (**a**) The sweep; (**b**) the lean.

With respect to the CGCT design, the groove number, width and depth of the grooves are fixed. Only the axial coverage of the CGCT is variable. The parameterization of the CGCT is shown in Figure 3. The groove number, along with the depth and width of each groove are given in the right side of the figure. The distance between any two adjacent grooves remains equal, to facilitate the manufacturing of the CGCT. In this figure, the abbreviation of "TG" means the tip gap. The depth of each groove is 4.04 times the tip gap size. The width of each groove is approximately equal to its depth. The dimension of the casing grooves refers to the previous research [24,25]. A few CFD tests, which are not provided in this paper, were also carried out to ensure the rationality of the parameter selection. Note that the width and depth design are not the interest of this paper; thus, only the axial coverage of the overall CGCT configuration installed on the casing can be controlled by two variables k and s, which correspond to the distance between the first groove and the blade's leading edge and the distance between two adjacent grooves, respectively. The variables k and s are non-dimensionalized by the axial chord length $C_{ax,\ tip}$.

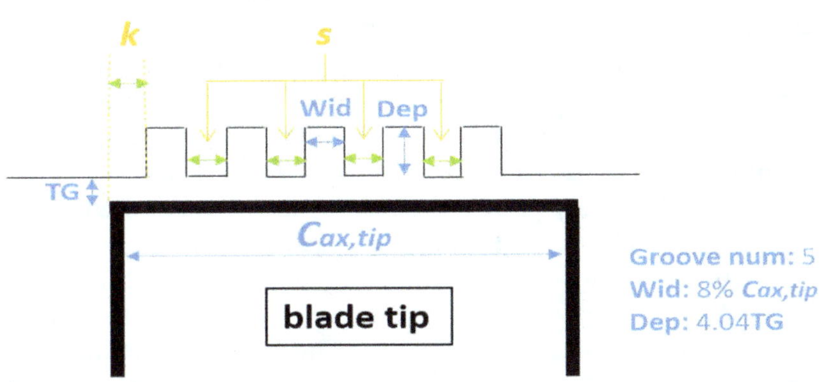

Figure 3. Illustration of the circumferential groove casing treatment (CGCT) design parameters.

3. Numerical Methods

An in-house CFD code called NSAWET (Navier-Stokes Analysis based on Window-Embedment Technique) is employed to solve the 3D compressible Reynold averaged Navier-Stokes (RANS) equation. An improved version of the HLL-Riemann solver by restoring the contact surface (HLLC) is selected to compute the convective flux [26]. The third-order monotone upstream-centred scheme for conservation laws (MUSCL) [27] with a Van Albada limiter [28] is adopted to ensure accuracy on non-uniform and skew cells. The k-ω shear stress transport (SST) turbulence model [29] is used to compute the turbulent viscosity. Moreover, the time marching is achieved by the Lower-Upper Symmetric-Gauss-Seidel (LU-SGS) implicit algorithm [30]. When the CGCT is adopted, the flux exchange between the groove-casing interfaces are dealt with using an overlap-area weighted reconstruction method [31]. The accuracy of NSAWET as applied to turbomachinery has been validated by previous studies [24,32,33].

The rotor is simulated in the single passage that is discretized by the multi-block structured mesh. The total pressure, total temperature, and axial flow angles are uniformly specified at the inlet boundary. With a given back pressure at the hub, a simplified radial equilibrium equation is employed to compute the radial distribution of the static pressure on the outlet plane. With a steady RANS, the stall point is approached by increasing the back pressure until a further 50 Pa increment will lead to an abrupt mass flow drop and the divergence of the computation. An automatic proportion-integration-differentiation (PID) conditioner [34] is embedded in the NSAWET code to adjust the back pressure until the mass flow converges to the target value at both the design point and the near-stall point.

The grid convergence study is conducted with three sets of meshes, corresponding to the coarse mesh, medium mesh, and fine mesh. The total grid numbers of the three sets of meshes are 73.6 million, 1.45 million, and 2.96 million, respectively. Figure 4 shows the details of the medium grid. The grid of circumferential grooves is shown in red, and the grid in the passage is shown in black. This set of mesh has 65 cells distributed along the whole blade span, and another 25 cells inside the tip gap in the radial direction. The wall unit y^+ of the first grid layer that is normal to the wall boundary, is strictly maintained below 1, in accordance with the requirements of the SST turbulence model. Each circumferential groove is discretized by a straight H block that has 40, 36, and 50 cells in the stream-, span- and pith-wise directions, respectively.

Figure 5 gives a comparison between the CFD results and the experimental data. The uncertainty in the measured total pressure ratio ($P_{t0,\ rat}$), the measured total temperature ratio ($T_{t0,\ rat}$), and the measured mass flow are 0.5%, 0.5%, and 1.0%, respectively. Figure 5a,b show the speed curve predicted by the three sets of meshes. All of the three meshes are capable of achieving a good agreement with the experimental data at the design point. The coarse mesh predicts a higher stall mass flow than the experimental data, due to the very sparse grid distribution in the passage. The speed curve obtained by the medium mesh is very close to the curve obtained by the fine mesh over the entire operating mass flow. The differences in the $P_{t0,\ rat}$ at the design point and the near-stall point between these two meshes are less than 0.15%, and this value reduces to 0.04% for the $T_{t0,\ rat}$ differences. The results of the medium and fine mesh show that the deviation of $P_{t0,\ rat}$ between CFD and experimental data occurs near the surge boundary. The present computation also slightly underestimates $T_{t0,\ rat}$. These numerical errors may be caused by the drawbacks of the turbulence model and the insufficient ability of the steady RANS for precisely capturing the large separation under near-stall conditions. Despite the numerical errors, the whole tendency of the speed curve predicted by the medium mesh and fine mesh coincides with the experimental data from choke to stall. In Figure 5c,d, the spanwise distributions of $P_{t0,\ rat}$ and $T_{t0,\ rat}$ predicted by the medium mesh at the downstream plane also show an acceptable agreement with the experimental data. To sum up, CFD results obtained by the medium mesh are close to the experimental data at the design point and the near-stall point. Because the optimization objectives are set at the design point, and the goal of the optimization is to gain a relative improvement based on the baseline blade, it is suitable to use the medium mesh to undertake CFD investigations to achieve a favorable balance between the accuracy and the computational cost.

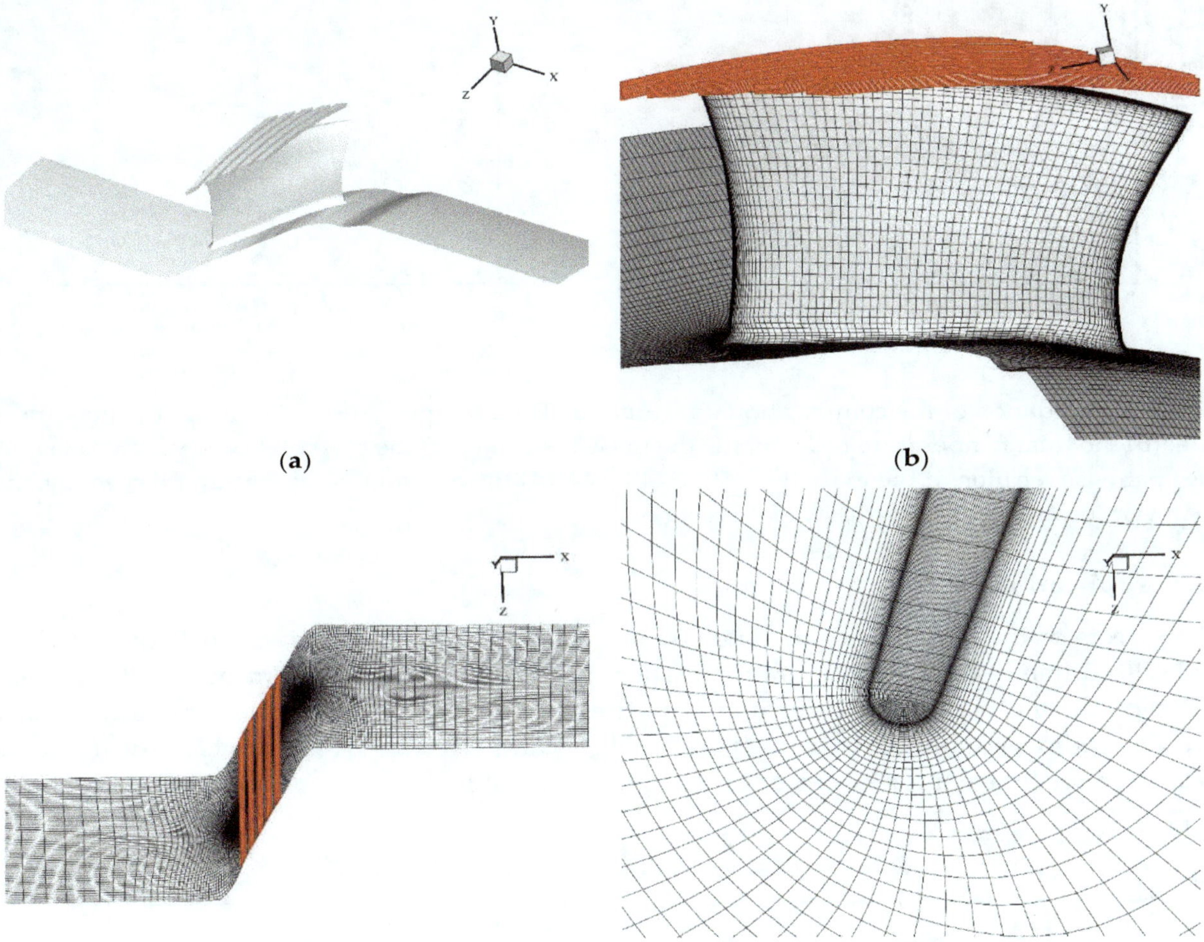

Figure 4. Structured mesh in the computational domain. (**a**) The geometry of the baseline rotor blade with five circumferential grooves installed on the casing; (**b**) the side view of the grid distribution on the blade surface; (**c**) the top view of the grid distribution on the casing; (**d**) the grid distribution near the leading edge of the blade tip.

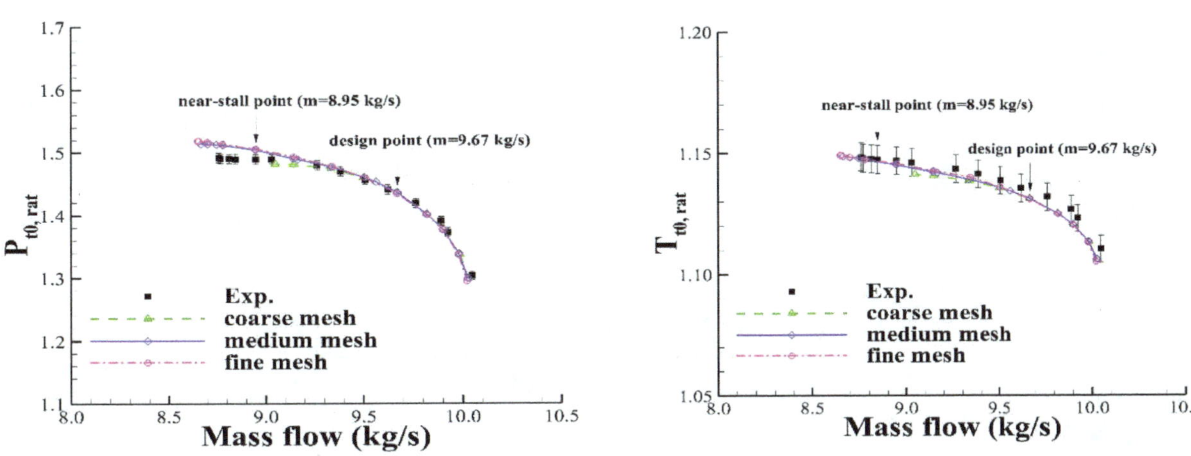

Figure 5. *Cont.*

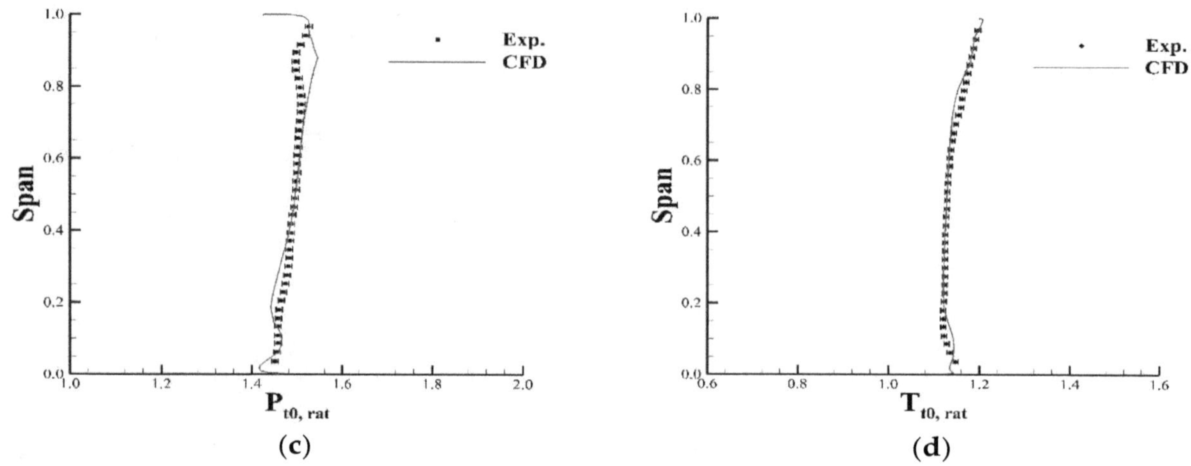

Figure 5. Validation of the computational results. (**a**) The total pressure ratio versus the mass flow rate; (**b**) the total temperature ratio versus the mass flow rate; (**c**) the exit total pressure ratio versus the spanwise location at the mass flow of 9.344 kg/s; (**d**) the exit total temperature ratio versus the spanwise location at the mass flow of 9.344 kg/s.

4. CGCT Application Tests

The effects of the CGCT and its axial coverage are first analyzed on the baseline blade that is installed with different CGCT configurations. Then, the sweep and lean are introduced to the blade tip. Influences of 3D blading on the performance of the CGCT are discussed. The mass flow rates of the design point and near-stall point are taken as 9.67 kg/s and 8.95 kg/s, respectively. The abbreviations adopted hereafter for each type of rotor are explained in Table 2.

Table 2. Definitions of the individual abbreviations.

Name	Definition
SW_*	solid casing rotor, including SW_0, SW_FS, SW_BS, SW_PL and SW_NL
CT_*	CGCT configuration, including CT_a, CT_b, CT_c, CT_d and CT_e
0	baseline ND-TAC rotor
FS	re-stacked rotor with forward swept tip
BS	re-stacked rotor with backward swept tip
PL	re-stacked rotor with positive leaned tip
NL	re-stacked rotor with negative leaned tip

4.1. Effects of the CGCT on the Performance of the Baseline Blade

By increasing the parameter k from 0 to 40% of $C_{ax,tip}$, five CGCT configurations with different axial coverage are generated. The parameter s is fixed as 5% of $C_{ax,tip}$ in all five of the CGCT configurations. Table 3 lists the axial coverages of these five CGCT configurations which are installed on the baseline blade. From CT_a to CT_e, the position of the CGCT is moved from the leading edge area to the trailing edge area.

Table 3. Five circumferential groove casing treatment configurations for testing.

Name	Axial Coverage (% $C_{ax,tip}$)
CT_a	0–60%
CT_b	10–70%
CT_c	20–80%
CT_d	30–90%
CT_e	40–100%

The speed curves of the ND-TAC rotor with and without installing the CGCT are shown in Figure 6. At both the design point and the near-stall point, all of the CGCT configurations improve both the total pressure ratio and the adiabatic efficiency (η). The stall mass flows of all of the grooved casing rotor are lower than that of the solid casing rotor. Specifically, the zoom-in of the stall point in Figure 6a shows that the stall mass flows of CT_a, CT_b and CT_c are very close. When the CGCT is moved to the trailing edge area, the ability of the CGCT to delay a stall dramatically decreases, which is reflected by the increased stall mass flows of CT_d and CT_e.

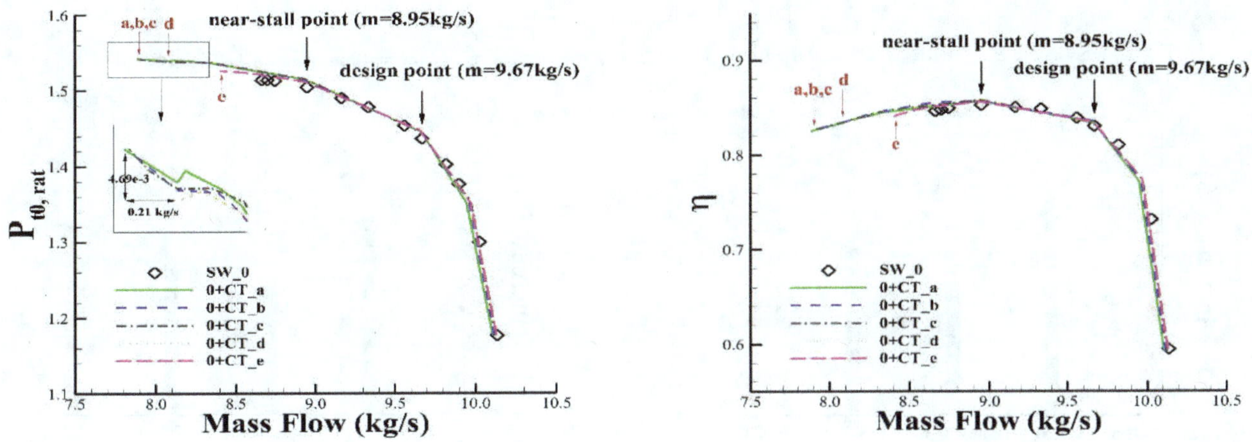

(a) (b)

Figure 6. Speed curves of the baseline blade installed with five CGCT configurations. (**a**) The total pressure ratio versus the mass flow rate; (**b**) the adiabatic efficiency versus the mass flow rate.

At the design point, the spanwise distribution of the total pressure ratio, as well as the adiabatic efficiency at 25% of the tip chord length downstream of the trailing edge are shown in Figure 7. In Figure 7a, the CGCT decreases the total pressure ratio above 95% of the span but enhances the working capacity of the rest of the span, which demonstrates that the CGCT redistributes the spanwise loading. In Figure 7b, the increase of the efficiency concentrated in the blade tip indicates that the CGCT reduces the aerodynamic loss in the tip region. The baseline rotor installed with CT_e is found to have the highest total pressure ratio below 80% of the span and the highest efficiency above 90% of the span.

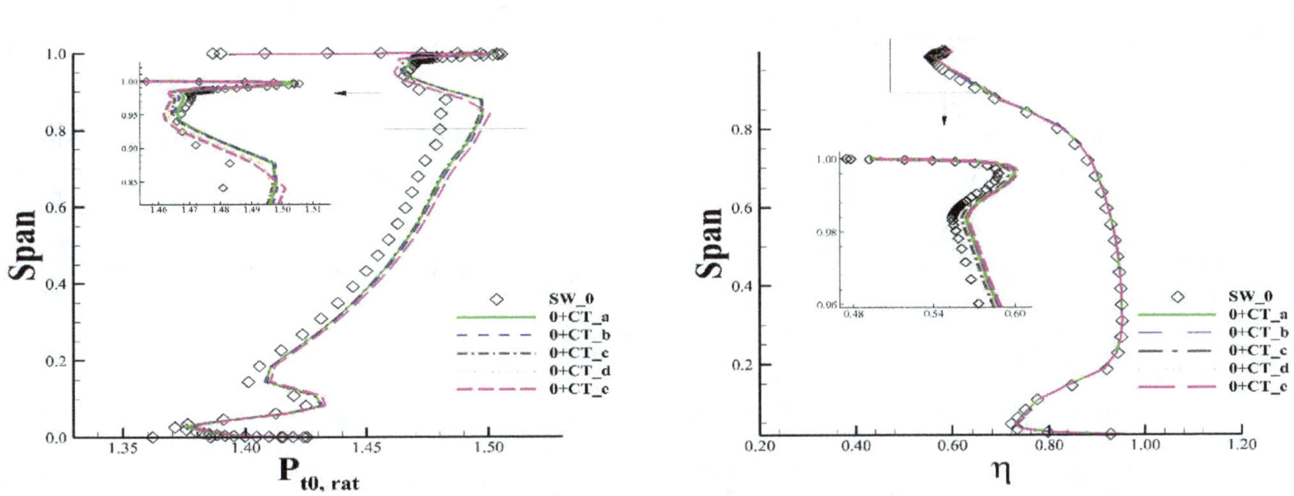

(a) (b)

Figure 7. Spanwise distribution of (**a**) the total pressure ratio, and (**b**) the adiabatic efficiency at the design point.

The design point performance and the stall mass flow of the ND-TAC rotor with different CGCTs installed are summarized in Table 4. The performance improvement and the decrease of the stall mass flow due to the installation of different CGCTs are shown in Figure 8. Although the grooved casing rotor 0 + CT_e has the most significant improvement in the design performance, the stall mass flow of 0 + CT_e is obviously higher than the corresponding stall mass flows of the other CGCT cases. The greatest difference between the five grooved casing rotors is more than 0.1% for the design point performance, and is more than 6% for the stall mass flow. In the present CGCT tests, the optimal performance at the design point and near-stall point are not obtained by the same CGCT configuration. The effectiveness of the CGCT varies with its axial coverage.

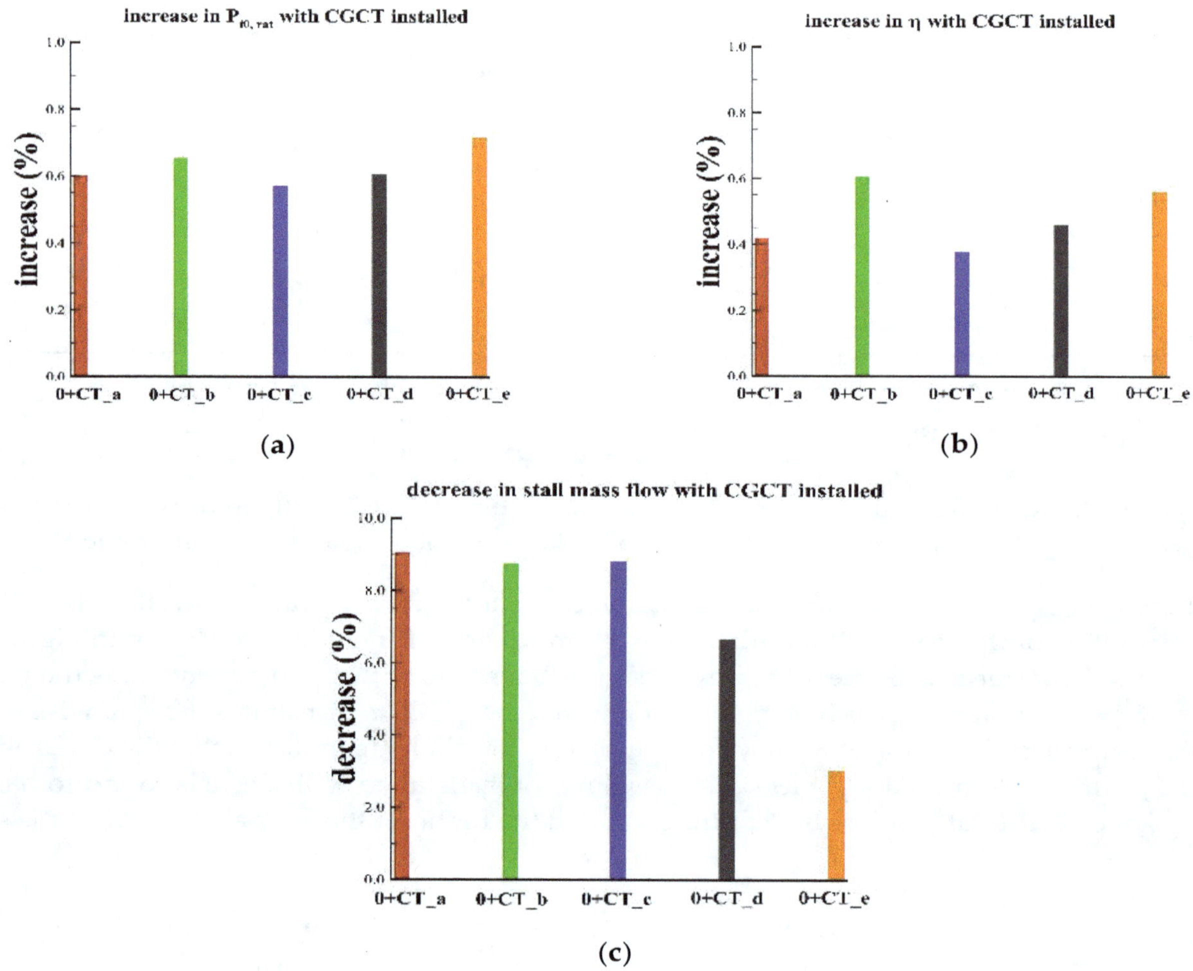

Figure 8. Variation of the (**a**) total pressure ratio, (**b**) the adiabatic efficiency, and (**c**) the stall mass flow when different CGCTs are installed on the baseline blade at the design point.

Table 4. Effects of the CGCT on the rotor's design point performance ($P_{t0,\ rat}$, η) and the stall mass flow rate (M_s).

Name	$P_{t0,\ rat}$	Relative to SW_0	η	Relative to SW_0	$M_s\ (kg/s)$	Relative to SW_0
SW_0	1.43655	-	0.83008	-	8.666	-
0 + CT_a	1.44519	0.602%	0.83355	0.418%	7.882	−9.044%
0 + CT_b	1.44596	0.655%	0.83511	0.606%	7.907	−8.759%
0 + CT_c	1.44475	0.571%	0.83323	0.379%	7.901	−8.820%
0 + CT_d	1.44525	0.606%	0.83391	0.461%	8.089	−6.663%
0 + CT_e	1.44685	0.717%	0.83474	0.561%	8.403	−3.033%

As indicated by Figure 8, the differences between CT_c and CT_e in the design point performance and the stall mass flow are the most obvious. Thus, the flow fields of 0 + CT_c and 0 + CT_e are extracted for comparison and analysis.

In Figure 9, the blockage at 93% of the span formed by the TLV after passing the shockwave is characterized by the relative Mach number contours. Common effects of CT_c and CT_e on the flow fields in the tip region are alleviations of blockages at both the design point and the near-stall point. When the mass flow decreases with the variation of the operating point, the behaviors of the low Mach number region in 0 + CT_c are basically unchanged (see Figure 9c,d). By contrast, from the design point to the near-stall point, the low Mach number area in 0 + CT_e expands rapidly, and the average speed in this blockage region is also reduced (see Figure 9e,f). The rapid growth of the blockage in 0 + CT_e implies that CT_e that locates between 40% of $C_{ax,tip}$ and 100% of $C_{ax,tip}$ is less effective at delaying the stall than CT_c that locates between 20% of $C_{ax,tip}$ and 80% of $C_{ax,tip}$. However, at the design point, the blockage in 0 + CT_e is smaller than that in 0 + CT_c (see Figure 9c,e), which indicates that CT_e is more helpful at relieving the blockage under a high-mass-flow condition than CT_c is. Therefore, the Mach number contours plotted in Figure 9 demonstrates that 0 + CT_e has a stronger working ability at the design point, but a smaller stall margin than 0 + CT_c has. In addition, the blockage that originates downstream of the shockwave and then extends to the aft-part of the pressure side (PS) is supposed to affect the blade's aft-loading. If the blade's chordwise loading is redistributed due to the 3D blading, the effectiveness of CGCT can be changed.

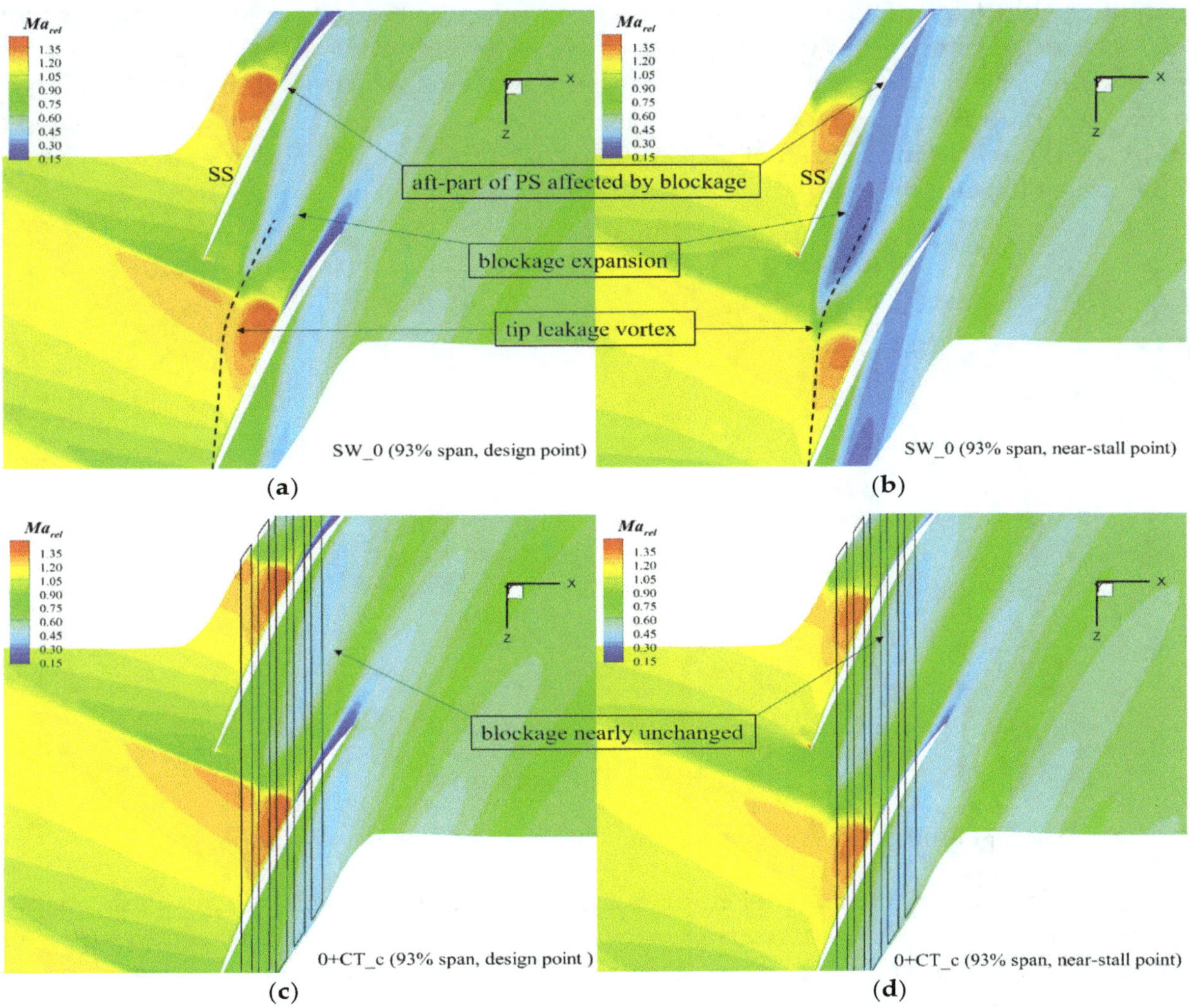

Figure 9. *Cont.*

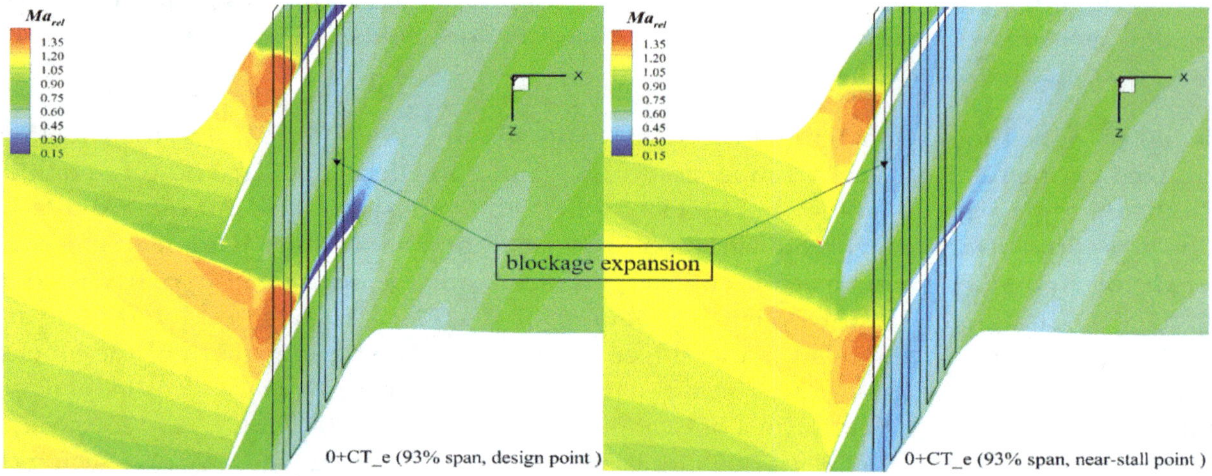

(e) (f)

Figure 9. The relative Mach number contours on the S1 plane for (**a**) SW_0 at the design point; (**b**) SW_0 at the near-stall point; (**c**) 0 + CT_c at the design point; (**d**) 0 + CT_c at the near-stall point, (**e**) 0 + CT_e at the design point, and (**f**) 0 + CT_e at the near-stall point.

Figure 10 quantitatively compares the mass flow distribution along the span at the near-stall condition. It shows that both CT_c and CT_e increase the mass flow near the blade tip due to the reduction of the blockage. The mass flow of 0 + CT_c near the blade tip region is higher than that of 0 + CT_e, which demonstrates the conclusion drawn from Figure 9 that CT_c is more effective to alleviate the blockage at a low-mass-flow condition. It explains why 0 + CT_c presents a larger stall margin than 0 and 0 + CT_e. Moreover, Figure 10 also shows that the mass flow near the blade root is reduced by CGCT, which indicates that CGCT not only alters the tip flow field, but also changes the flow behaviors at the non-tip region.

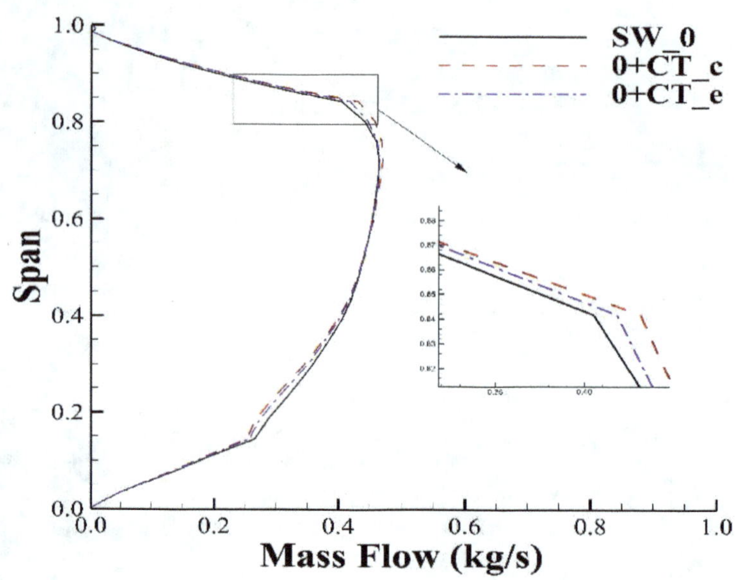

Figure 10. The spanwise mass flow distribution at the near-stall point.

Along the trajectory of the TLV, seven planes perpendicular to the Z-axis are extracted, and the entropy contours on them at the design point are shown in Figure 11. The yellow square in the figure encloses the cross-section where the core of the TLV interacts with the shockwave. Apparently, the entropy in this area of 0 + CT_e is less than the corresponding entropy of 0 + CT_c, which indicates that the strength of the shock-TLV interaction in 0 + CT_e is weaker than that in 0 + CT_c. Consequently, the high-entropy region marked by the dashed black square in the low-speed area of 0 + CT_e is smaller. The variations of the entropy distributions demonstrate that at the design point, CT_e is more effective at increasing the efficiency than CT_c is.

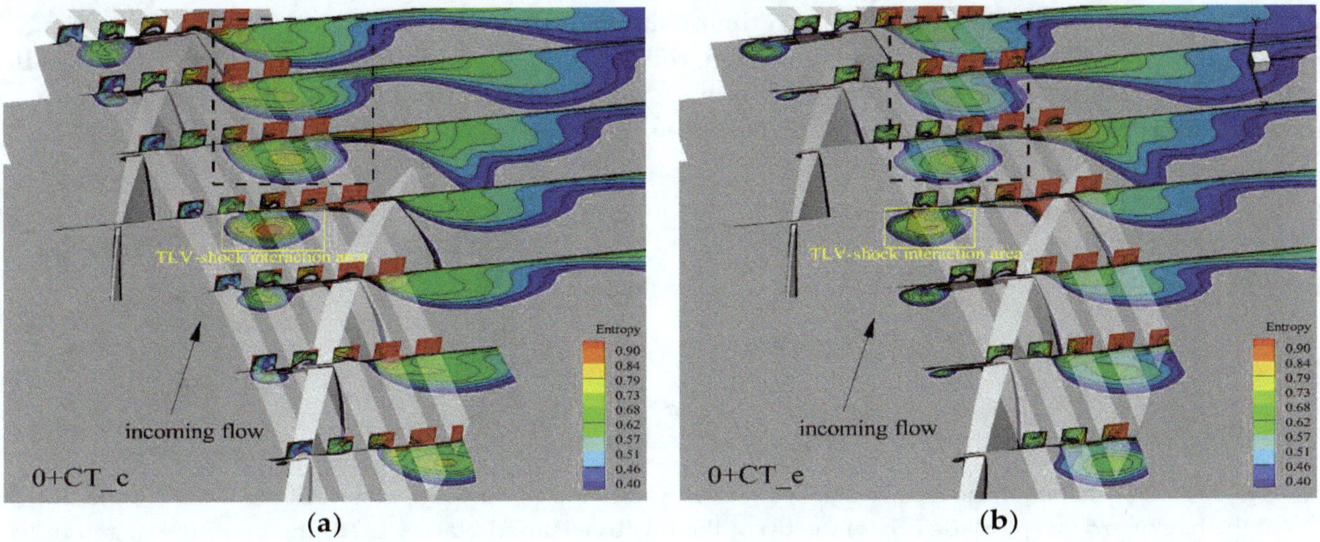

Figure 11. The entropy contours on the Z-planes at the design point in (**a**) 0 + CT_c and (**b**) 0 + CT_e.

The distributions of the relative total pressure ($P_{t0,\ rel}$), non-dimensionalized by the inlet total pressure in the tip clearance, are plotted in Figure 12. Figure 12a shows that the improvement of $P_{t0,\ rel}$ induced by CT_e after the 40% point of $C_{ax,\ tip}$ is higher than the corresponding improvement for CT_c at the design point; however, at the near-stall point, Figure 12b shows that CT_c is more effective at increasing $P_{t0,\ rel}$ after the 30% point of $C_{ax,\ tip}$ than CT_e is. The variation of $P_{t0,\ rel}$ in the tip clearance proves that the effectiveness of different CGCT configurations varies with the operating conditions.

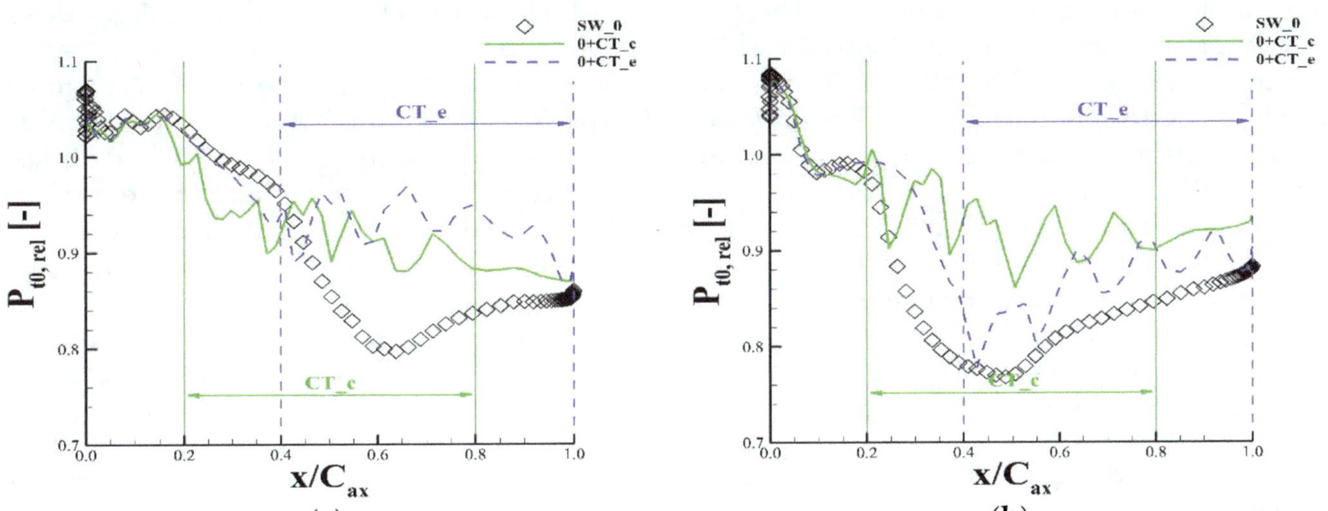

Figure 12. Distribution of the non-dimensional relative total pressure in the tip gap at (**a**) the design point and (**b**) the near-stall point.

4.2. Effects of the Sweep and Lean on the Effectiveness of the CGCT

As shown in Table 4, in terms of both the design point performance and the stall margin, the performance of CT_b is more satisfactory than the other CGCT configurations. In this section, CT_b is applied to four re-stacked rotor blades named FS, BS, PL, and NL to test the effects of the sweep and lean. The sweep and lean distributions adopted in this study are given in Figure 2. The 3D shapes of the four blades are plotted in Figure 13. Compared with the baseline ND-TAC blade, the control point of the sweep at 100% span is forward and backward shifted by 12% of the mean chord length to form the blades FS and BS, respectively. Moreover, the control point of the lean at 100% of the span is positively and negatively shifted by 3.5% of the mean leading-trailing edge deviation to produce the blades PL and NL, respectively. Although the position of the blade tip airfoil in the flow passage is changed in accordance with the sweep and lean, the geometry parameters of the 2D airfoil, including the thickness, camber and chord length, are unchanged. Additionally, the relative position of the circumferential grooves to the blade tip is fixed, regardless of how the blade is re-stacked.

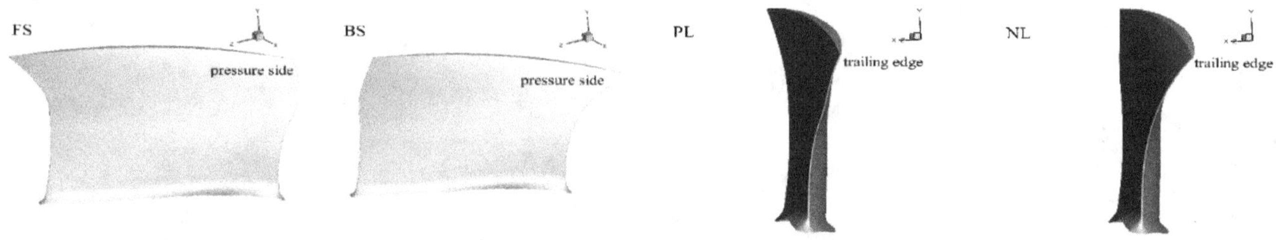

(a) (b) (c) (d)

Figure 13. 3D shapes of the four re-stacked blades: (**a**) the tip of the forward swept blade FS; (**b**) the tip of the backward swept blade BS; (**c**) the tip of the positive leaned blade PL; (**d**) the tip of the negative leaned blade NL.

The CFD results show that when CT_b is installed, all four of the re-stacked blades are able to operate at the mass flow rate of 8.10 kg/s, which is approximately 0.60 kg/s lower than the stall mass flow rate of the baseline blade. Therefore, the stall margin of the re-stacked blades with CT_b installed is sufficient. Consequently, we focus on the performance of these blades instead of the stall margin.

Figure 14 shows the performance variations of re-stacked blades at both the design point and the near-stall point induced by CT_b. The forward and backward sweep result in a slight design performance deterioration, but the backward sweep leads to greater loss in $P_{t0,\ rat}$ and η than the forward sweep (see symbol Δ and ∇ in Figure 14a). After installing CT_b on the swept rotor's casing, the performance is improved significantly. The performance improvement induced by CT_b in blade FS is more than the corresponding improvement in the baseline blade 0 and blade BS (see Figure 14b). In addition, after introducing the sweep to the tip, the improvement of the near-stall performance induced by CT_b in FS and BS is more than twice the corresponding improvement in blade 0 (see Figure 14d). The positive-leaned blade PL has a slightly higher design performance than blade 0. By contrast, the negative lean causes a performance deterioration (see symbol $\diamond$ and $\circ$ in Figure 14a). The performance improvement after installing CT_b in blade PL is less than the corresponding improvement in blade 0; however, the performance improvement of NL induced by CT_b is more than what we find in blade 0 (see Figure 14b,d).

From the above quantitative comparisons, it can be concluded that the sweep and lean introduced to the blade tip affect both the working capacity and the efficiency of the blade. The sweep and negative lean slightly decrease the solid casing rotor's performance. Moreover, the effectiveness of the CGCT is also altered by the sweep and lean. In these tests, the effectiveness of the CGCT at improving the solid casing rotor's performance is magnified by the forward sweep and negative lean in the tip.

Figure 15 shows the variations of the spanwise total pressure ratio distribution caused by the sweep and lean in the blade tip. In Figure 15a, whereas the forward sweep in the tip reduces the working ability of the blade tip, the backward sweep reduces the working ability of the rest of the blade span without changing the tip performance. When CT_b is installed on the casing, the total pressure ratio of the forward swept blade FS is greatly improved from the root to 90% of the span.

The performance improvement is even larger than the corresponding improvement to the baseline blade with CT_b installed; however, for the backward swept blade BS, the performance improvement induced by CT_b is less than the corresponding improvement in FS at the same spanwise position. In Figure 15b, the negative lean reduces the total pressure ratio along the entire span, whereas the positive lean slightly increases the total pressure ratio in the tip. The differences between the grooved leaned blades PL + CT_b and NL_CT_b are not as obvious as the differences between the grooved swept blades FS + CT_b and BS + CT_b.

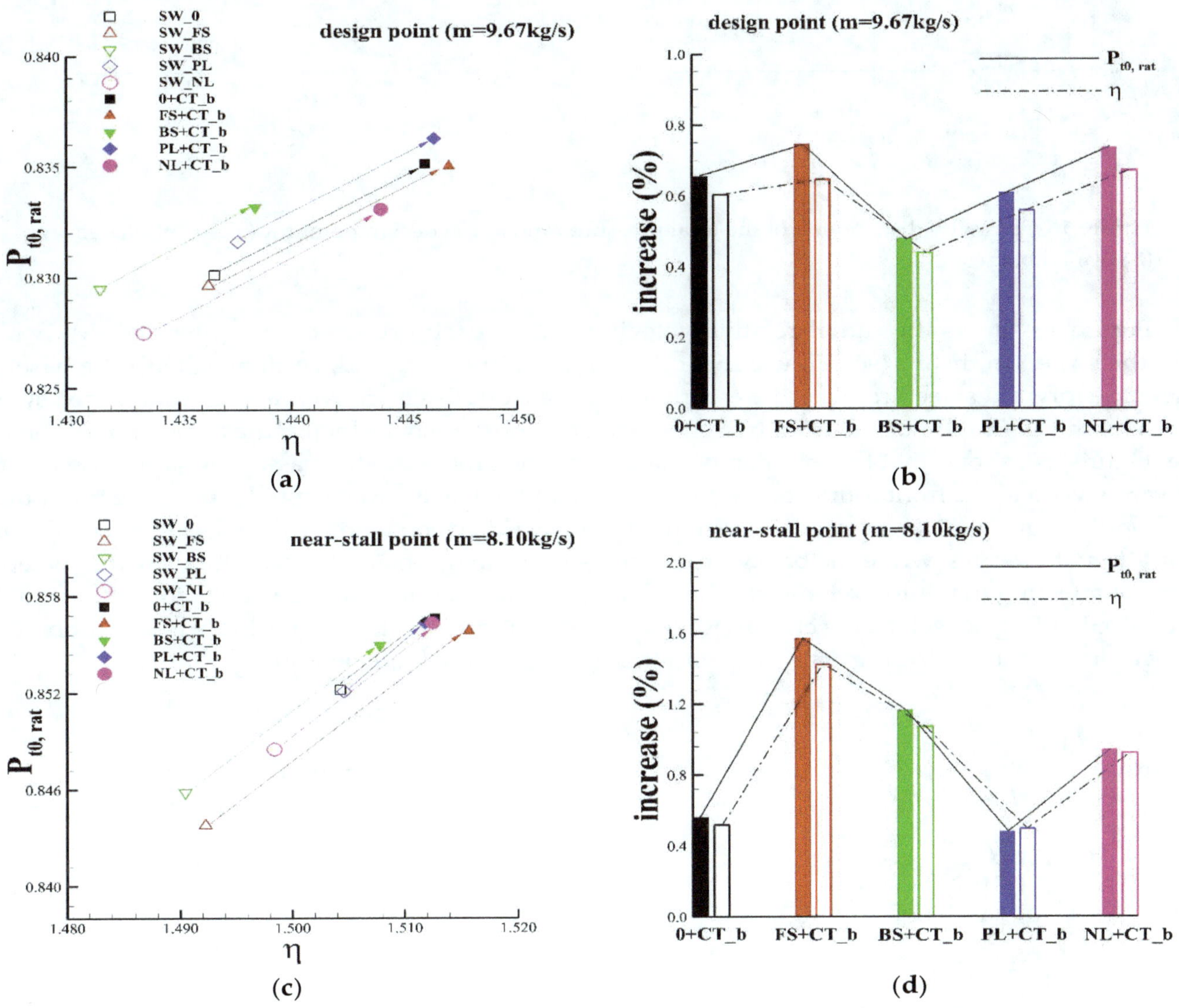

Figure 14. Performance variations of re-stacked blades induced by CT_b at the design point and the near-stall point: (**a**) $P_{t0,\text{ rat}}$ and η at the design point; (**b**) the improvements in $P_{t0,\text{ rat}}$ and η of the grooved casing rotors relative to the respective solid casing rotors at the design point; (**c**) $P_{t0,\text{ rat}}$ and η at the near-stall point; (**d**) the improvements in $P_{t0,\text{ rat}}$ and η of the grooved casing rotors relative to the respective solid casing rotors at the near-stall point.

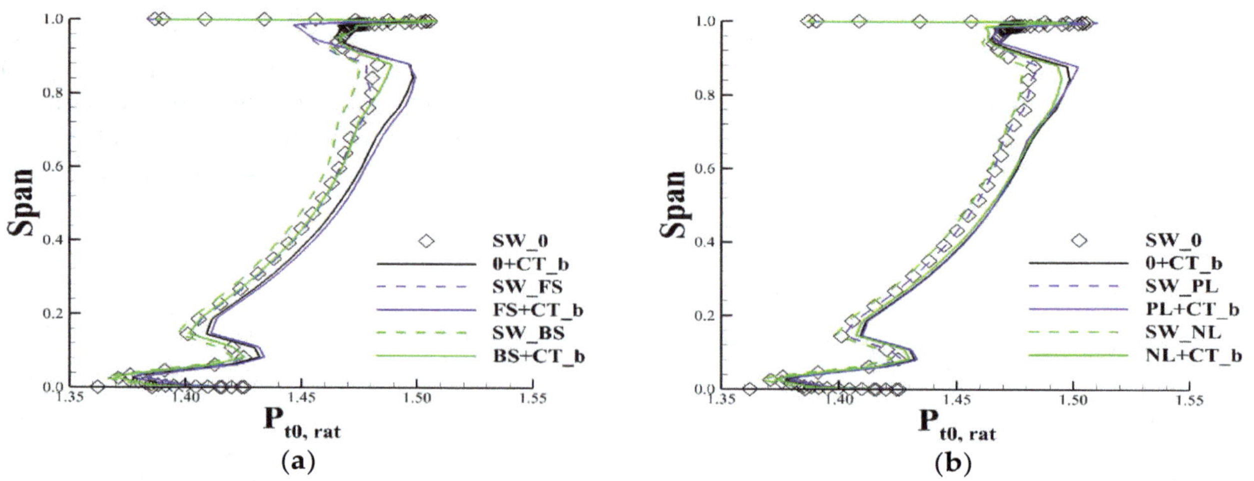

Figure 15. Spanwise distribution of the total pressure ratio at the design point for (**a**) swept blades and (**b**) leaned blades.

Figure 16 shows the radial relative velocities W_r at each groove-casing interface. W_r varies with the sweep in the tip (see Figure 16a–c). As shown in Figure 16d, compared with the baseline blade, the forward sweep reduces the aspiration of the front three grooves, and the backward sweep increases the aspiration of the front two grooves while obviously reducing the aspiration of the last one. Figure 16e) shows that the injection of fluids from the grooves to the passage is suppressed by the forward sweep in the front four grooves. By contrast, in the backward swept blade BS, the injection of the front four grooves is enhanced. To sum up, the forward sweep tends to decrease the injection and aspiration of grooves, while the backward sweep shows an opposite trend to the forward sweep in terms of affecting the fluid exchange at the groove–casing interfaces. The sweep is found to shift the spanwise loading as well as the chordwise loading [35]; therefore, the aspiration and injection of the grooves in the corresponding area vary with the changes of the blade loading.

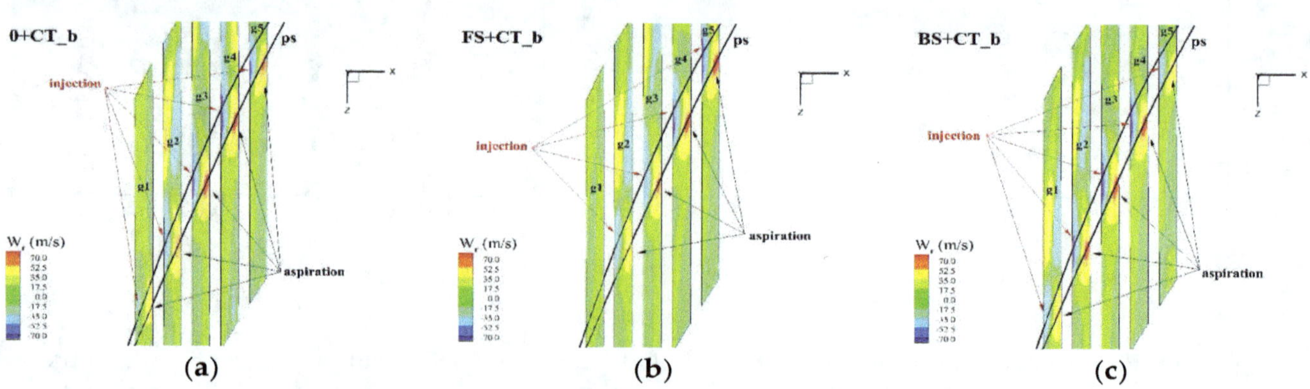

Figure 16. *Cont.*

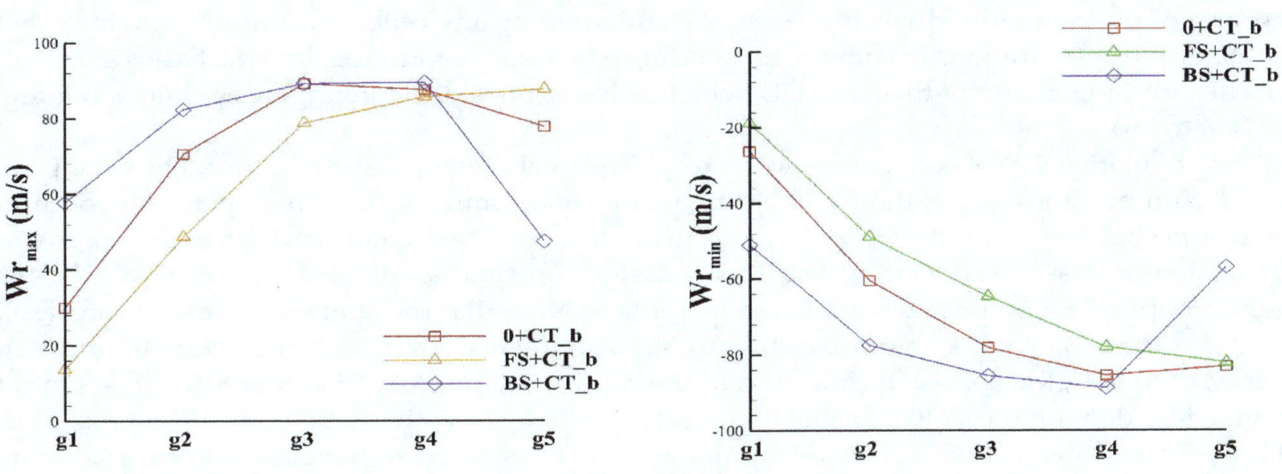

(d) (e)

Figure 16. Comparison of the radial relative velocity at the casing-groove interfaces (**a**) 0 + CT_b; (**b**) FS + CT_b; (**c**) BS + CT_b; (**d**) maximal radial velocity at each interface (aspiration); (**e**) minimal radial velocity at each interface (injection).

Although the relative position between the circumferential grooves and the tip airfoil is unchanged when the sweep and lean are introduced, the interactions between the blade and CGCT are altered, which impacts the performance of the blade and the effectiveness of the grooves. The effects of the 3D blading on the CGCT suggest that it is promising to achieve better design point performance while ensuring an adequate stable operating range by considering 3D blading in the process of the CGCT design.

5. CGCT-Blade Integrated Optimization

Because the CGCT-blade interactions are found to play a significant role in altering the rotor's performance and stability, we undertake an integrated optimization of the CGCT-blade combination. The goal is to verify the effectiveness of the CGCT design strategy that introduces 3D blading to the blade tip during the optimization of the grooves' axial distribution.

5.1. Optimization Setup

The sweep and lean of the blade tip and the axial distribution of the circumferential grooves are optimized together. The total number of design variables is four. The variation ranges of the sweep and lean at the blade tip are [−12%, +12%] and [−3.5%, +3.5%], respectively. The ranges of k and s controlling the grooves' location are [−0.8%, +20%] and [1.6%, 12%], respectively. A negative value of k means that the first groove is upstream of the leading edge. The variation range of s equals 0.2–1.5 times the width of the groove. Since the width of each groove is 8% of $C_{ax,\ tip}$, the start position of groove N can be expressed by Equation (3):

$$L_{start} = k + (8\% + s)\cdot(N-1) \tag{3}$$

where $N \in \{1, 2, 3, 4, 5\}$.

A hybrid aerodynamic optimization algorithm HSADE [36] is used in this paper to search a Pareto Front. HSADE combines the differential evolution (DE) algorithm with the radial basis function (RBF) response surface. Specifically, DE is a stochastic evolutionary optimization algorithm, well-known for its robustness, its strong global searching ability, and its suitability for high-dimensional problems [36]. RBF evaluates the similarity between the candidate design variables and known samples in the design space to predict their corresponding objective functions by interpolation. The results are used to construct the response surface for searching for potential optimal individuals. The Pareto Front is a group of optimal individuals that are not dominated by any other individuals obtained by the multi-objective optimization. To sum up, embedding RBF response surfaces into the basic DE improves the local searching ability of the basic DE, which helps to boost the convergence while obtaining an ideal Pareto Front.

The population size of each generation is nine. The total pressure ratio and the adiabatic efficiency at the design point are set as the two objectives for the optimization. Three operating conditions, namely, the choking point, design point, and near-stall point, are computed for each design in the optimization process. The operating conditions, design constraints, and objectives of the CGCT-blade integrated optimization problem are listed in Table 5. Note that the stall mass flow of the baseline ND-TAC rotor is 8.67 kg/s, and a baseline CGCT can decrease the stall mass flow by more than 0.7 kg/s; therefore, the near-stall condition of the rotor with the CGCT is set at 8.10 kg/s. Once the optimized blade with the CGCT is able to operate at 8.10 kg/s with a total pressure ratio of at least 1.511, the newly obtained CGCT-blade combination is considered to have a sufficient stall margin. The purpose of the CGCT-blade optimization is to obtain improvements to $P_{t0,\,rat}$ and η at the design point, as well as a favorable stall margin.

Table 5. Definition of the CGCT-blade integrated optimization problem.

Working Conditions	Constraints	Objectives
choking point	$10.08\ \text{kg/s} \leq m \leq 10.34\ \text{kg/s}$	-
design point (m = 9.67 kg/s)	-	max $P_{t0,\,rat}$ and max η
near-stall point (m = 8.10 kg/s)	$P_{t0,\,rat} \geq 1.511$	-

5.2. Results and Analysis

The optimization ran for 3400 CPU hours on a cluster with 300 cores in total, and it converged after 32 generations. The optimization histories of the two objectives are plotted in Figure 17. The performance differences between the individuals in the last few generations are small, and the objective values are concentrated on a narrow strip, which indicates that the optimization has converged.

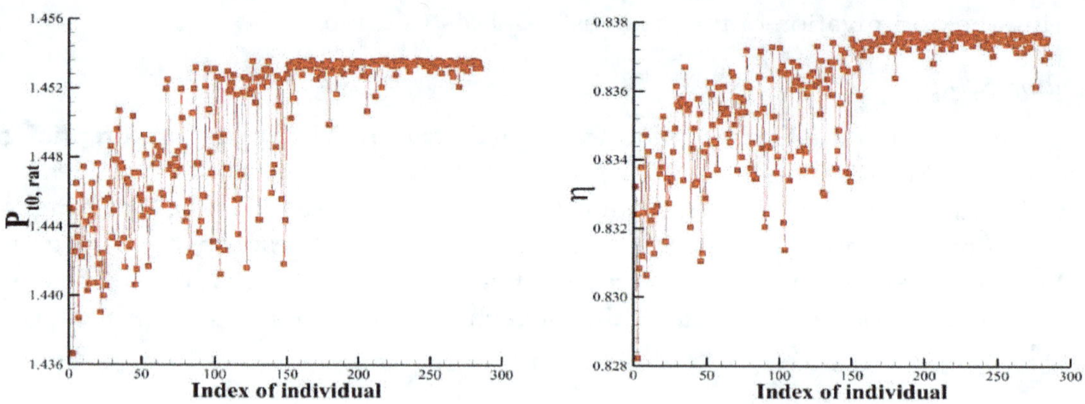

(a) (b)

Figure 17. Convergence histories of the two design objectives: (**a**) the total pressure ratio, and (**b**) the adiabatic efficiency.

The values of the two design objectives of the individuals produced in the CGCT-blade optimization are plotted in Figure 18a. Performances of baseline blade 0, swept blade FS, BS, and leaned blade PL and NL, which have been discussed in Section 4.2, are also indicated in the figure with blue deltas. Compared with the tested blades, the CGCT-blade integrated optimization further improves the total pressure ratio $P_{t0,\,rat}$ and the adiabatic efficiency η. The result verifies the effectiveness of the CGCT-blade integrated optimization strategy. The Pareto Front of the optimization is shown in Figure 18b. The difference between these optimal individuals is less than 0.02% for $P_{t0,\,rat}$ and no greater than 0.05% for η.

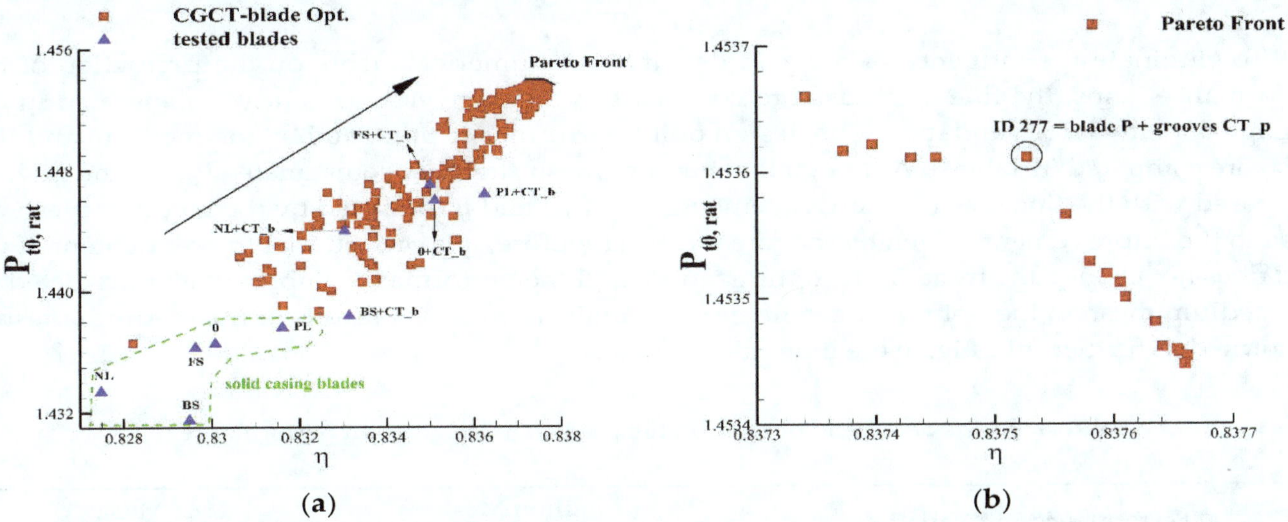

Figure 18. Design objectives of the CGCT-blade integrated optimization results for (**a**) all qualifying individuals, and (**b**) 18 optimal individuals on the Pareto Front.

Table 6 shows the average design performance of the grooved tested blades and the grooved optimized blades on the Pareto Front. The average performance of all of the grooved swept and leaned blades manually generated for the test is slightly lower than the performance of grooved baseline blade 0 + CT_b. Compared with 0 + CT_b, the CGCT-blade integrated optimization improves $P_{t0,\,rat}$ and η by 0.53% and 0.29%, respectively. When the optimization results are compared with the solid-casing baseline blade, the above two values increase to 1.19% and 0.90%, respectively.

Table 6. The average performance at the design point.

Type	$P_{t0,\,rat}$	η
SW_0	1.43655	0.83008
0 + CT_b	1.44596	0.83511
tested blades with CT_b	1.44393	0.83437
CGCT-blade integrated opt.	1.45354	0.83755

For the optimal individuals on the Pareto Front, the average values and variances of the four design variables are shown in Table 7. The differences between these optimal individuals are very small. The optimal blades are characterized by a forward swept and positive leaned tip. The distance between two adjacent grooves is approximately 7.7% of $C_{ax,tip}$, which is approximately 96% of the grooves' width. By substituting the average values of k and s into Equation (3), we find that the axial coverage of the CGCT obtained in the optimization is located between 11.7% and 82.5% of $C_{ax,tip}$.

Table 7. Averages and variances of the values of design variables of the individuals on the Pareto Front.

Statistics	3D Blading		CGCT Distribution	
	Sweep (%*L*)	**Lean (%(*r*·`))**	**k (%$C_{ax,tip}$)**	**s (%$C_{ax,tip}$)**
average	11.9669	2.9675	11.6883	7.7067
variance	0.0419	0.3449	0.0028	0.0673

The CGCT-blade combination ID 277 marked by the black circle in Figure 18b is taken as the representative of the optimization for further analysis. The values of the sweep, lean, k, and s of ID 277 are 12.0000%, 2.8466%, 11.7790%, and 7.6247%, respectively. For the sake of simplicity, the blade of ID 277 is named "P", and the CGCT of ID 277 is named "CT_p" in the following discussion.

To eliminate the influence of the grid density and numerical errors on the prediction of the performance gains, the fine mesh used to conduct the grid convergence study is generated in the passage of blade P for validation. The design point performance predicted by the medium and fine mesh are compared in Table 8. Although the medium mesh slightly overestimates $P_{t0,\ rat}$ and η when compared with the fine mesh, the improvements in $P_{t0,\ rat}$ and η predicted by the medium mesh are close to the improvements given by the fine mesh. Therefore, it is proved that the refinement of the grid does not change the trend of the optimization, and the performance improvement predicted by the medium mesh in the optimizer is reliable. The analysis based on results of the medium mesh is performed to further investigate the inner mechanisms.

Table 8. Effects of the grid density on the predicted design point performance.

Performance and Variations	Medium Mesh (Used in the Optimization)	Fine Mesh (Used for Validation)
$P_{t0,\ rat}$ of baseline blade 0	1.43655	1.43449
$P_{t0,\ rat}$ of grooved blade 0 + CT_b	1.44596	1.44101
$P_{t0,\ rat}$ of optimized grooved blade P + CT_p	1.45361	1.45009
Improvement in $P_{t0,\ rat}$ (P+CT_p versus 0)	1.1876%	1.0875%
Improvement in $P_{t0,\ rat}$ obtained by optimization (P+CT_p versus 0+CT_b)	0.5291%	0.6301%
η of baseline blade 0	0.83008	0.82853
η of grooved blade 0 + CT_b	0.83511	0.83178
η of optimized grooved blade P +CT_p	0.83753	0.83626
Improvement in η (P+CT_p versus 0)	0.8975%	0.9330%
Improvement in η (P+CT_p versus 0+CT_b)	0.2898%	0.5386%

The speed curves of ID 277 with and without the CGCT are shown in Figure 19. The baseline blade installed with CT_p is also simulated for comparison. As shown by the blue solid squares and the black solid diamonds, the design point performance of the solid casing blade P is almost identical with the performance of the baseline blade, which indicates that the benefits of CGCT-blade integrated optimization are not obtained by improving the solid casing blade's performance. After installing CGCT, the stall mass flow is reduced below 8 kg/s in both rotors, and the near-stall performance of both are very similar, which shows that the ability of optimized grooves CT_p to extend the stall margin does not vary with the re-shape of the baseline blade. The zoom-in views at the design point show that the performance improvement induced by optimized grooves CT_p in the optimized blade P is larger than the corresponding improvement in the baseline blade. This result demonstrates that the CGCT-blade integrated optimization which alters the 3D shape of the blade and the axial distribution of the CGCT simultaneously, improves the CGCT-blade interactions to pursue an optimal CGCT-blade combination.

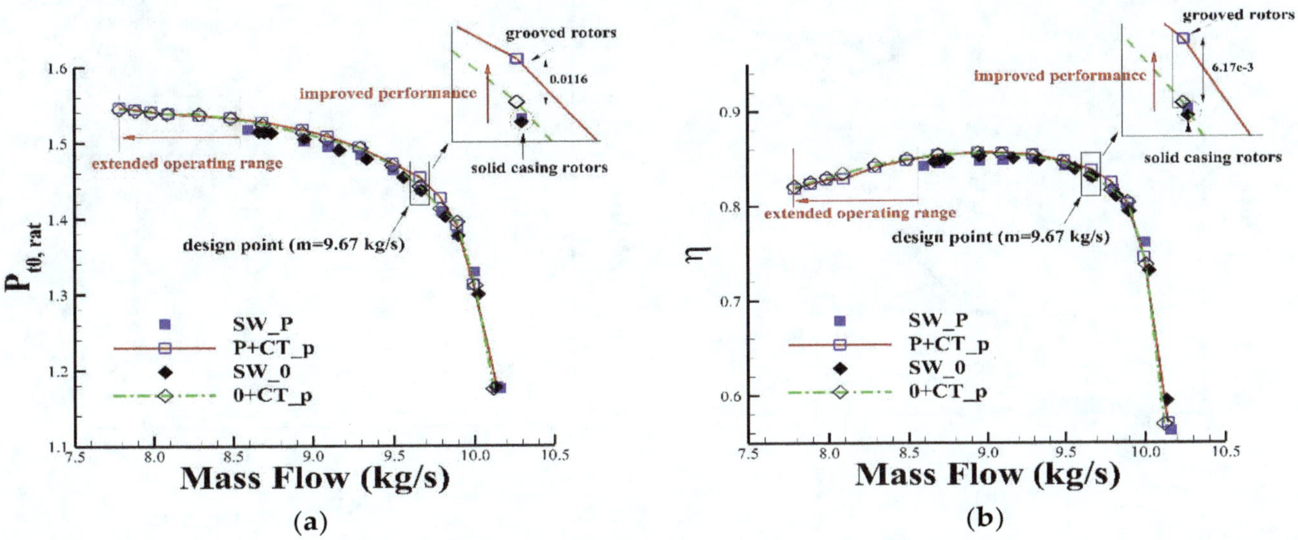

Figure 19. Speed curves of optimal blade ID 277 for (**a**) the total pressure ratio versus the mass flow rate; (**b**) the adiabatic efficiency versus the mass flow rate.

The working capacities and efficiencies of the baseline blade and optimized blade with and without installing CT_p are compared along the spanwise direction in Figure 20. In Figure 20a, the total pressure ratio of optimized blade P is reduced only in the tip region by the forward sweep and positive lean. When CT_p is installed, the total pressure ratio continues to decrease above 90% of the span, however, it increases below 90% of the span. Moreover, the total pressure ratio improvement induced by CT_p below 90% of the span in optimized blade P is obviously larger than the improvement in the baseline blade. Figure 20b shows that the efficiency improvement induced by CT_p is concentrated in the region above 85% of the span. The improvement of the efficiency in optimized blade P is also larger than the corresponding improvement in the baseline blade. The comparison of the diffusion factors in the spanwise direction is shown in Figure 20c. In the solid casing rotor, the forward sweep and positive lean in optimized blade P slightly reduce the tip loading; however, when CT_p is installed, optimized blade P has a more significant tip loading reduction than what the baseline blade has. Meanwhile, the enhancement of the loading in the non-tip region induced by CT_p is larger in optimized blade P than that in the baseline blade 0, which indicates that the forward swept and positive leaned blade tip improves the effectiveness of CT_p. These three plots demonstrate that the good performance of the optimal CGCT-blade combination is attributable to the redistributed spanwise loading caused by the CGCT-blade interactions in the tip region. Because the optimization does not change the design point performance of the solid casing rotor, the spanwise loading redistribution works to improve the CGCT-blade interactions, which is helpful to achieve a better overall performance of the CGCT-blade combination.

Figure 20a shows that the difference in the total pressure ratio between 0 + CT_p and P + CT_p is remarkable at 85% of the span; thus, the static pressure (P_S) distributions at the corresponding position are plotted in Figure 21 for comparison. The forward sweep and positive lean in the tip of blade P push the shockwave downstream, leading to a larger aft-loading in chordwise direction. After installing CT_p, P_S on the pressure side is increased. Compared with CT_p in the baseline blade, the zoom-in view in this figure shows that CT_p in the optimized blade is more effective at increasing P_S on the pressure side between 50% and 95% of the chord length. Note that P_S in this section of the optimized blade with a solid casing equals P_S of the baseline blade with a solid casing, which means the increase of P_S induced by CGCT is further improved after the CGCT-blade optimization.

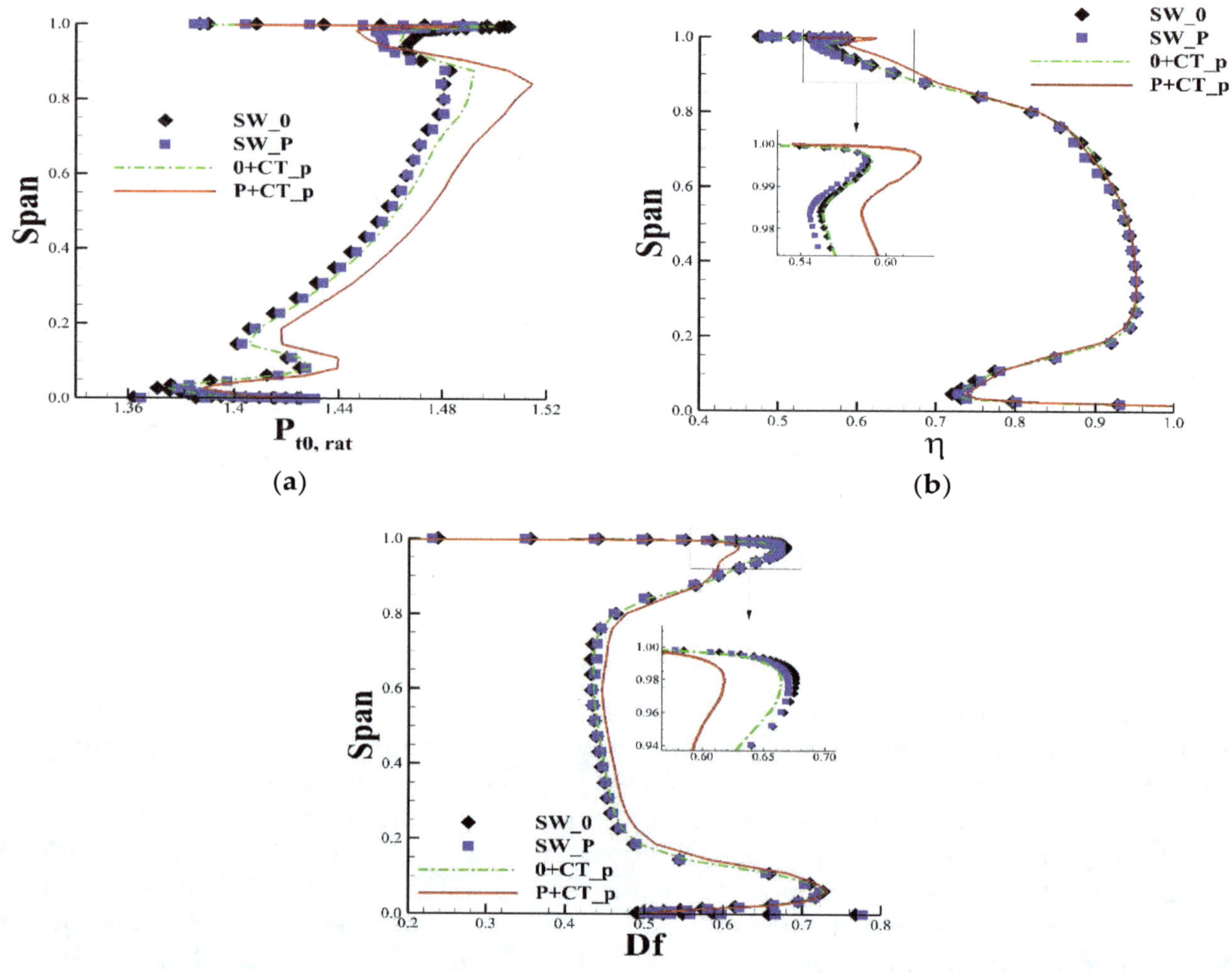

Figure 20. Spanwise distribution of (**a**) the total pressure ratio, (**b**) the adiabatic efficiency, and (**c**) the diffusion factor at the design point for optimal blade P and baseline blade 0 with and without CT_p.

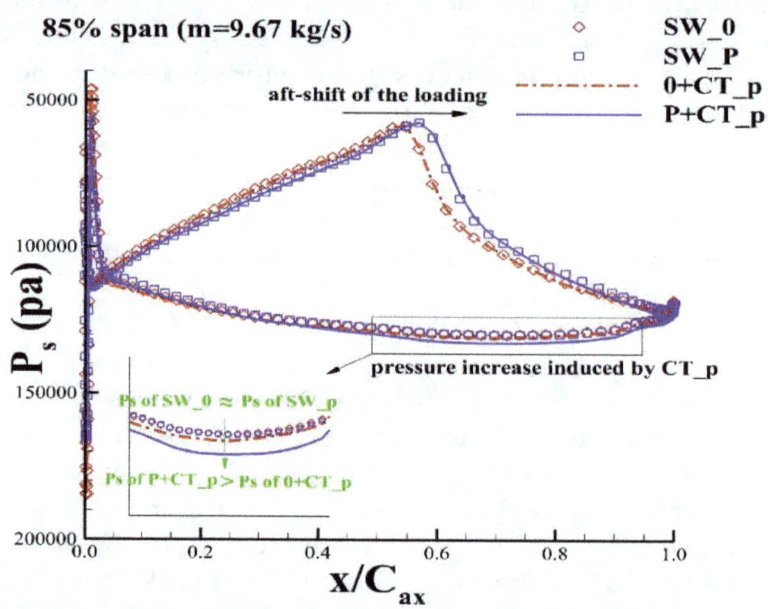

Figure 21. Static pressure distribution at 85% of the span in baseline blade 0 and optimized blade "P" with and without CT_p.

At 93% of the span, flow structures of 0 + CT_p and P + CT_p are shown in Figure 22. The main differences between the two flow fields are illustrated in this figure. The detached bow shock in the flow passage of blade P that has a forward swept and positive leaned tip is closer to the downstream. The area of the shock-induced separation on the suction side (SS) of blade P is also smaller. Moreover, because CGCT in blade P is more effective at relieving the blockage than CGCT in the baseline blade is, the flow tube sandwiched between PS and the blockage expands, which decelerates the near-wall flow inside the tube and increases the static pressure in the aft-part of the blade tip (see Figure 21). To sum up, when the airfoil is more aft-loaded in chordwise direction due to the forward sweep and positive lean, the effectiveness of CGCT to enhance the working capacity of the aft-part of the blade is improved.

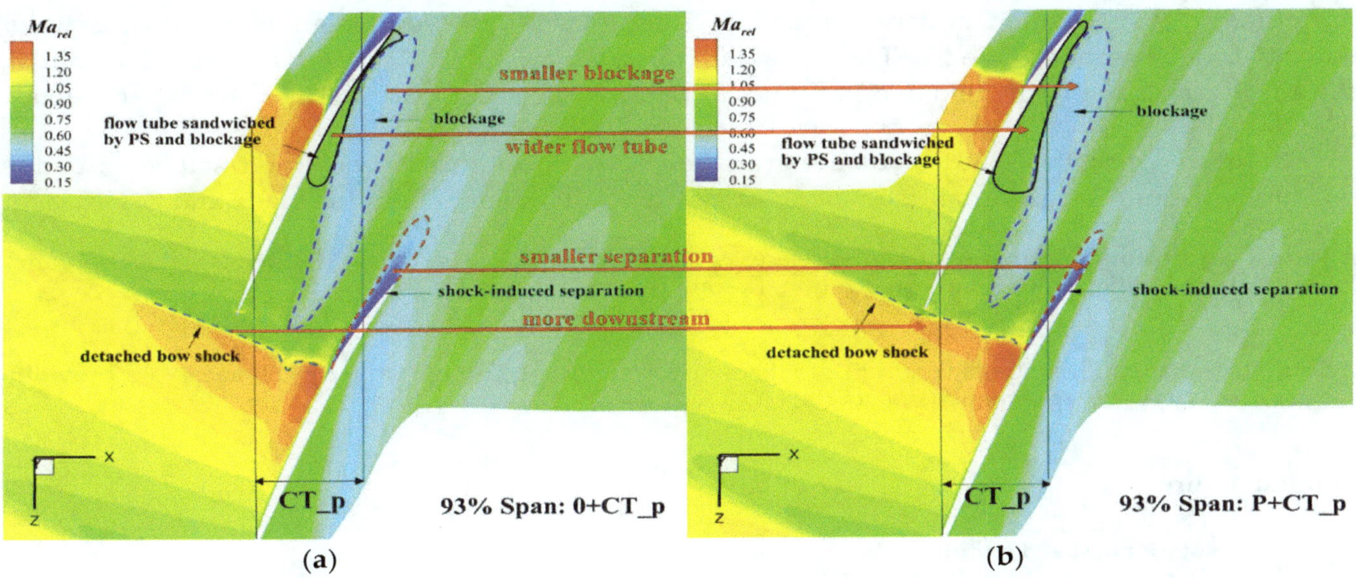

Figure 22. Comparison of flow structures between the baseline blade with CT_p and the optimized blade with CT_p at the design point: (**a**) baseline blade with optimized CGCT: 0 + CT_p; (**b**) optimized CGCT-blade combination: P + CT_p.

6. Conclusions

This paper studies multiple circumferential casing grooves' axial distribution design and optimization. The conventional CGCT design with a given blade is developed into a CGCT-blade integrated optimization. In the process of the optimization, the performance of CGCT-blade combinations is evaluated at the choking point, design point and near-stall point, which fulfills the constraints while improving the objectives. The work can be summarized as follows:

(1) A CGCT-blade combination with good design point performance does not necessarily results in good near-stall performance. For the ND-TAC rotor, the most effective CGCT that leads to the greatest working capacity improvement in the blade at the design point is located between 40% and 100% of the tip chord length; however, this CGCT configuration results in the smallest stall margin enhancement, because the grooves near the trailing edge are ineffective at suppressing the growth of the blockage in the tip region. Therefore, it is suggested that the CGCT design should coordinate the effects of CGCT under both the design and near-stall conditions.

(2) The sweep and lean introduced to the tip alter the CGCT-blade interactions, leading to the performance variations. The working ability of the blade at different span, as well as the fluid exchange between the passage and the grooves, are affected by the 3D blading, thus, it is promising to obtain a better CGCT-blade combination if the sweep and lean are introduced in the process of the CGCT design.

(3) The CGCT-blade integrated optimization improves the coupling between the CGCT and the blade tip, which strengthens the ability of the CGCT in improving the rotor's design point

performance. Compared with the baseline blade and all of the tested re-stacked blades with CGCTs, the CGCT-blade integrated optimization achieves the highest total pressure ratio and adiabatic efficiency while maintaining a sufficient stall margin. In the optimal CGCT-blade combination, a forward swept and positive leaned tip are generated, together with a CGCT located between 11.7% and 82.5% of $C_{ax,tip}$. In the spanwise direction, the blade loading is reduced in the tip region. In the chordwise direction, the detached bow shock is pushed downstream and the blade is more aft-loaded. These variations in loading are good for the CGCT to reduce the blockage and enhance the working capacity of the non-tip span and aft-part of the blade.

The newly developed CGCT-blade integration optimization method is helpful for advancing the understanding of CGCT-blade interactions. The combination of the CGCT and the 3D blading technique breaks the boundary between the CGCT design and the blade design. However, this design strategy is only verified in the ND-TAC rotor with the assistance of CFD. Efforts are needed to validate the effectiveness of this method in other types of rotors by both CFD calculations and experiments.

In future work, the CGCT-blade integrated design and optimization method will be refined by introducing more design variables to increase the design freedom of both the blade and CGCT.

Author Contributions: Formal Analysis, Software, Writing–Review & Editing, W.S.; Methodology, Data Curation, Project Administration, Y.Z.; Conceptualization, Resources, Supervision, H.C.

Acknowledgments: The authors would like to thank the Notre Dame Turbomachinery Laboratory for providing the geometry and the experimental data of the ND-TAC.

Nomenclature

N_c	Rotational speed (rpm)
U_{tip}	Blade tip velocity (m/s)
N_b	Blade number
R_{casing}	Radius of casing (mm)
R_{hub}	Radius of hub (mm)
τ	Tip gap size (mm)
m	Mass flow (kg/s)
M_s	Stall mass flow (kg/s)
$P_{t0,\ rat}$	Total pressure ratio
$T_{t0,\ rat}$	Total temperature ratio
η	Adiabatic efficiency
Ma_{rel}	Relative Mach number
$P_{t0,\ rel}$	Non-dimensional relative total pressure
Df	Diffusion factor
W_r	Relative radial velocity (m/s)
C_{ax}	Axial chord length (mm)
P_s	Static pressure (Pa)
PS	Pressure side
SS	Suction side

References

1. Broichhausen, K.D.; Ziegler, K.U. Supersonic and Transonic Compressors: Past, Status and Technology Trends. In Proceedings of the ASME Turbo Expo 2005: Power for Land, Sea, and Air, Reno, NV, USA, 6–9 June 2005; pp. 63–74. [CrossRef]
2. Fujita, H.; Takata, H. A Study on Configurations of Casing Treatment for Axial Flow Compressors. *Bull. JSME* **1984**, *27*, 1675–1681. [CrossRef]

3. Hathaway, M.D. *Passive Endwall Treatments for Enhancing Stability*; NASA Rep. No. TM-2007-214409; NASA Glenn Research Center: Cleveland, OH, USA, 2007.
4. Rabe, D.C.; Hah, C. Application of Casing Circumferential Grooves for Improved Stall Margin in a Transonic Axial Compressor. In Proceedings of the ASME Turbo Expo 2002: Power for Land, Sea, and Air, Amsterdam, The Netherlands, 3–6 June 2002; pp. 1141–1153. [CrossRef]
5. Müller, M.W.; Schiffer, H.-P.; Hah, C. Effect of Circumferential Grooves on the Aerodynamic Performance of an Axial Single-Stage Transonic Compressor. In Proceedings of the ASME Turbo Expo 2007: Power for Land, Sea, and Air, Montreal, QC, Canada, 14–17 May 2007; pp. 115–124. [CrossRef]
6. Sakuma, Y.; Watanabe, T.; Himeno, T.; Kato, D.; Murooka, T.; Shuto, Y. Numerical Analysis of Flow in a Transonic Compressor with a Single Circumferential Casing Groove: Influence of Groove Location and Depth on Flow Instability. *J. Turbomach.* **2013**, *136*, 031017. [CrossRef]
7. Houghton, T.; Day, I. Stability Enhancement by Casing Grooves: The Importance of Stall Inception Mechanism and Solidity. *J. Turbomach.* **2012**, *134*, 021003. [CrossRef]
8. Du, J.; Li, J.; Gao, L.; Lin, F.; Chen, J. The Impact of Casing Groove Location on Stall Margin and Tip Clearance Flow in a Low-Speed Axial Compressor. *J. Turbomach.* **2016**, *138*, 121007. [CrossRef]
9. Minsuk Choi Effects of circumferential casing grooves on the performance of a transonic axial compressor. *Int. J. Turbo Jet-Engines* **2010**, *32*, 361–371. [CrossRef]
10. Mao, X.; Liu, B.; Tang, T.; Zhao, H. The impact of casing groove location on the flow instability in a counter-rotating axial flow compressor. *Aerosp. Sci. Technol.* **2018**, *76*, 250–259. [CrossRef]
11. Kim, J.H.; Choi, K.J.; Kim, K.Y. Aerodynamic analysis and optimization of a transonic axial compressor with casing grooves to improve operating stability. *Aerosp. Sci. Technol.* **2013**, *29*, 81–91. [CrossRef]
12. Zhao, Q.; Zhou, X.; Xiang, X. Multi-objective optimization of groove casing treatment in a transonic compressor. *Proc. Inst. Mech. Eng. Part A J. Power Energy* **2014**, *228*, 626–637. [CrossRef]
13. Houghton, T.O. *Axial Compressor Stability Enhancement*; University of Cambridge: Cambridge, UK, 2010.
14. Denton, J.D.; Xu, L. The Effects of Lean and Sweep on Transonic Fan Performance. In Proceedings of the ASME Turbo Expo 2002: Power for Land, Sea, and Air, Amsterdam, The Netherlands, 3–6 June 2002; Volume 1, pp. 23–32. [CrossRef]
15. Denton, J.D.; Xu, L. The exploitation of three-dimensional flow in turbomachinery design. *Proc. Inst. Mech. Eng. Part C J. Mech. Eng. Sci.* **1998**, *213*, 125–137. [CrossRef]
16. Wadia, A.R.; Szucs, P.N.; Crall, D.W. Inner Workings of Aerodynamic Sweep. *J. Turbomach.* **1998**, *120*, 671. [CrossRef]
17. Hah, C.; Puterbaugh, S.L.; Wadia, A. Control of Shock Structure and Secondary Flow Field inside Transonic Compressor Rotors through Aerodynamic Sweep. In Proceedings of the ASME 1998 International Gas Turbine and Aeroengine Congress and Exhibition, Stockholm, Sweden, 2–5 June 1998; pp. 1–15. [CrossRef]
18. Gallimore, S.J.; Bolger, J.J.; Cumpsty, N.A.; Taylor, M.J.; Wright, P.I.; Place, J.M.M. The Use of Sweep and Dihedral in Multistage Axial Flow Compressor Blading—Part I: University Research andMethods Development. *J. Turbomach.* **2002**, *124*, 521–532. [CrossRef]
19. Hah, C. The Inner Workings of Axial Casing Grooves in a One and a Half Stage Axial Compressor with a Large Rotor Tip Gap: Changes in Stall Margin and Efficiency. *J. Turbomach.* **2018**, accepted.
20. Goinis, G.; Nicke, E. Optimizing surge margin and efficiency of a transonic compressor. In Proceedings of the ASME Turbo Expo 2016: Turbomachinery Technical Conference and Exposition, Seoul, Korea, 13–17 June 2016; pp. 1–12. [CrossRef]
21. Cameron, J.D.; Cameron, J.D.; Morris, S.C.; Morris, S.C.; Corke, T.C.; Corke, T.C. A Transonic Axial Compressor Facility for Fundamental Research and Flow Control Development. *Mech. Eng.* **2006**. [CrossRef]
22. Kelly, R.; Hickman, A.R.; Shi, K.; Morris, S.C.; Jemcov, A. Very Large Eddy Simulation of a Transonic Axial Compressor Stage. In Proceedings of the 52nd AIAA/SAE/ASEE Joint Propulsion Conference, Salt Lake City, UT, USA, 25–27 July 2016; pp. 1–13. [CrossRef]
23. Cameron, J.D.; Morris, S.C. Spatial Correlation Based Stall Inception Analysis. In Proceedings of the ASME Turbo Expo 2007: Power for Land, Sea, and Air, Montreal, QC, Canada, 14–17 May 2007; pp. 433–444. [CrossRef]
24. Chen, H.; Huang, X.; Shi, K.; Fu, S.; Ross, M.; Bennington, M.A.; Cameron, J.D.; Morris, S.C.; McNulty, S.; Wadia, A. A Computational Fluid Dynamics Study of Circumferential Groove Casing Treatment in a Transonic Axial Compressor. *J. Turbomach.* **2013**, *136*, 031003. [CrossRef]

25. Shi, K.; Chen, H.X.; Fu, S. Numerical investigation of the casing treatment mechanism with a single circumferential groove. *Sci. China Phys. Mech. Astron.* **2013**, *56*, 353–365. [CrossRef]
26. Toro, E.F.; Spruce, M.; Speares, W. Restoration of the contact surface in the HLL-Riemann solver. *Shock Waves* **1994**, *4*, 25–34. [CrossRef]
27. Van Leer, B. Towards the ultimate conservative difference scheme III. Upstream-centered finite-difference schemes for ideal compressible flow. *J. Comput. Phys.* **1977**, *23*, 263–275. [CrossRef]
28. Van Albada, G.D.; Van Leer, B.; Roberts, W.W. A comparative study of computational methods in cosmic gas dynamics. *Astron. Astrophys.* **1982**, *108*, 76–84. [CrossRef]
29. Menter, F.R. Improved two-equation k-omega turbulence models for aerodynamic flows. *NASA Tech. Memo.* **1992**, 1–31. [CrossRef]
30. Yoon, S.; Jameson, A. Lower-upper Symmetric-Gauss-Seidel method for the Euler and Navier-Stokes equations. *AIAA J.* **1988**, *26*, 1025–1026. [CrossRef]
31. Chen, H.X.; Fu, S.; Li, F.W. Navier-Stokes simulations for transport aircraft wing-body combinations with deployed high-lift systems. *J. Aircr.* **2003**, *40*, 883–890. [CrossRef]
32. Huang, X.; Chen, H.; Fu, S. CFD Investigation on the Circumferential Grooves Casing Treatment of Transonic Compressor. In Proceedings of the ASME Turbo Expo 2008: Power for Land, Sea and Air, Berlin, Germany, 9–13 June 2008; pp. 1–9.
33. Chen, H.; Huang, X.; Fu, S. CFD Investigation on Stall Mechanisms and Casing Treatment of a Transonic Compressor. In Proceedings of the 42nd AIAA/ASME/SAE/ASEE Joint Propulsion Conference & Exhibit, Sacramento, CA, USA, 9–12 July 2006. [CrossRef]
34. Siller, U.; Voß, C.; Nicke, E. Automated Multidisciplinary Optimization of a Transonic Axial Compressor. In Proceedings of the 47th AIAA Aerospace Sciences Meeting Including The New Horizons Forum and Aerospace Exposition, Orlando, FL, USA, 5–8 January 2009; pp. 1–12. [CrossRef]
35. Ji, L.; Chen, J.; Lin, F. Review and Understanding on Sweep in Axial Compressor Design. In Proceedings of the ASME Turbo Expo 2005: Power for Land, Sea, and Air, Reno, NV, USA, 6–9 June 2005; pp. 1–9. [CrossRef]
36. Deng, K.; Chen, H. A Hybrid Aerodynamic Optimization Algorithm Based on Differential Evolution and RBF Response Surface. In Proceedings of the 17th AIAA/ISSMO Multidisciplinary Analysis and Optimization Conference, Washington, DC, USA, 13–17 June 2016. [CrossRef]

4

Impeller Optimized Design of the Centrifugal Pump: A Numerical and Experimental Investigation

Xiangdong Han [1,2,3], Yong Kang [1,2,3,4,*], Deng Li [1,2,3] and Weiguo Zhao [5]

1 Key Laboratory of Hydraulic Machinery Transients, Ministry of Education, Wuhan University, Wuhan 430072, China; hanxiangdong@whu.edu.cn (X.H.); 2008lee@whu.edu.cn (D.L.)
2 Hubei Key Laboratory of Waterjet Theory and New Technology, Wuhan University, Wuhan 430072, China
3 School of Power and Mechanical Engineering, Wuhan University, Wuhan 430072, China
4 Collaborative Innovation Center of Geospatial Technology, Wuhan 430079, China
5 School of Energy and Power Engineering, Lanzhou University of Technology, Lanzhou 730050, China; hanromeolut@163.com
* Correspondence: kangyong@whu.edu.cn.

Abstract: Combined numerical simulation with experiment, blade wrap angle, and blade exit angle are varied to investigate the optimized design of the impeller of centrifugal pump. Blade wrap angles are 122°, 126°, and 130°. Blade exit angles are 24°, 26°, and 28°. Based on numerical simulation, internal flow of the centrifugal pump with five different impellers under 0.6, 0.8, 1.0, 1.2, and 1.5 Q_d are simulated. Variations of static pressure, relative velocity, streamline, and turbulent kinetic energy are analyzed. The impeller with blade wrap angle 126° and blade exit angle 24° are optimal. Distribution of static pressure is the most uniform and relative velocity sudden changes do not exist. Streamlines are the smoothest. Distribution scope of turbulent kinetic energy is the smallest. Based on performance experiments, head and efficiency of the centrifugal pump with the best impeller are tested. The values of head and efficiency are higher than that of the original pump. Centrifugal pump with the best impeller has better hydraulic performance than the original centrifugal pump.

Keywords: centrifugal pump; impeller; blade wrap angle; blade exit angle; optimized design

---

1. Introduction

A centrifugal pump is one kind of general machinery [1] which is widely and fully utilized in the industrial and agricultural fields [2], such as irrigation and water supply. It normally includes four different parts: suction pipe, impeller, volute, and exit pipe. The impeller is the core part and it converts the mechanical energy into pressure energy [3], which directly determines the transport capacity and the hydraulic performances of centrifugal pump. So, optimized design of the impeller is essential and significant for the efficient operation of a centrifugal pump [4,5].

One-dimensional flow theory, two-dimensional flow theory, and three-dimensional flow theory are three basic theories to optimally design the centrifugal pump [6,7]. Two-dimensional flow theory regards the distribution of meridian velocity across the cross-section as symmetrical. The flow is one kind of potential flow. Optimized design of hydraulic machinery in engineering fields, which is based on two-dimensional flow theory, is rare. It is only employed to optimally design the higher specific speed impellers of mixed-flow pump and runners of mixed-flow turbine. Three-dimensional flow theory, proposed by Wu [8], is also called S_1 and S_2 relative flow surface theory. In this method, the three-dimensional flow is converted into two two-dimensional flows in the surfaces S_1 and S_2. Only in the ideal conditions of flow being inviscid, incompressible, and unsteady, can this method be applied to design the impeller successfully. This method is difficult and complex. Thus, it is not always employed to optimally design an impeller for a centrifugal pump in the engineering field. However, one-dimensional flow theory, which is based on Euler equations, is convenient and simple. Flow in the meridional surface is symmetrical. Engineers coming from home and abroad factories widely employ one-dimensional flow theory to optimally design an impeller for a centrifugal pump [9–11].

Blade exit angle (Φ) and blade wrap angle (β_2) are two significant parameters in the impeller optimized design process according to one-dimensional flow theory [12–15]. The hydraulic performance of a centrifugal pump is mainly determined by these parameters. In the actual optimized design process, blade exit angle often varies from 15° to 40° and blade wrap angle often varies from 90° to 130° [16]. Some scholars studied the effects of blade exit angle and blade wrap angle on the hydraulic performances of the centrifugal pump [17–19]. The selection of blade wrap angle and blade exit angle mainly depend on the specific speed of the centrifugal pump and the actual optimized design experience. If the blade wrap angle is too large, the impeller friction area becomes large, which is bad for the improvement of efficiency. If the blade wrap angle is too small, the impeller cannot control the flow of water effectively and the impeller cannot operate stably. For the blade exit angle, a moderate value could effectively improve the hydraulic performances of the centrifugal pump. Comprehensively considering the specific speed of the target centrifugal pump (n_s = 112) and the effects of blade wrap angle and blade exit angle on the hydraulic performances of the centrifugal pump, the variation scope of the blade wrap angle and blade exit angle was selected carefully. The blade wrap angle was tested in the range of 120° to 130° and the values were Φ = 122°, 126°, and 130°, respectively. The blade exit angle was tested in the range of 20° to 30° and the values were β_2 = 24°, 26°, and 28°, separately.

2. Basic Parameters of the Impeller

The inlet diameter of the impeller D_1 = 200 mm and the outlet diameter D_2 = 420 mm. The exit width b_2 = 34 mm. The original blade wrap angle Φ = 120°. The original blade exit angle β_2 = 26°. The basic parameters of the impeller are shown in Figure 1.

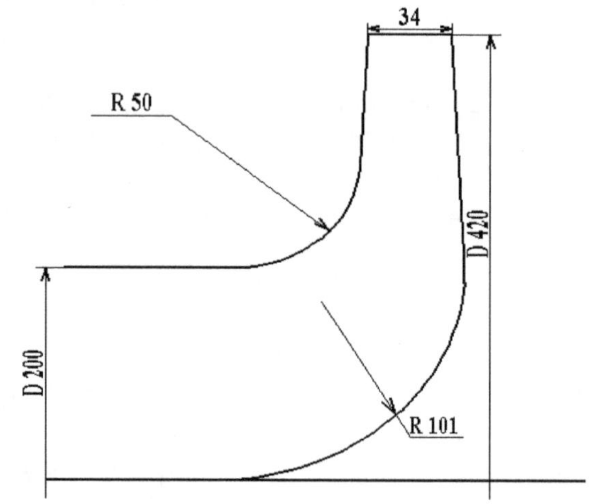

Figure 1. Two-dimensional model of the impeller.

3. Method to Modify the Impeller

The impeller shape is directly determined by the blade profiles. In the impeller optimization design process, all blade profiles must be smooth. The bend of the blade profiles should be in one single direction. An 'S' shape should not exist [16,20,21]. All blade profiles should be symmetrical. The blade profile differential equation is

$$d\theta = \frac{ds}{r \tan \beta} \tag{1}$$

Here, s is the axial streamline. θ denotes the axial angle. r represents the radius of one arbitrary point in the blade profile. β is the blade angle.

The speed triangle is shown in Figure 2. Here, c represents absolute velocity, w is the relative velocity, and u is the following velocity. c_m and w_m are the component of absolute velocity and relative velocity in the meridional plane, respectively. c_u is the component in the circumferential plane.

Based on the speed triangle, tanβ could be described as

$$\tan\beta = \frac{c_m}{u - c_u} \tag{2}$$

Thus,

$$\frac{ds}{rd\theta} = \frac{c_m}{u - c_u} \tag{3}$$

So,

$$ds = \frac{c_m}{u - c_u} rd\theta \tag{4}$$

According to the above equations, blade profile optimized design has a closed relation with blade inlet angle (β_1), blade exit angle (β_2), and blade wrap angle (Φ). Three different methods are usually employed to modify the blade profiles. They are that

(1) β_1 is the constant. β_2 and Φ are changed.
(2) β_1 and β_2 are fixed. Φ is varied.
(3) β_2 and Φ are invariant. β_1 is altered.

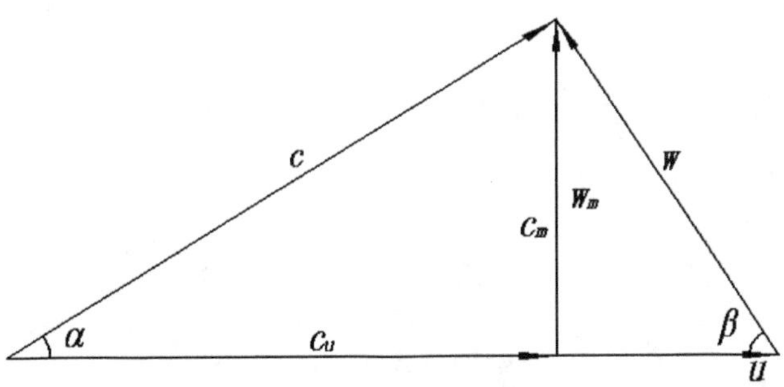

Figure 2. Speed triangle.

Method 1 (β_1 is the constant. β_2 and Φ are changed.) was always employed to optimally design the impeller. As shown in Table 1, the blade exit angle of Schemes 1–3 was identical and the blade wrap angles varied from 122° to 130°, which were employed to investigate the effects of different blade wrap angles on the optimized design of the centrifugal pump. For Schemes 2, 4, and 5, the blade wrap angle was the same and the blade exit angle was changed, which was used to discuss the effects of diverse blade exit angles on the optimized design of the centrifugal pump.

Table 1. Optimized schemes of the blade profile.

Scheme	1	2	3	4	5
Φ	122°	126°	130°	126°	126°
β_2	26°	26°	26°	28°	24°

The impeller optimized design procedure is that, via the analysis of effects of blade wrap angle (Schemes 1–3) on the hydraulic performances of the centrifugal pump, one optimal scheme could be determined via computational fluid dynamics (CFD). According to the discussion of effects of blade exit angle (Schemes 2, 4, and 5) on the variations of head and efficiency, one optimal scheme could be got by numerical simulation. Then, these two different schemes are compared and the best one could be obtained. The optimized procedure includes two main parts, as shown in Figure 3.

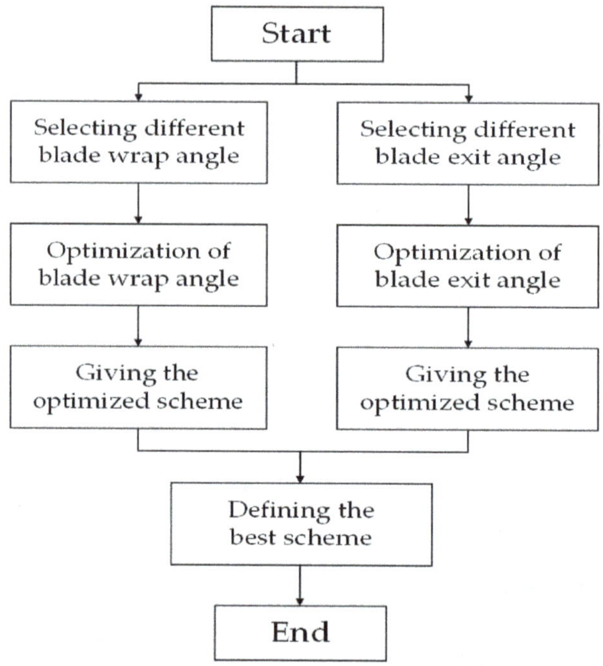

Figure 3. Flow chart of the optimized design of the impeller.

The physical models of the blades under different blade wrap angle and blade exit angle conditions are given in Figure 4.

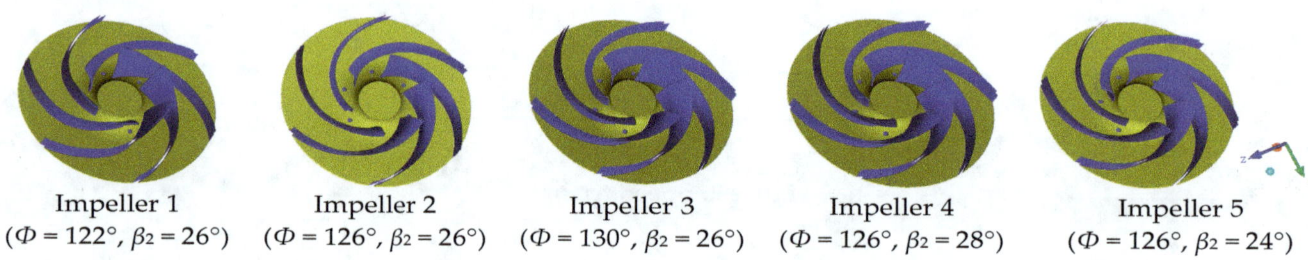

Figure 4. Physical model of the blades.

4. Numerical Method

4.1. Fundamental Equations

The fundamental equations [22] are employed to describe the flow characteristics in the centrifugal pump, which include two main parts: continuity equation and motion equation, corresponding with mass conservation law and momentum conservation law.

$$\frac{\partial \rho}{\partial t} + \frac{\partial (\rho u_i)}{\partial x_i} = 0 \tag{5}$$

$$\rho \frac{\partial u_i}{\partial t} + \rho u_j \frac{\partial u_i}{\partial x_j} = \rho f_i - \frac{\partial p}{\partial x_i} + \frac{\partial}{\partial x_j}\left[\mu\left(\frac{\partial u_i}{\partial x_j} + \frac{\partial u_j}{\partial x_i}\right)\right] - \frac{2}{3}\frac{\partial}{\partial x_i}\left(\mu \frac{\partial u_j}{\partial x_j}\right) \tag{6}$$

Here, ρ is the density of water. t represents the time. u denotes the velocity. x is the Cartesian coordinate. f_i is the body force vector. p is the pressure. μ is the dynamic viscosity.

4.2. Turbulent Model

RNG k-ε turbulent model proposed by Yakhot and Orzag [23] was employed to deal with turbulent flow. In the centrifugal pump, the impeller is the rotating part whose rotation effects could be fully dealt by the turbulent dissipation rate (ε) equation in this turbulent model. On the other hand, the RNG k-ε turbulent model has high-precision, which could guarantee the accuracy of the numerical results.

$$\frac{\partial(\rho k)}{\partial t} + \frac{\partial(\rho k u_i)}{\partial x_i} = \frac{\partial}{\partial x_j} \times \left[(\alpha_k(\mu + \mu_t))\frac{\partial k}{\partial x_j}\right] + G_k + \rho\varepsilon \tag{7}$$

$$\frac{\partial(\rho \varepsilon)}{\partial t} + \frac{\partial(\rho \varepsilon u_i)}{\partial x_i} = C_{1\varepsilon}\frac{\varepsilon}{k}G_k C_{2\varepsilon}\rho\frac{\varepsilon^2}{k} + \frac{\partial}{\partial x_j}\left[\alpha_\varepsilon(\mu + \mu_t)\frac{\partial \varepsilon}{\partial x_j}\right] \tag{8}$$

$$\mu_t = \rho c_\mu \frac{k^2}{\varepsilon} \tag{9}$$

Here, k is the turbulent kinetic energy. ε is the turbulent dissipation rate. μ_t is the turbulent viscosity. The five terms, $C_{1\varepsilon}$, $C_{2\varepsilon}$, α_k, α_ε, and c_μ are empirical coefficients and the values are 1.42, 1.68, 1.39, 1.39, and 0.09, separately. G_k is one generation term of turbulent kinetic energy which is caused by the mean velocity gradient.

5. Numerical Simulation Setup

5.1. Physical Model

Three-dimensional (3D) single-stage and single-suction centrifugal pump was employed, as shown in Figure 5. The suction pipe was employed to keep the uniform of the flow. Exit pipe was utilized to avoid the backflow. The performance parameters were designed flow rate Q_d = 550 m^3/h; designed head H_d = 50 m; rated motor power P = 110 KW; and rated rotational speed n = 1480 rpm. The geometric parameters of the impeller and the volute were shown in Table 2.

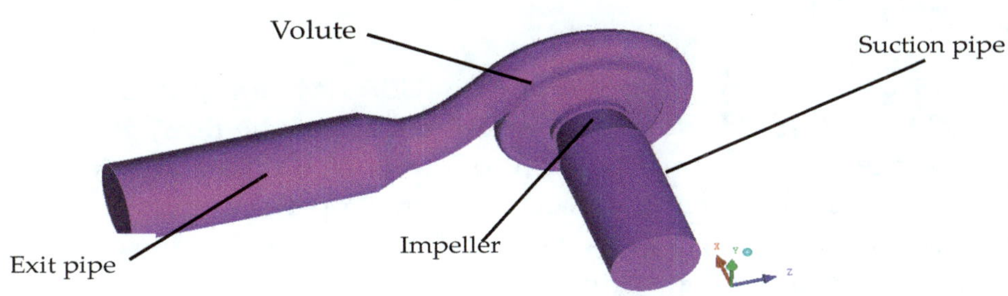

Figure 5. Physical model of the centrifugal pump.

Table 2. Geometric parameters of the impeller and volute

Parameter	Value
Impeller inlet diameter D_1, mm	200
Impeller outlet diameter D_2, mm	420
Impeller exit width b_2, mm	34
Number of blade Z	6
Base diameter of the volute D_3, mm	435
Volute inlet width b_3, mm	72
Volute outlet diameter D_4, mm	250

To get accurate numerical simulation results a balance hole, front chamber, and back chamber were added to the 3D physical model of the centrifugal pump, as displayed in Figures 6 and 7. Diameter of the balance hole was 10mm and it was mainly employed to balance the axial force [24].

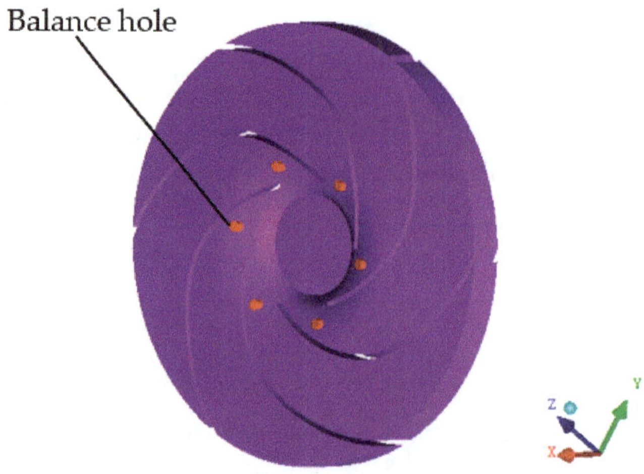

Figure 6. Physical model of the impeller.

Figure 7. Chamber of the centrifugal pump.

5.2. Mesh Generation

ANSYS-ICEM (15.0, ANSYS, Inc., Canonsburg, PA, USA) is employed to discrete the centrifugal pump computational domains. To guarantee the uniformity of the flow, structured meshes were employed to discretize the computational domains of the suction pipe and exit pipe, as displayed in Figure 8d–e. Fully considering the complex structure of the impeller and the volute, to get the better adaptability of the flow, unstructured meshes were employed to discrete the computational domains of impeller, front and back chamber, and volute, as shown in Figure 8a–c. Meshes in the leading edges of the impeller and balance hole were refined.

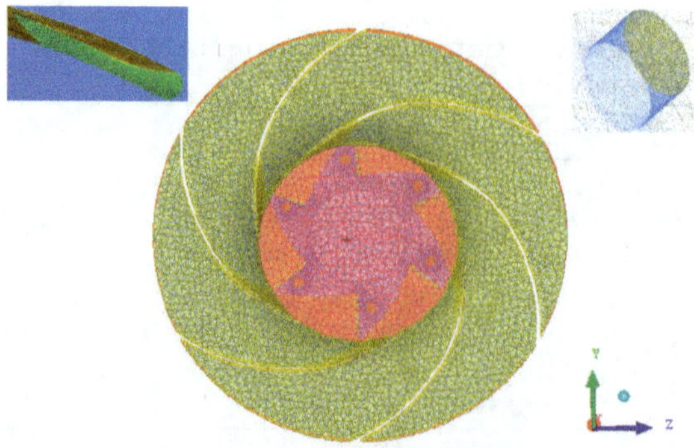

(**a**) Mesh of the impeller

Figure 8. *Cont.*

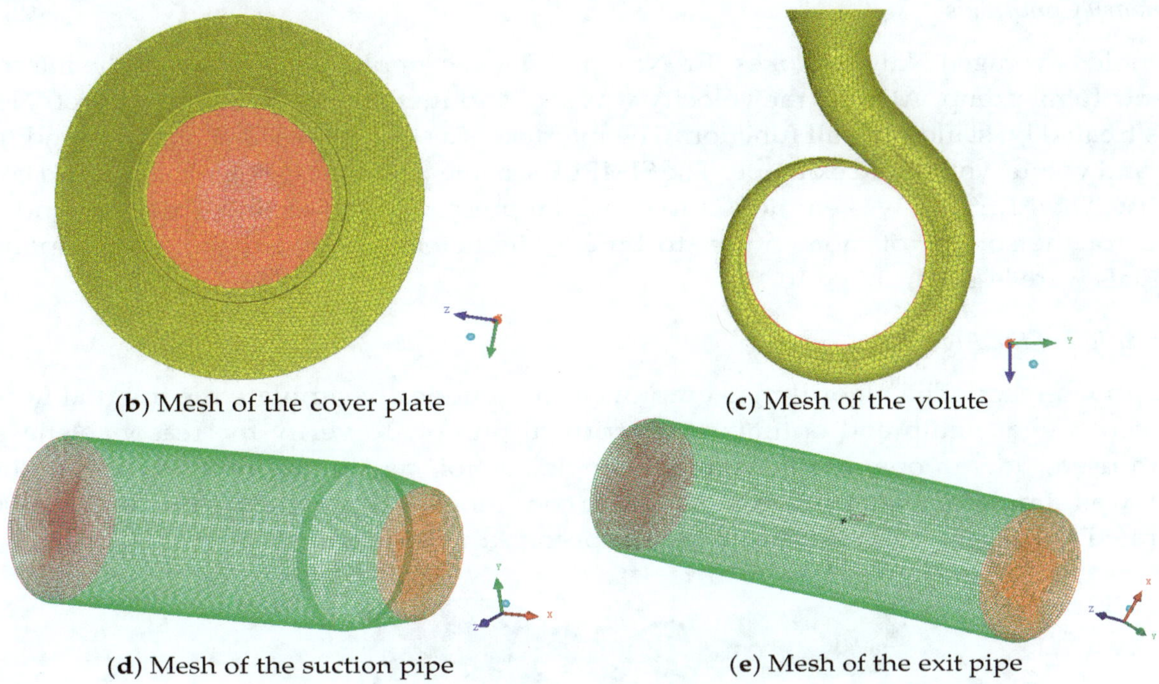

Figure 8. Meshes of the computational domain of the centrifugal pump.

Mesh numbers of the five different impellers were 3,472,923, 3,473,052, 3,471,896, 3,472,816, and 3,473,011, respectively. All mesh quality was higher than 0.3. Mesh number and quality of front and back chamber, volute, suction pipe, and exit pipe were displayed in Table 3.

Table 3. Mesh number and quality

	Front and Back Chamber	Volute	Suction Pipe	Exit Pipe
Mesh number	934,607	2,238,179	498,506	327,624
Mesh quality	>0.4	>0.4	>0.8	>0.8

Head under designed flow rate condition was calculated to verify mesh independence. Results indicated that with the increase of mesh number, head increased firstly. Then, the differences of head were slight, as shown in Figure 9. Total mesh numbers of the five different centrifugal pumps were 7,139,826, 7,146,497, 7,143,016, 7,145,682, and 7,147,024, separately.

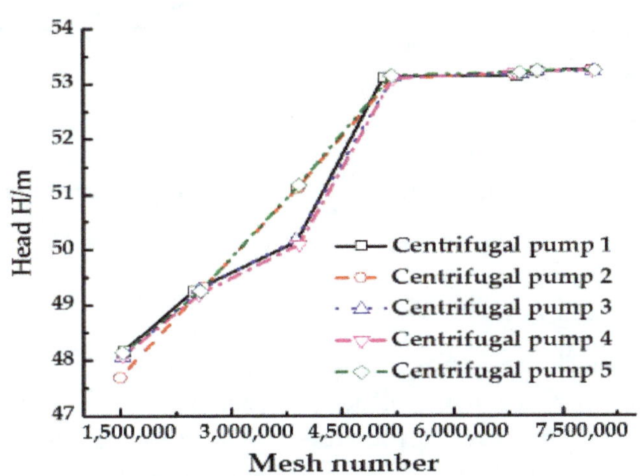

Figure 9. Meshes independence.

5.3. Boundary Conditions

Reynolds averaged Naiver–Stokes (RANS) method was employed to simulate the internal flow of the centrifugal pump. At inlet, the velocity was set. At outlet, the free outflow was set. Near wall flow was treated by standard wall function. The interface was set between suction pipe and impeller, impeller and volute, volute and exit pipe. The SIMPLE scheme was used to solve the coupled equations of velocity. The PRESTO! was employed to compute pressure. The second order upwind scheme was used for the solving of momentum, turbulent kinetic energy, and turbulent dissipation rate. All residuals were less than 1.0×10^{-5}.

5.4. Verification of the Algorithm

The above-mentioned algorithm is employed to numerically simulate the internal flow of the original centrifugal pump and optimized centrifugal pump. To verify the reasonableness of the algorithm used, flow in one single-stage and single suction centrifugal pump which is shown in Figure 10 was simulated under different flow rate conditions. The rated flow rate is $Q_d = 500$ m^3/h and the rated head is $H_d = 53$ m. Rated rotating speed is $n = 1480$ rpm.

Figure 10. Experimental centrifugal pump.

The numerical results of head and efficiency had a good agreement with the experimental results, as displayed in Figure 11. The maximum relative error of the head was less than 3.7%. The largest fractional error of the efficiency was less than 2.2%. The algorithm designed was reasonable.

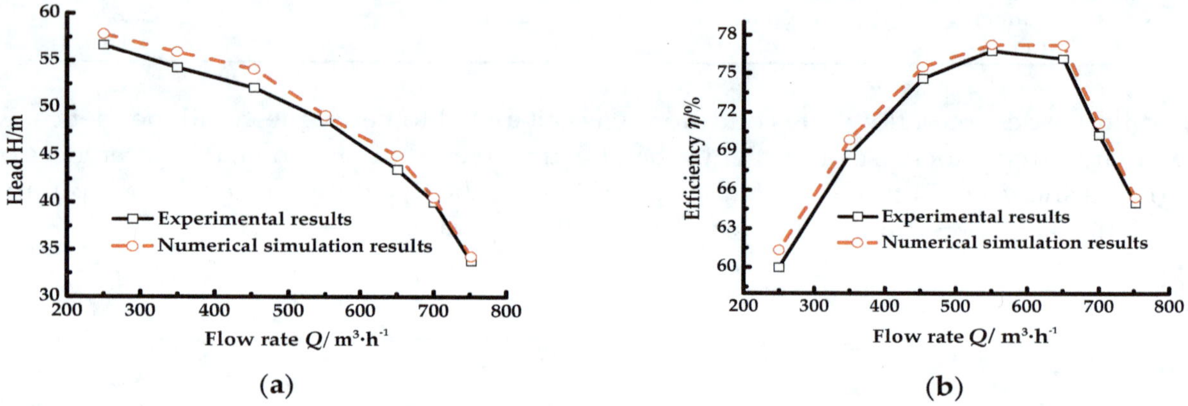

Figure 11. Comparison of the hydraulic performances of centrifugal pump. (**a**) Head (H)-flow rate (Q); (**b**) Efficiency (η)-flow rate (Q).

6. Numerical Simulation Results Analysis

To measure the effects of blade wrap angle and blade exit angle on the performances of centrifugal pump, variations of static pressure, relative velocity, streamlines, and turbulent kinetic energy in the middle span of the impeller were analyzed under the typical flow rate conditions (low flow rate 0.6 Q_d and 0.8 Q_d, rated flow rate 1.0 Q_d, and high flow rate 1.2 Q_d and 1.5 Q_d). Also, the variations of head and efficiency under the above flow rate conditions were compared.

6.1. *Variation of Static Pressure*

The overall static pressure variation law is that static pressure in different passages distributed uniformly. Differences of static pressure in all passages were relatively small, as shown in Figure 12. At the inlet of the impeller, static pressure was the lowest. Static pressure at the outlet of the impeller was the highest. For Impellers 1–5, static pressure increased overall with the growing of flow rate. Under 0.8 Q_d condition, the increase of static pressure was more manifest. At outlet, static pressure attained the maximum under 1.5 Q_d condition.

0.6 Q_d 0.8 Q_d 1.0 Q_d 1.2 Q_d 1.5 Q_d

(**a**) Impeller 1 ($\Phi = 122°$, $\beta_2 = 26°$)

0.6 Q_d 0.8 Q_d 1.0 Q_d 1.2 Q_d 1.5 Q_d

(**b**) Impeller 2 ($\Phi = 126°$, $\beta_2 = 26°$)

0.6 Q_d 0.8 Q_d 1.0 Q_d 1.2 Q_d 1.5 Q_d

(**c**) Impeller 3 ($\Phi = 130°$, $\beta_2 = 26°$)

0.6 Q_d 0.8 Q_d 1.0 Q_d 1.2 Q_d 1.5 Q_d

(**d**) Impeller 4 ($\Phi = 126°$, $\beta_2 = 28°$)

0.6 Q_d 0.8 Q_d 1.0 Q_d 1.2 Q_d 1.5 Q_d

(**e**) Impeller 5 ($\Phi = 126°$, $\beta_2 = 24°$)

Figure 12. *Cont.*

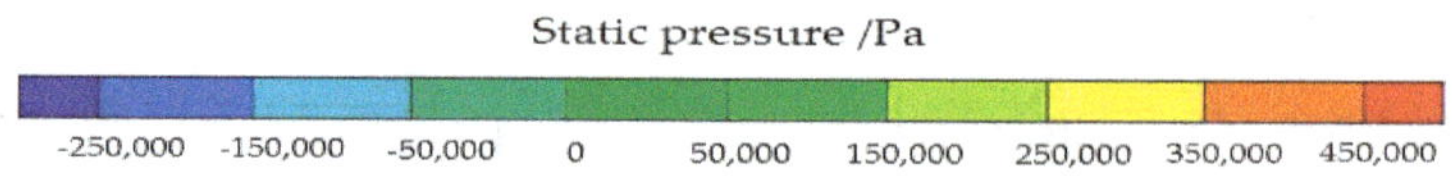

Figure 12. Variation of static pressure in different impellers.

Static pressure in Impellers 1, 2, and 4 fluctuated more obviously and the pressure gradient was higher, which was caused by the intense rotor-stator interaction of the impeller and the volute [25]. Volumetric loss of the centrifugal pump is severe. Thus, the centrifugal pump could not operate efficiently. For Impeller 2, under low flow rate conditions, the distribution scope of lower pressure was much larger than that of other impellers. Centrifugal pumps operating under these conditions were unstable. Compared with Impellers 1, 2, 4, and 5, the lower pressure region in Impeller 3 was much larger under 0.8 Q_d and 1.2 Q_d conditions, mainly caused by secondary flow. Cavitation in Impeller 3 could occur easily, which could make the performances of the centrifugal pump decrease sharply [26]. In Impeller 5, the pressure gradient was the smallest and the lower pressure region was the smallest under all flow rate conditions, proving that the optimization of Impeller 5 is the best. Impeller with $\Phi = 126°$ and blade exit angles $\beta_2 = 24°$ could guarantee the safety operation of the centrifugal pump.

6.2. Variation of Relative Velocity

At the inlet of the impeller, the relative velocity was small. At the outlet of the impeller, the relative velocity attained the maximum, which was in agreement with the experimental and theoretical results [27]. Relative velocity in the pressure surface was much smaller than that of suction pressure. With the increase of flow rate, the low relative speed zone became smaller, as shown in Figure 13.

Under low flow rate conditions, low relative velocity zone in Impeller 5 was smaller than that of other impellers, which could guarantee the stable operation of the centrifugal pump. Velocity gradient in Impellers 1–4 was larger and the flow in these impellers was disorderly. Under rated flow rate conditions, the low speed velocity zone in Impeller 1 was the most obvious. Although differences of low speed zones of Impellers 2–5 were smaller, the zone of Impeller 5 was smaller than that of Impellers 2–4. Under high flow rate conditions, the low velocity speed zone of Impeller 5 was the smallest. Velocity sudden change did not appear. The distribution of relative velocity was uniform, which reflected that blade wrap angle $\Phi = 126°$ and blade exit angle $\beta_2 = 24°$ could let the centrifugal pump operate stably. The hydraulic performances of Impeller 5 were better than others.

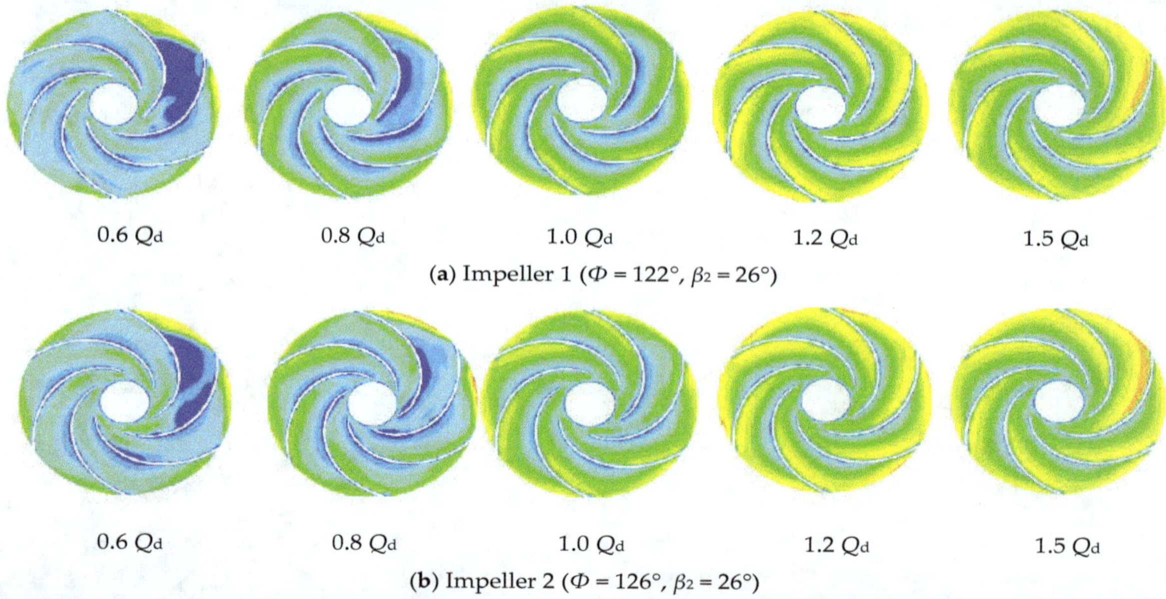

Figure 13. *Cont.*

0.6 Q_d 0.8 Q_d 1.0 Q_d 1.2 Q_d 1.5 Q_d

(c) Impeller 3 ($\Phi = 130°$, $\beta_2 = 26°$)

0.6 Q_d 0.8 Q_d 1.0 Q_d 1.2 Q_d 1.5 Q_d

(d) Impeller 4 ($\Phi = 126°$, $\beta_2 = 28°$)

0.6 Q_d 0.8 Q_d 1.0 Q_d 1.2 Q_d 1.5 Q_d

(e) Impeller 5 ($\Phi = 126°$, $\beta_2 = 24°$)

Relative velocity / m/s

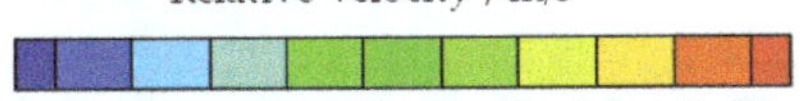

Figure 13. Variation of absolute velocity in different impellers.

6.3. Variation of Streamlines

Under low flow rate conditions, streamlines in Impellers 1–4 were disorder and there were manifest vortices, especially for 0.6 Q_d condition, which could cause severe energy loss and secondary flow. For Impeller 2, the vortices were the most obvious, as shown in Figure 14b. The centrifugal pumps under these two flow rate conditions operated unstably, which could induce severe vibrations and noises [28] and the efficiency of the centrifugal pumps decreased. However, the streamlines in Impeller 5 under these two conditions were smooth. The secondary flow could be avoided successfully. Compared with low flow rate conditions, streamlines in rated and high flow rate conditions became smooth. However, streamlines of Impellers 1–4 were more disorder than that of Impeller 5. In Impellers 1, 2, and 4, the low speed region on the pressure surface was dramatic, which could let the flow in suction surface was faster. Energy loss in this region could be caused easily. Overall, the streamlines in Impeller 5 were the smoothest, as displayed in Figure 14e, which could result in lower energy loss. The flow separation and back flow could be controlled. The centrifugal pump could operate stably.

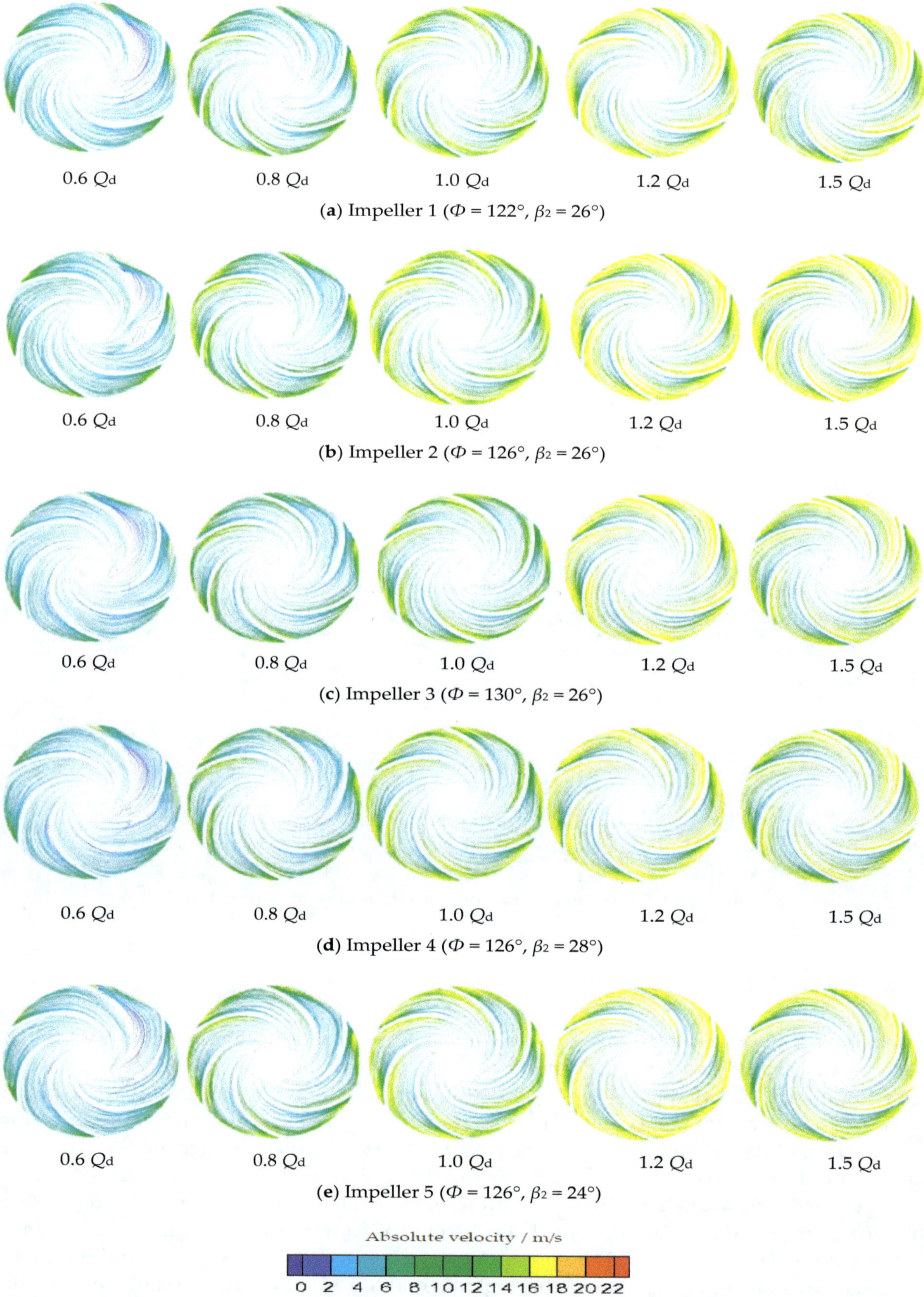

Figure 14. Variation of streamlines in different impeller.

6.4. Variation of Turbulent Kinetic Energy

The distribution characteristics of turbulent kinetic energy in the middle span of the impeller were sharply different under diverse flow rate conditions, which were shown in Figure 15. The distribution scope became smaller gradually with the increase of flow rates. Under the 0.6 Q_d condition, the distribution range of turbulent kinetic energy was the largest and under the 1.5 Q_d condition, the scope was the smallest for all the impellers.

Under low flow rate conditions, the distribution scope of turbulent kinetic energy in Impeller 2 was larger than that of other impellers. Velocity vector had obvious change when water flows from the impeller to the volute. Centrifugal pump with Impeller 2 could not operate stably. Under rated flow rate conditions, the turbulent kinetic energy distribution scopes of Impellers 1, 3, and 4 were similar and they were slightly larger than that of Impeller 5. For high flow rate conditions, the distribution scope of turbulent kinetic energy was almost identical for all the impellers.

Based on the above analysis of turbulent kinetic energy and according to Equation (7), turbulent intensity l [29] of Impeller 5 was less intense than that of Impellers 1–4 overall. Thus, the energy loss in Impeller 5 was less than other impellers, which could guarantee the high transportation capacity of the centrifugal pump.

$$k = \frac{3}{2}(v_{\text{in}} l)^2 \tag{10}$$

Here, v_{in} is the inlet velocity. l is the turbulent intensity.

0.6 Q_d 0.8 Q_d 1.0 Q_d 1.2 Q_d 1.5 Q_d

(**a**) Impeller 1 (Φ = 122°, β_2 = 26°)

0.6 Q_d 0.8 Q_d 1.0 Q_d 1.2 Q_d 1.5 Q_d

(**b**) Impeller 2 (Φ = 126°, β_2 = 26°)

0.6 Q_d 0.8 Q_d 1.0 Q_d 1.2 Q_d 1.5 Q_d

(**c**) Impeller 3 (Φ = 130°, β_2 = 26°)

Figure 15. *Cont.*

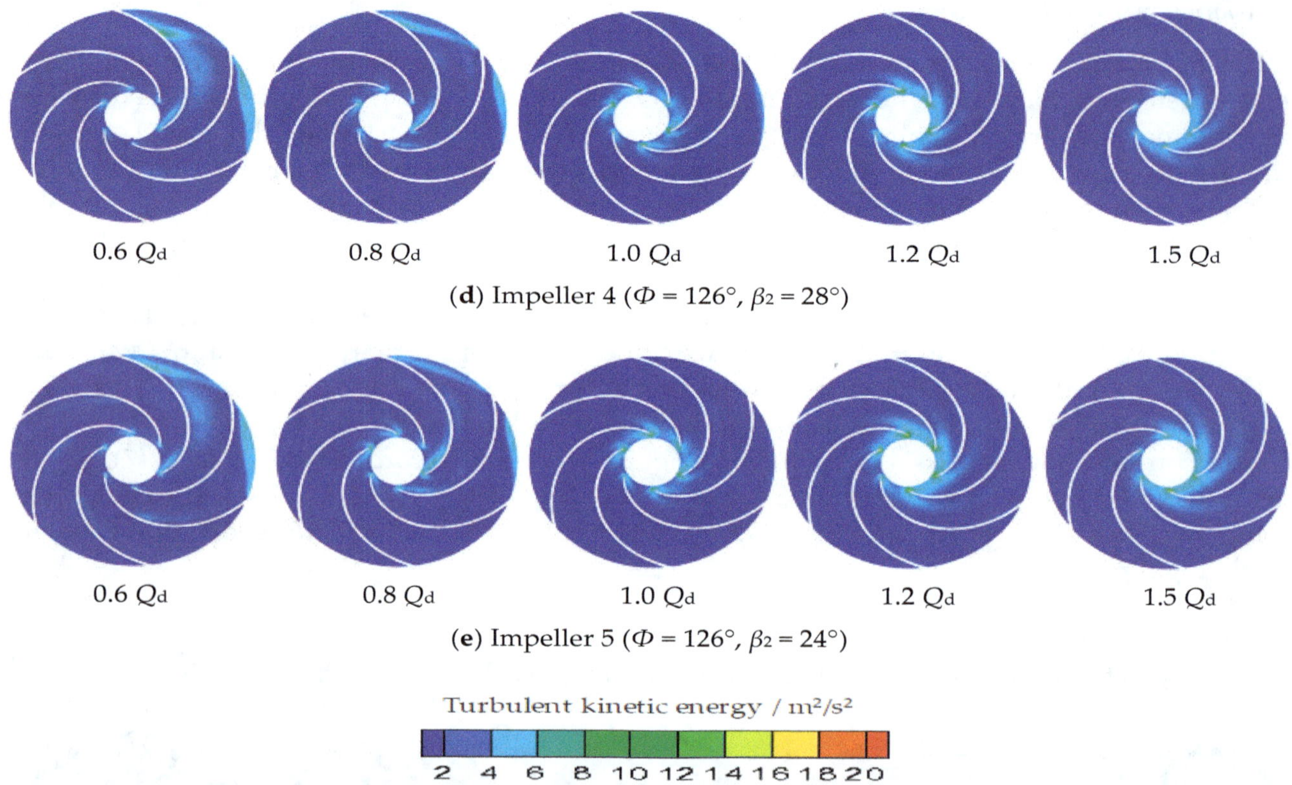

Figure 15. Variation of turbulent kinetic energy in different impellers.

6.5. Variations of Head and Efficiency

With the increase of flow rate, head decreased gradually and efficiency increased firstly then decreased, as exhibited in Figure 16. Differences of head were relatively slight. A head with $\Phi = 122°$ was higher than that of $\Phi = 126°$. The head of the centrifugal pump with $\Phi = 130°$ was the highest one. Efficiency attained the maximum under designed flow rate. Efficiency with $\Phi = 130°$ was higher than that of $\Phi = 122°$ and $126°$. Compared with Schemes 1 and 2, Scheme 3 was the optimal one.

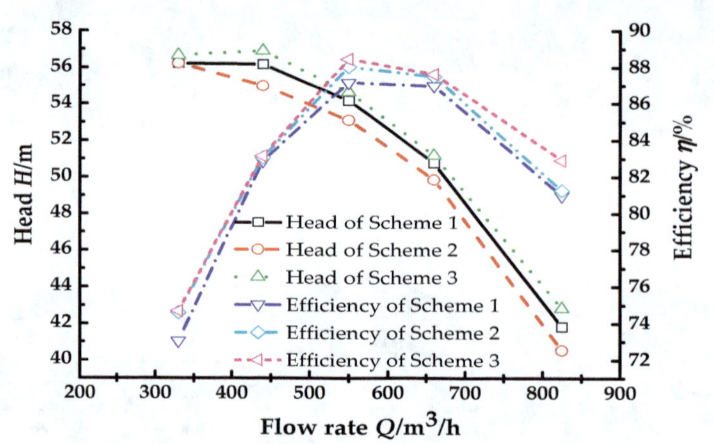

Figure 16. Hydraulic performances of the centrifugal pump with different blade wrap angles.

Schemes 2, 4, and 5 reflected the effects of blade exit angles on hydraulic performances of the centrifugal pump, as displayed in Figure 17. Head decreased constantly and efficiency increased firstly then decreased with the increase of flow rate. The efficiency attained the maximum under the rated flow rate. Head with $\beta_2 = 24°$ was slightly higher than that of $\beta_2 = 28°$. Differences of head with

β_2 = 26° and 28° were sharp. The distinctions of efficiency were smaller. Efficiency of centrifugal pump with β_2 = 24° was the highest. Compared with Schemes 2, 4, and 5, the centrifugal pump with β_2 = 24° had the best hydraulic performances.

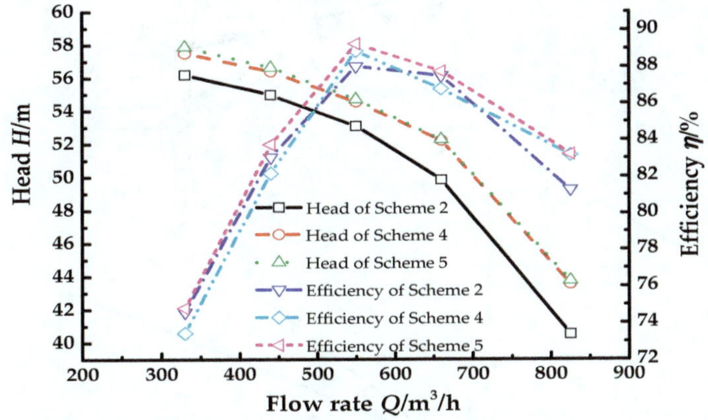

Figure 17. Hydraulic performances of the centrifugal pump with different blade exit angles.

In Scheme 3, blade wrap angle Φ = 130° was the optimal one. The corresponding blade exit angle β2 = 26°. In Scheme 5, blade exit angle β2 = 24° was the best one. The corresponding blade wrap angle Φ = 126°. Head and efficiency of Scheme 5 were higher than that of Scheme 3, as shown in Figure 18. So, Scheme 5 was selected to improve the hydraulic performances of the centrifugal pump.

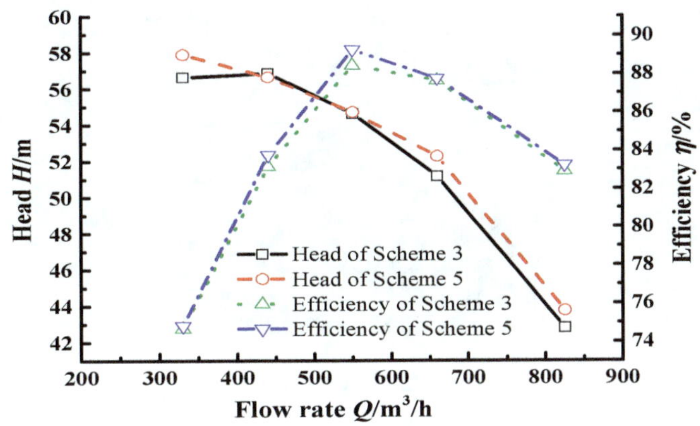

Figure 18. Hydraulic performances of Schemes 3 and 5.

7. Experimental Analysis

Via numerical simulation in Part 6, Scheme 5 was determined to be the best one. The corresponding impeller was machined. Then, the performance experiments were done in the closed performance experiment rig of Shanghai Kaiquan Pump Group to verify that head and efficiency of the centrifugal pump with the Impeller 5 showing obvious improvement.

Experiment Setup

The impeller machining process was displayed in Figure 19. At the outlet of the impeller, the allowance was 5 mm, as shown in region A. Allowance of the impeller front shroud and back shroud was 3 mm, as exhibited in region B. At the inlet of the impeller, the allowance was 3 mm too, as displayed in region C. Allowance of the hub was 3 mm, as shown in region D.

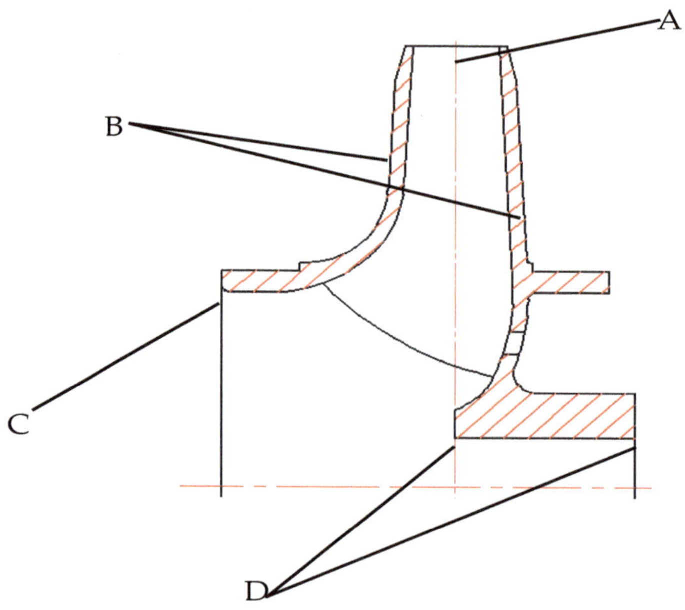

Figure 19. The machining process of the impeller.

To measure the optimized design of the impeller of Scheme 5, 3D wax pattern impeller, given in Figure 20, was machined firstly in the HRPS-V rapid prototyping machine, as shown in Figure 21. Laminated object manufacturing (LOM) in this machine—which is based on the data coming from CAD laying model, CO_2 laser beam, and hot pressing device—was employed to machine the 3D wax pattern impeller.

Figure 20. 3D wax pattern.

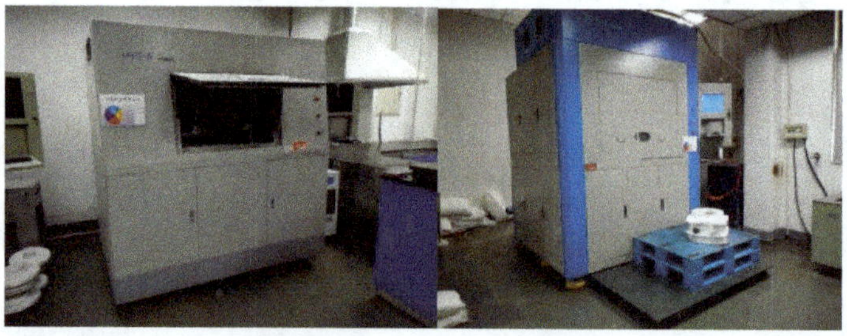

Figure 21. HRPS-V rapid prototyping machine.

Then, the casted impeller of Scheme 5 was machined, which was displayed in Figure 22. The material was 2Cr13. To guarantee that the passages of the impeller were smooth and clean, silt particles and cast joint flash in the passages were cleaned carefully.

Figure 22. The casted impeller.

The performance experiments of the centrifugal pump were done at the closed experiment rig of Shanghai Kaiquan Pump Group. The experiment rig, given in Figure 23, includes the model centrifugal pump, suction pipes, exit pipes, pressure gauges, valves, and flow meters. The type of pressure sensor is XU12087105 (1.0, Shanghai Automation Instrumentation Co. Ltd., Shanghai, China). The type of flow meter is DN300 (1.0, Tianjin Flow Meter Co. Ltd., Tianjin, China). The type of valve is ZA2.T. All testing precisions are national grade 1 (GB3216-2005 and ISO 9906-1999).

Figure 23. Performance experiment rig.

Figure 24 is the typical components of the closed performance experiment rig. It includes 16 different parts.

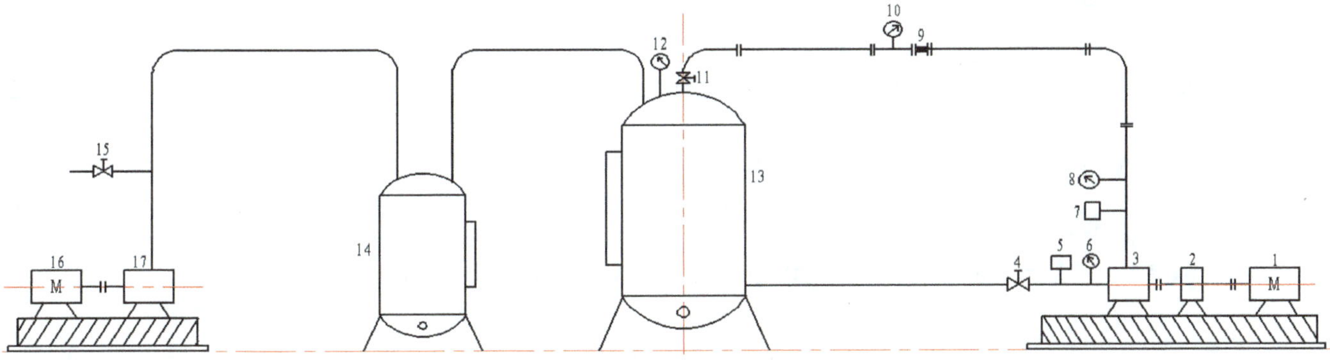

Figure 24. Components of the closed performance experiment rig. 1–electric motor; 2–torque and speed sensor; 3–centrifugal pump; 4–sluice valve; 5–inlet pressure sensor; 6–turbine flow meter; 7–outlet pressure sensor; 8–turbine flow meter; 9–expansion joint; 10–turbine flow meters; 11–regulating valve; 12–turbine flow meters; 13–cavitation tank; 14–separation tank of vapor and water; 15–ball valve; 16–electric motor; 17–vacuum pump.

Via the inlet pressure Sensor 5 and the outlet pressure Sensor 7, M_1 and M_2 were determined. According to Equations (11) and (12), the inlet pressure and the outlet pressure of the centrifugal pump could be got.

The inlet pressure is calculated by

$$\frac{p_1}{\rho g} = 102M_1 + h_1 \tag{11}$$

The outlet pressure is determined by

$$\frac{p_2}{\rho g} = 102M_2 - h_3 \tag{12}$$

where h_1 is the vertical height between the center of inlet pressure sensor with axial lead of the centrifugal pump. h_3 is the vertical height between the center of the outlet pressure sensor with the outlet pressure port.

The inlet flow rate (Q_1) and outlet flow rate (Q_2) of the centrifugal pump could be measured by turbine flow meter 6 and 8, respectively.

Thus, the inlet velocity of the centrifugal pump v_1 could be calculated by

$$v_1 = \frac{Q_1}{A_1} \tag{13}$$

The outlet velocity of the centrifugal pump v_2 could be got by

$$v_2 = \frac{Q_2}{A_2} \tag{14}$$

Here, A_1 is the inlet cross-section area. A_2 is the outlet cross-section area.

Head (H) could be calculated by Equation (15)

$$H = \left(\frac{p_2}{\rho g} - \frac{p_1}{\rho g}\right) + \left(\frac{v_2^2}{2g} - \frac{v_1^2}{2g}\right) + (Z_2 - Z_1) \tag{15}$$

where Z_2 and Z_1 are the static head.

In the performance experiment rig,

$$Z_1 = 0 \tag{16}$$

For Z_2, it could be obtained by

$$Z_2 = h_2 + h_3 \tag{17}$$

Here, h_2 is the vertical height between the center of outlet pressure sensor with axial lead of the centrifugal pump.

In the performance experiment rig,

$$h_2 = h_1 \tag{18}$$

The Equation (15) could be modified. The new form was shown in Equation (20).

$$H = 102(M_2 - M_1) + \left(\frac{v_2^2}{2g} - \frac{v_1^2}{2g}\right) \tag{19}$$

Efficiency (η) could be calculated by Equation (16)

$$\eta = \frac{\rho g Q H}{1000P} \times 100\% \tag{20}$$

Here, P is the inlet power, which could be measured by torque and speed sensor 2.

Hydraulic performance curves of the improved centrifugal pump and the original centrifugal pump were shown in Figure 25. Experimental results indicated that head and efficiency of the improved centrifugal pump were higher than that of the original centrifugal pump. Under low flow rate condition (0.8 Q_d), the head increased by 3.76 m and the efficiency increased by 3.84%. With rated flow rate condition (1.0 Q_d), the head increased by 2.74 m and the efficiency increased by 5.77%, separately. For high flow rate condition (1.2 Q_d), the head and efficiency increased by 2.67 m and 5.0%, respectively. The experimental results indicated that the optimized design is successful.

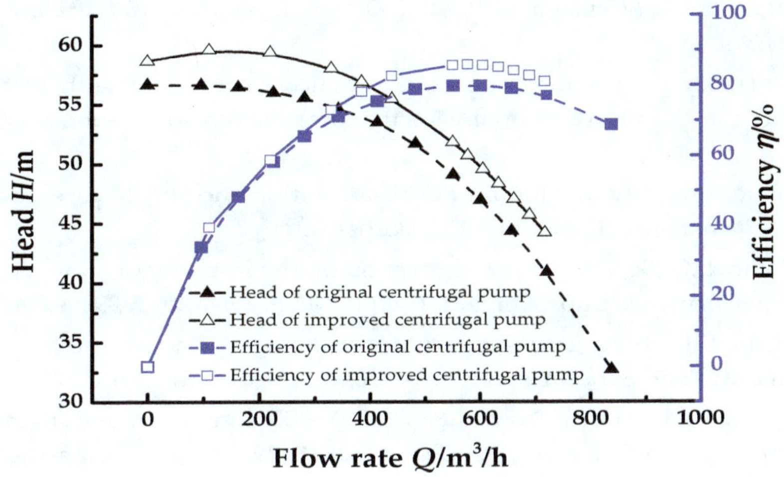

Figure 25. Hydraulic performance of improved and original centrifugal pump.

8. Conclusions

In this paper, effects of blade exit angle and blade wrap angle on the optimized design of the impeller were comprehensively investigated. Flows in the centrifugal pump with five different impellers under low flow rate, rated flow rate, and high flow rate were numerically simulated. Variations of static pressure, relative velocity, streamlines, and turbulent kinetic energy were analyzed. Head and efficiency of the centrifugal pump of different schemes were compared. The experimental head and efficiency of the centrifugal pump with the best impeller were compared with that of the original impeller. The main conclusions are as follows:

(1) The impeller with blade wrap angle 126° and blade exit angle 24° was the best one.

(2) For the best impeller, static pressure and relative velocity was the most uniform distribution. Streamlines were the smoothest and vortices did not exist. Compared with other impellers, the distribution scope of turbulent kinetic energy in the best impeller under all flow rate conditions was the smallest.

(3) For the centrifugal pump with optimized impeller, head and efficiency were higher than that of the original pump. With low flow rate (0.8 Q_d), the head and efficiency increased by 3.76 m and 3.84%. With rated flow rate, the head and efficiency increased by 2.74 m and 5.77%. With high flow rate (1.2 Q_d), the head and efficiency increased by 2.67 m and 5.0%.

Author Contributions: X.H. and Y.K. presented the optimal scheme and designed the experiments. X.H. and D.L. made the numerical simulation and performed the experiment. W.Z. analyzed the data. X.H. wrote the paper.

Acknowledgments: This research is financially supported by the National Key Basic Research Program of China (no. 2014CB239203), the National Natural Science Foundation of China (no. 51474158), and the Hubei Provincial Natural Science Foundation of China (no. 2016CFA088). We deeply acknowledge the help of Hubei Key Laboratory of Waterjet Theory and New Technology, School of Energy and Power Engineering of Lanzhou University of Technology, and Shanghai Kaiquan Group.

References

1. Yedidiah, S. *Centrifugal Pump User's Guidebook: Problems and Solutions*; Springer Group: Berlin, Germany, 1996.
2. Gülich, J.F. *Centrifugal Pumps*; Springer Group: Berlin, Germany, 2008.
3. Karassik, I.J. *Centrifugal Pump Clinic*; Taylor Francis Inc.: Oxford, UK, 1989.
4. Tuzson, J. *Centrifugal Pump Design*; Wiley-Interscience: Washington, DC, USA, 2000.
5. Grist, E. *Cavitation and the Centrifugal Pump: A Guide for Pump Users*; Taylor Francis Inc.: Oxford, UK, 1998.
6. Lobanoff, V.S.; Ross, R.R. *Centrifugal Pump: Design and Application*; Gulf Professional Publishing: Houston, TX, USA, 1992.
7. Girdhar, P.; Moniz, O. *Practical Centrifugal Pumps. Design, Operation, Maintenance*; Newnes: Amsterdam, The Netherlands, 2005.
8. Wu, Z.H. Three-dimensional turbomachine flow equations expressed with respect to non-orthogonal curvilinear coordinates and non-orthogonal velocity components and methods of solution. *J. Mech. Eng.* **1979**, *15*, 1–23.
9. Zhang, Z.H. Streamline similarity method for flow distribution and shock losses at the impeller inlet of the centrifugal pump. *J. Hydrodyn.* **2018**, *30*, 140–152. [CrossRef]
10. Pei, J.; Wang, W.J.; Yuan, S.Q.; Zhang, J.F. Optimization on the impeller of a low-specific-speed centrifugal pump for hydraulic performance improvement. *Chin. J. Mech. Eng.* **2017**, *29*, 992–1002. [CrossRef]
11. Tan, L.; Zhu, B.S.; Cao, S.L.; Bing, H.; Wang, Y.M. Influence of blade wrap angel on centrifugal pump performance by numerical and experimental study. *Chin. J. Mech. Eng.* **2014**, *27*, 171–177. [CrossRef]
12. Kim, J.H.; Lee, H.C.; Kim, J.H.; Kim, S.; Yoon, J.Y.; Choi, Y.S. Design techniques to improve the performance of a centrifugal pump using CFD. *J. Mech. Sci. Technol.* **2015**, *29*, 215–225. [CrossRef]
13. Stepanoff, A.J. *Centrifugal and Axial Pump: Theory, Design and Application*; John Wiley Sons Inc.: Hoboken, NJ, USA, 1986.
14. Yang, A.L.; Lang, D.P.; Li, G.P.; Chen, E.Y.; Dai, R. Numerical research about influence of blade outlet angel on flow-induced noise and vibration for centrifugal pump. *Adv. Mech. Eng.* **2015**, *6*, 1–11.
15. Heo, M.W.; Kim, J.H.; Seo, T.W.; Kim, K.Y. Aerodynamic and aeroacoustic optimization for design of a forward-curved blades centrifugal fan. *Proc. Inst. Mech. Eng. Part A J. Power Energy* **2017**, *230*, 154–174. [CrossRef]
16. Guan, X.F. *Modern Pumps Theory and Design*; China Astronautic Publishing House: Beijing, China, 2011. (In Chinese)
17. Bai, Y.X.; Kong, F.Y.; Yang, S.S.; Chen, K.; Dai, T. Effect of blade wrap angel in hydraulic turbine with forward-curved blades. *Int. J. Hydrogen Energy* **2017**, *42*, 18709–18717. [CrossRef]
18. Wang, W.J.; Pei, J.; Yuan, S.Q.; Zhang, J.F.; Yuan, J.P.; Xu, C.Z. Application of different surrogate models on the optimization of centrifugal pump. *J. Mech. Sci. Technol.* **2016**, *30*, 567–574. [CrossRef]
19. Zhou, L.; Shi, W.D.; Wu, S.Q. Performance optimization in a centrifugal pump impeller by orthogonal experiment and numerical simulation. *Adv. Mech. Eng.* **2013**, *6*, 1–11. [CrossRef]
20. Nelik, L. *Centrifugal and Rotary Pumps: Fundamentals with Applications*; CRC Press: London, UK, 1999.
21. Li, W.G.; Su, F.Z.; Ye, Z.M.; Xia, D.L. Experiment on effect of blade pattern on performance of centrifugal oil pumps. *Chin. J. Appl. Mech.* **2002**, *19*, 31–34. [CrossRef]
22. Lohner, R. *Applied Computational Fluid Dynamics Techniques: An Introduction Based on Finite Element Methods*; John Wiley & Sons: Hoboken, NJ, USA, 2008.
23. Yakhot, V.; Orzag, S.A. Renormalization group analysis of turbulence: Basic theory. *J. Sci. Comput.* **1986**, *1*, 3–51. [CrossRef]
24. Zhou, L.; Shi, W.D.; Li, W.; Agarwal, R. Numerical and experimental study of axial force and hydraulic performance in a deep-well centrifugal pump with different impeller rear shroud radius. *J. Fluid Eng.* **2013**, *135*, 749–760. [CrossRef]
25. Posa, A.; Lippolis, A. A LES investigation of off-design performance of a centrifugal pump with variable-geometry diffuser. *Int. J. Heat Fluid Flow* **2018**, *70*, 299–314. [CrossRef]
26. Chen, H.X.; He, J.W.; Liu, C. Design and experiment of the centrifugal pump impellers with twisted inlet vice blades. *J. Hydrodyn.* **2017**, *29*, 1085–1088. [CrossRef]

27. Yang, S.L.; Kong, F.Y.; Chen, H.; Su, X.H. Effects of blade wrap angel influencing a pump as turbine. *J. Fluid Eng.* **2012**, *134*, 1–8. [CrossRef]
28. Cheah, K.W.; Lee, T.S.; Winoto, S.H.; Zhao, Z.M. Numerical flow simulation in a centrifugal pump at design and off-design conditions. *Int. J. Rotating Mach.* **2007**, *2*, 1–9. [CrossRef]
29. Ferziger, J.H.; Peric, M. *Computational Methods for Fluid Dynamics*; Springer Group: Berlin, Germany, 2002.

5

Computational Study of the Noise Radiation in a Centrifugal Pump When Flow Rate Changes

Ming Gao [1,*], Peixin Dong [1,2], Shenghui Lei [3] and Ali Turan [4]

1 School of Energy and Power Engineering, Shandong University, Jinan 250061, China; paytonsdu@hotmail.com
2 School of Mechanical and Mining Engineering, University of Queensland, Brisbane 4067, Australia
3 Thermal Management Research Group, Efficient Energy Transfer (ηET) Department, Bell Labs Ireland, Nokia, Dublin D15 Y6NT, Ireland; shenghui.lei@nokia-bell-labs.com
4 School of Mechanical, Aerospace and Civil Engineering, University of Manchester, Manchester M60 1QD, UK; a.turan@manchester.ac.uk
* Correspondence: gm@sdu.edu.cn.

Academic Editor: Bjørn H. Hjertager

Abstract: Noise radiation is of importance for the performance of centrifugal pumps. Aiming at exploring noise radiation patterns of a typical centrifugal pump at different flow rates, a three-dimensional unsteady hydro/aero acoustic model with large eddy simulation (LES) closure is developed. Specifically, the Ffowcs Williams-Hawkings model (FW-H) is employed to predict noise generation by the impeller and volute. The simulated flow fields reveal that the interactions of the blades with the volute induce root mean square (RMS) pressure and further lead to noise radiation. Moreover, it is found that the profiles of total sound pressure level (*TSPL*) regarding the directivity field for the impeller-generated noise demonstrate a typical dipole characteristic behavior, whereas strictly the volute-generated noise exhibits an apparently asymmetric behavior. Additionally, the design operation (Here, 1 Q represents the design operation) generates the lowest *TSPL* vis-a-vis the off-design operations for all the flow rates studied. In general, as the flow rates decrease from 1 Q to 0.25 Q, *TSPL* initially increases significantly before 0.75 Q and then levels off afterwards. A similar trend appears for cases having the larger flow rates (1–1.25 Q). The *TSPL* deviates with the radiation directivity and the maximum is about 50%. It is also found that *TSPL* by the volute and the blades can reach ~87 dB and ~70 dB at most, respectively. The study may offer a priori guidance for the experimental set up and the actual design layout.

Keywords: centrifugal pump; 3D flow field; varying flow rate; impeller and volute radiation noise; total sound pressure level (*TSPL*)

---

1. Introduction

Centrifugal pumps as turbomachinery components have been extensively used to transport fluids by converting the rotational kinetic energy consumed to the hydrodynamic energy of the flow [1]. In general, the centrifugal pump routinely operates under various conditions imposed via different operational requirements; hence, a wide frequency range regarding the noise radiation typically occurs and needs to be managed. It has been realized that the noise generation within a centrifugal pump can significantly degrade the pump performance, shift the working point away from the optimal design, reduce the pump efficiency, and further consume extra external energy [2]. In addition, it may give rise to some issues such as safety and reliability. Therefore, it is extremely important to have a better understanding of the hydro/aero acoustics behavior of centrifugal pumps to incorporate in comprehensive pump designs.

In the past, considerable research has been carried out both experimentally and numerically in terms of noise generation in centrifugal pumps. It is generally assumed that the noise sources in the pump mainly comprise of mechanical, (structure borne) noise and flow-induced noise [3]. Choi et al. [4] reported that the noise generation is caused by the large scale flow field instability in a pump impeller. Recently, computational fluid dynamics (CFD) were widely used to elucidate and predict the hydro/aero acoustic generation [5,6]. Chu et al. [7,8] numerically studied a centrifugal pump and reported that the vibration and radiation of noise is mainly determined by the unsteady flow characteristics in centrifugal pumps, which made a great contribution to settle flow-induced noise by CFD methods. More recently, Langthjem et al. [9,10] employed Lighthill's acoustic analogy to predict the radiation of noise for a centrifugal pump based on the 2D flow field. Kato et al. [11,12] reported pressure fluctuation levels on the volute wall using large eddy simulation (LES) in a finite element formulation. Huang [13] also performed numerical simulations on a centrifugal pump via LES and the Ffowcs Williams-Hawkings (FW-H) acoustic model to study the influence of different blade shapes on the efficiency of noise radiation in a centrifugal pump.

Notwithstanding, all the above previous studies concerned radiation of noise in centrifugal pumps, most authors regarded the centrifugal pump as an all-encompassing composite source of sound. However, in this paper, attention is first focused specifically on noise generation by the impeller and/or the volute to map out individual/separate contributions and interactions via each source. In addition, it is rather practical and common for pumps to work with varying flow rates, thus, further emphasis has been put on elucidating the changing rules of radiation noise from both impeller and volute with varying flow rates. To that end a numerical model incorporating LES in conjunction with the FW-H approach in a finite volume environment (FLUENT) is developed. Detailed comparisons with experimental/analytical results on a 2D/3D case are performed to validate the model. The instantaneous flow fields on the blade and volute surfaces are discussed in detail to highlight the nature of the field profiles for the total sound pressure level (*TSPL*, defined using Eqation (8)) obtained to quantitatively reveal the noise distribution induced by different sound sources at varying flow rates.

2. Materials and Methods

2.1. Flow Field Solver

The turbulent flow field within a centrifugal pump is very complex and, in particular, it is challenging to adequately capture the pressure fluctuations accompanying the flow required in the acoustic analysis. To carry out hydro/aero acoustics predictions for a centrifugal pump, three-dimensional transient Navier-Stokes equations are solved for an incompressible fluid using FLUENT [14]. Large eddy simulation is employed to close the governing equations. In the study, a widely accepted sub-grid scale, SGS, model is used to compute the SGS Reynolds stress field via:

$$\tau_{ij} = \frac{1}{3}\delta_{ij}\tau_{kk} - 2\upsilon_{SGS}\overline{S_{ij}}, \tag{1}$$

where:

$$\upsilon_{SGS} = (C_S\Delta)^2|\overline{S}|, \quad |\overline{S}| = \sqrt{2\overline{S_{ij}}\overline{S_{ij}}}, \quad \overline{S_{ij}} = \frac{1}{2}\left(\frac{\partial\overline{u_i}}{\partial x_j} + \frac{\partial\overline{u_j}}{\partial x_i}\right), \tag{2}$$

where C_s is the Smagorinsky constant, and Δ is the size of the grid filter.

The governing equations with the LES formulation are discretized using a second-order scheme for the spatial terms and a second-order implicit scheme for the temporal terms. The pressure implicit with splitting of operator (PISO) algorithm is used to solve the pressure-velocity coupling equation [15].

2.2. The Acoustic Submodel

To further capture the noise radiation in a centrifugal pump induced by the inner flow field encompassing the impellers and/or volute, the FW-H acoustic model is employed, in line with the built in assumptions that the flow speed is low, noise is primarily produced by the unsteady pressure field and the chosen noise sources sit in an open region when calculating the level of radiation noise which means the interaction between solid and fluid is neglected. For this model, it is inherently required that all the receivers are located far away from the primary sources of sound. Based on the transient flow simulation discussed above, the FW-H model can be expressed as [16]:

$$\frac{1}{a_0^2}\frac{\partial^2 p'}{\partial t^2} - \nabla^2 p' = \frac{\partial}{\partial t}\{[\rho_0 v_n + \rho(u_n - v_n)]\delta(f)\} - \frac{\partial}{\partial x_i}\{[p_{ij}n_j + \rho u_i(u_n - v_n)]\delta(f)\} + \frac{\partial^2}{\partial x_i \partial x_j}\{T_{ij}H(f)\}, \quad (3)$$

where p' is the far-field acoustic pressure, a_0 is the speed of sound, ρ_0 is the reference density of the fluid, f is the surface, u_i and v_i are the flow velocity component and the surface velocity component in the x_i direction respectively, u_n and v_n is the flow velocity component and the surface velocity normal to the surface f respectively, n_j is the unit normal vector, $\delta(f)$ is Dirac delta function and $H(f)$ is Heaviside function. Here T_{ij} is the Lighthill stress tensor given by:

$$T_{ij} = \rho u_i u_j + p_{ij} - a_0^2(\rho - \rho_0)\delta_{ij}, \quad (4)$$

where p_{ij} is the stress tensor expressed as:

$$p_{ij} = p\delta_{ij} - \mu\left[\frac{\partial u_i}{\partial x_j} + \frac{\partial u_j}{\partial x_i} - \frac{2}{3}\frac{\partial u_k}{x_k}\delta_{ij}\right], \quad (5)$$

Physically, the three RHS terms appearing in Equation (3) correspond to monopole, dipole, and quadrupole acoustic sources respectively. In this study, the quadrupole noise source is ignored [17] in view of the flow having low speed compared with the local acoustic velocity. In addition, for a well-designed centrifugal pump, the monopole noise source is relatively weak and also neglected. Thus, the dipole noise induced by the wall surface encompassing the transient flow RMS pressure is the primary concern in this paper.

Furthermore, according to the GB/T29529-2013 standard [18], the volute and/or impeller wall surfaces chosen as monitoring stations are employed to study the impeller and/or volute noise radiation. As shown in Figure 1, the monitoring surface is circular, having a radius of 1000 mm; 24 receivers denoted by P_1–P_{24} are evenly arranged along the surface and, in particular, P_1 is placed strictly towards the volute tongue. A Fourier transformation involving the acoustic signals obtained at these receivers is applied to extract the relevant acoustic spectra. The sound pressure level (*SPL*) is calculated by:

$$SPL = 20\ \log_{10}\frac{P_e}{P_{ref}}, \mathrm{dB}, \quad (6)$$

where P_{ref} is the reference sound pressure (=2×10^{-5} Pa in air), and P_e is the effective value of acoustic pressure defined as:

$$P_e = \sqrt{\frac{1}{T}\int_0^T p'^2 dt}, \quad (7)$$

To reveal the intensity of noise radiation, it is necessary to derive a temporal intensity profile involving a superposition of acoustic pressures at each Fourier frequency. In this regard, the *TSPL* is introduced and expressed as:

$$TSPL = 10\lg\sum_{i=1}^{n} 10^{SPL_i/10}, \quad (8)$$

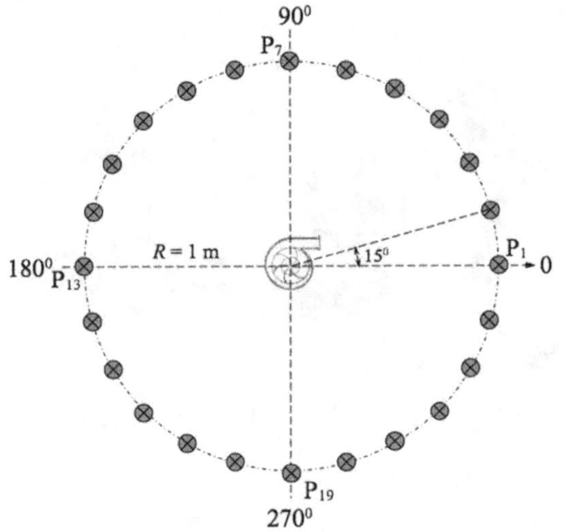

Figure 1. Arrangement of receivers outside the centrifugal pump.

2.3. Numerical Issue Concerning the Set-Up

In this study, a single-stage single-suction centrifugal pump (media, water) having a medium specific speed of 112 is selected. Table 1 lists all the geometric and performance parameters. The 3D computational domain for the simulations comprises of the inlet pipe, impeller, and the volute, as shown in Figure 2. It should be noted that the practical disk friction losses and leakage losses are neglected, but an account is taken of the potential backflow at the inlet and outlet. A hybrid mesh system is generated using GAMBIT (2.4.6, Fluent, Inc., New York, NY, USA) [19], as shown in Figure 3. Considering the complexity of the geometrical layout and the volute tongue, an unstructured grid is used for these domains and an appropriate mesh refinement is further applied, whereas, a structured mesh is used for the remaining regions. Here, particular attention is paid to the near wall regions and the unstructured mesh is automatically refined to have appropriate y^+ values ($y^+ \approx 1$) and thus to ensure the requirement of wall-resolved LES for the unsteady calculation [20]. It is observed that different levels of mesh refinement may lead to different average y^+ values, and when the grid number reaches almost 2.5 million (2.5 M) cells, the average y^+ values decrease to around 1, as shown in Figure 4. With the increase of mesh refinement, the average y^+ values gradually reduce. With a combining consideration of both the mesh accuracy and the limited computing resources, the grid of 2.5 M cells is chosen for the calculation. What is more, it is not difficult to find that the average y^+ values are correctly consistent with the sizes of the first boundary layer. Therefore, different grid systems shown are tested for the required grid-dependency based on the convergence pattern for the pump hydraulic head to the designed value (~28 m), as shown in Table 2. To improve the prediction accuracy while reducing the attendant computational costs, Grid-C is chosen in the study.

Table 1. Geometric and performance parameters used for a simulated centrifugal pump (water).

Parameter	Value	Unit
Impeller outlet diameter D_2	270	mm
Impeller outlet width D	30	mm
Design flow rate Q	80	m^3/h
Design head H	28	m
Design rotational speed n	1450	r/min
Blade number Z	6	1
Axis pass frequency f_a (APF)	24.2	Hz
Blade pass frequency f_b (BPF)	145	Hz

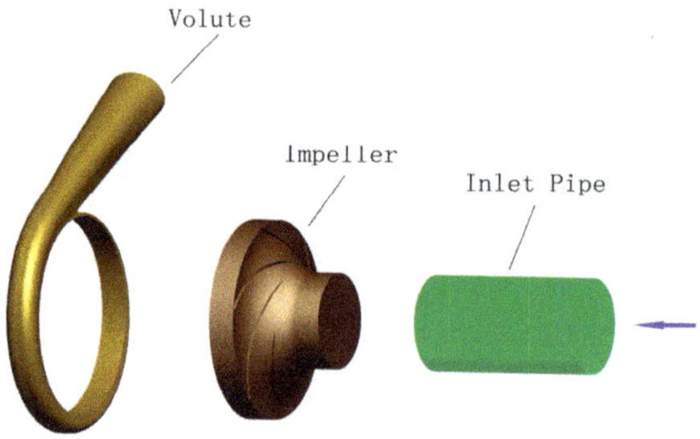

Figure 2. Computational geometry of the centrifugal pump for each part.

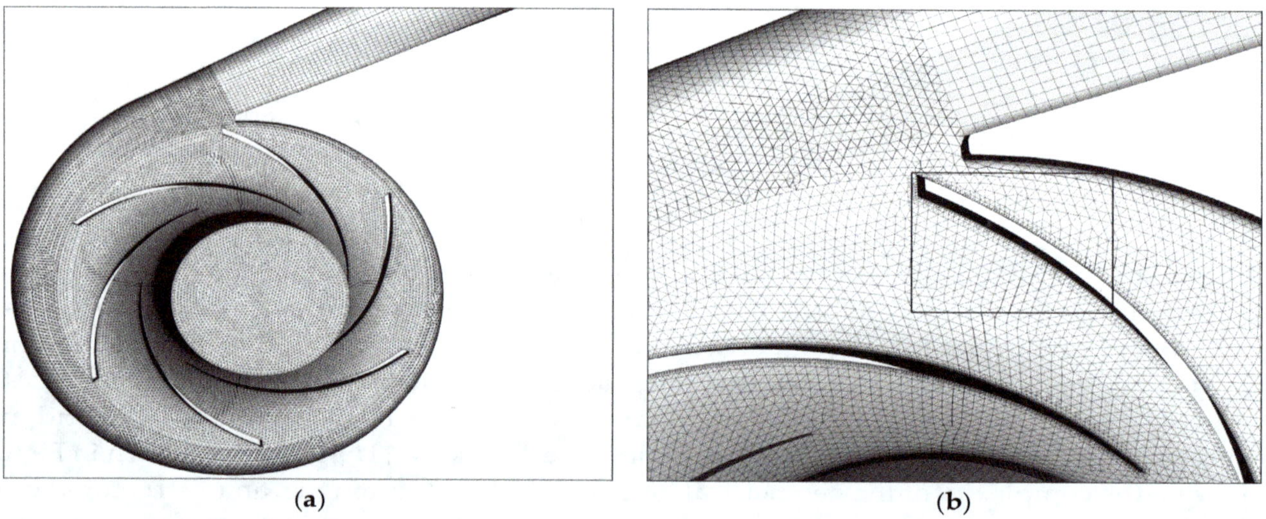

Figure 3. Grid System of the centrifugal pump. (**a**) Overall view; and (**b**) magnified view of the volute tongue.

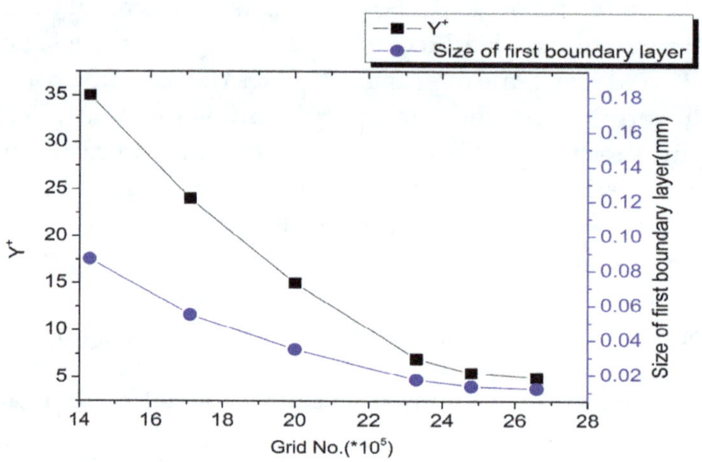

Figure 4. Y^+ values and corresponding sizes of first boundary layer.

Table 2. Grid independent Analysis.

Grid System	Grid Number	Head (m)
Grid-A	2,012,658	28.82
Grid-B	2,324,567	28.70
Grid-C	2,496,543	28.63
Grid-D	2,676,328	28.62

For the boundary conditions, a uniform axial velocity profile is imposed at the inlet while 'outlet' boundary conditions are adopted at the outlet. All the walls are treated as non-slip boundaries. The pre-converged steady flow field obtained via the k-ε is adopted as the initial condition for the subsequent transient wall-resolved LES simulations with a Subgrid-Scale Model of Smagorinsky-Lilly. The time step is specified by:

$$\Delta t = \frac{60}{nK}, \tag{9}$$

where K is the step number in a rotational period of the impeller (=360 in the study), and n is the rotational speed. With the designed values given in Table 1, Δt is 1.1494×10^{-4} s in the study.

3. Results and Discussion

3.1. Numerical Validaton

To verify the FW-H model used, a 2D cylinder case was first studied and its computational domain and the boundary conditions are shown in Figure 5. It should be pointed out that the receiver is placed at a distance of 1.3 m from the center of the cylinder to accord with the experimental set-up [21]. A constant velocity of 40 m/s is specified at all the inlets, whereas an ambient pressure of 1 atm is used at the outlet. A structured mesh system of about 90,000 cells is used and the finest cell has a minimum edge length of 0.005 mm. Figure 6a shows the predicted and measured sound pressure spectra. As can be seen, the prediction matches rather well with the measurements at the receiver. Furthermore, the two peak frequencies at 800 Hz and 1500 Hz are adequately captured. Figure 6b displays the pressure coefficients around the cylinder surface. The simulation displays the overall trend regarding the pressure coefficient obtained in the experiment with a maximum discrepancy less than 10%.

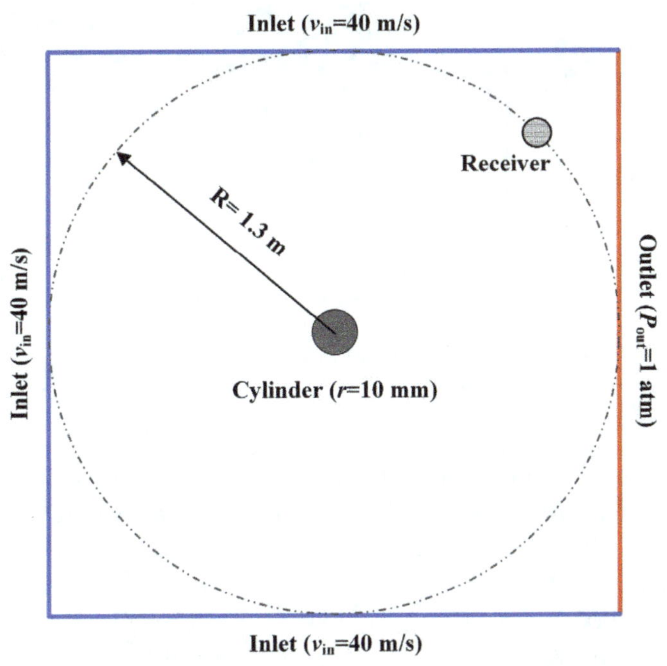

Figure 5. The validation case of a 2D cylinder (not to scale).

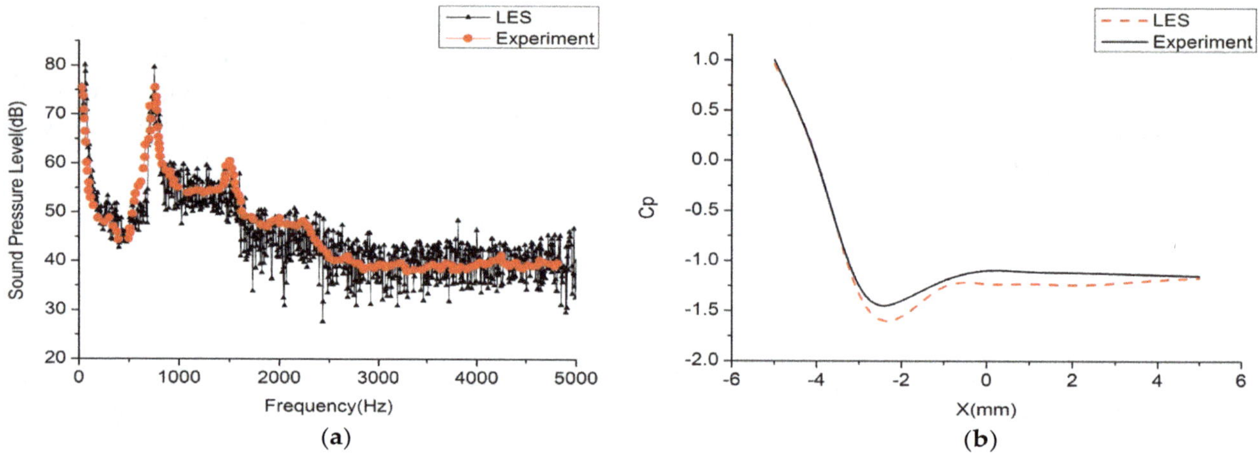

Figure 6. Comparison of numerical predictions with experimental measurements. (**a**) Sound pressure spectra; and (**b**) pressure coefficient (C_p).

Additionally, the numerical method adopted should also be verified. In terms of the 3D case shown in Figure 2, the hydraulic head predicted by the numerical model is compared with the nominal design head for the centrifugal pump. The simulation yields a head of 28.63 m. This prediction agrees well with the designed valued (28 m) and the relative error of the calculation is less than 2.3%. The discrepancy is essentially due to the fact that the disk friction losses and leakage losses appearing in practice are neglected in the simulations. The sound spectra for the receiver 1 in terms of the design flow rate are also shown in Figure 7. The predicted rotational frequency (APF) and blade pass frequency (BPF) are 24.1 Hz and 142.3 Hz respectively, very close to the design values of 24.2 Hz and 145 Hz marked in dashed lines. Clearly, the numerical model presented in the paper is able to yield reasonably accurate predictions on the flow field as well as the noise.

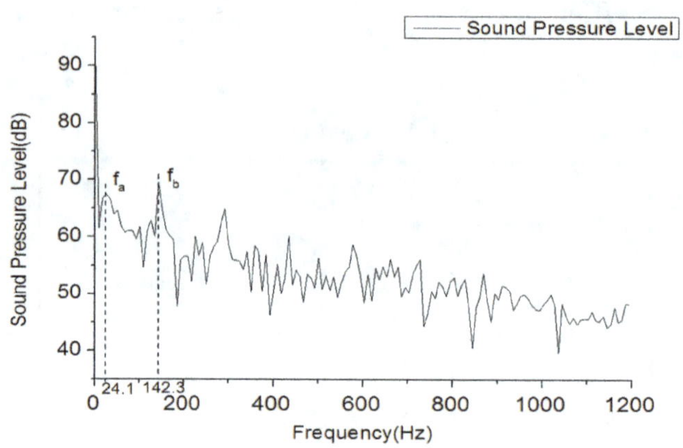

Figure 7. The sound spectra of receiver 1.

3.2. Flow Field

Throttle governing is a common control mode routinely employed in the operational envelope of the centrifugal pump system. By such means installation of the throttle employing valves or baffles and adjusting the opening of the throttle can change the local loss and achieve the desired goal of throttle governing. In this study, 13 flow rate conditions are studied to analyze the noise radiation, including 20 m^3/h, 30 m^3/h, 40 m^3/h, 50 m^3/h, 60 m^3/h, 70 m^3/h, 78 m^3/h, 80 m^3/h (which is the design flow rate), 82 m^3/h, 90 m^3/h, and 100 m^3/h, respectively.

To elucidate the noise generation by the flow field, an index employing a value of RMS pressure field is introduced to provide information for the pressure variations with time step on the wall of the impeller and volute, P_{rms} defined by:

$$P_{rms} = \sqrt{\frac{1}{N}\sum_{i=0}^{N-1} [P(t) - \overline{P}]^2}, \tag{10}$$

where $P(t)$ represents the time-dependent static pressure, $\overline{P}$ is the time-averaged static pressure in a circulation period ($N = 360$ in this study).

Figure 8 describes the distribution of RMS values on the surfaces of impeller and volute. When the flow rate is 0.75 Q, the RMS pressure field primarily resides on the side of inlet, displaying values much stronger vis-a-vis the opposite side, as is shown in Figure 8a. It can be inferred from such considerations that while the flow rate is less than the design value, the velocity field distribution on the impeller is substantially non-uniform, providing for extremely unsteady flow near the inlet side vis-a-vis the outlet. Thus, the primary reason for the impeller noise is the unsteady flow field at 0.75 Q. When the flow rate reaches 1 Q, the average RMS is a minimum and the flow field is also smoothly distributed, especially near the inlet side, yielding the impeller noise under the design flow rate condition to be substantially lower than other flow rate conditions. Additionally, the RMS pressure field displays extreme values near both the inlet and the outlet regions under the 1.25 Q flow rate condition to yield a maximum noise field generated.

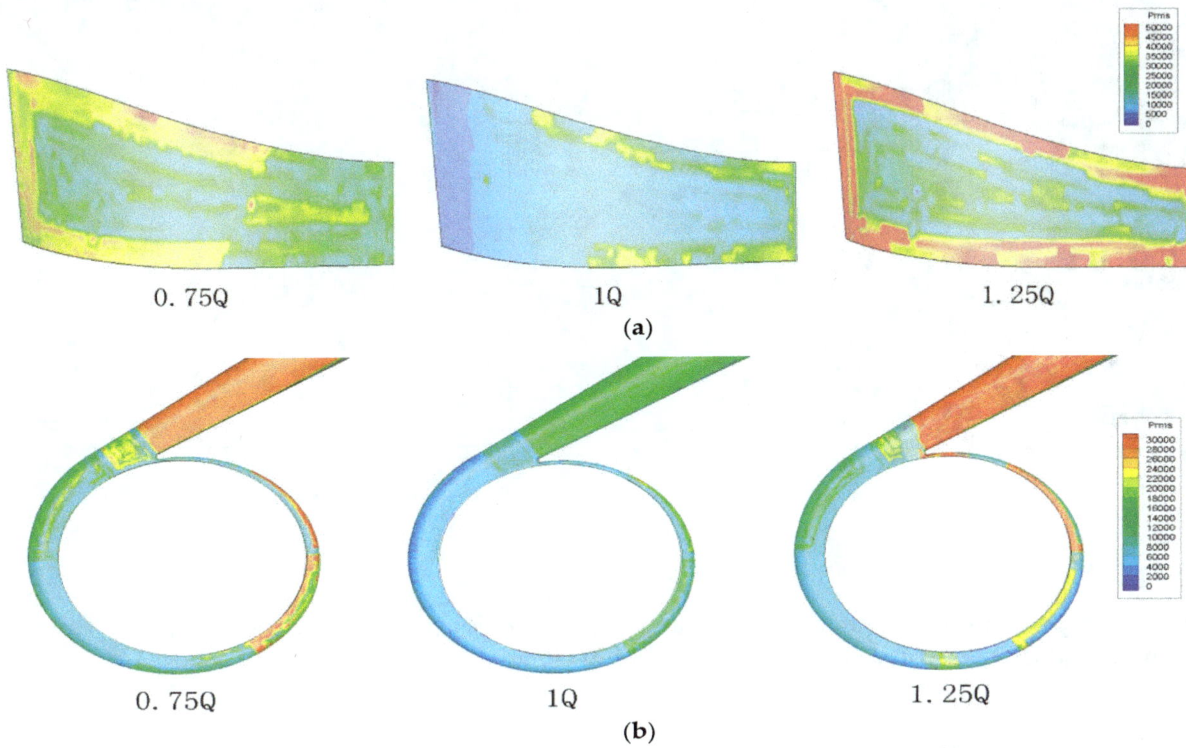

Figure 8. Distribution of root mean square (RMS) value on both impeller and volute under three separate flow rates, including 0.75 Q, 1 Q and 1.25 Q. (**a**) Distribution of RMS value on impeller; and (**b**) distribution of RMS value on volute.

It can also be inferred from a study of the RMS pressure field occurring in the volute at different flow rates that the resulting acoustic behavior is rather similar to that of the impeller (seen in Figure 8b). The overall distribution of RMS at the wall of the volute under the three flow rate conditions is that the RMS values in the diffuser are unmistakably the highest and, near the volute tongue, the fluctuation field is rather "stiff", particularly when the fluid is just transported past the tongue to flow into the volute, implying that the volute tongue is the primary source of the RMS pressure and even the noise. Although the RMS at 1 Q is smaller than those at other flow rates, the distribution resembles that under 0.75 Q and 1.25 Q flow rate conditions.

3.3. Noise Analysis

3.3.1. The Impeller Radiation Noise

The noise generated by the blades is originally studied by defining only the surfaces of the blades as the sources of sound. The distribution of impeller radiation noise for different flow rates calculated using the FW-H model is shown in Figure 9. The profiles of *TSPL* of directivity field shown in Figure 9a apparently demonstrate the dipole characteristic behavior. It is also found that two *TSPL* valleys appear at $\theta = 75°$ and 270°, but the weak asymmetric distribution occurs along these two points. Figure 9b reveals that the design operation (1 Q) generates the smallest *TSPL* than off-design operations. As the flow rates decrease from 1 Q to 0.25 Q, the *TSPL* initially increases significantly before 0.75 Q and then level off afterwards. A similar trend occurs for the cases for the larger flow rates (1–1.25 Q), but the *TSPL* increases very quickly. For example, the 1.25 Q case produces much more noise in comparison with 0.75 Q, even though they have the same off-design deviation. In general, the noise level deviates about 50% in different directions as demonstrated by the deviation bar in Figure 9b. Nevertheless, the *TSPL* is eventually saturated at ~70 dB with the increase/decrease of the flow rate.

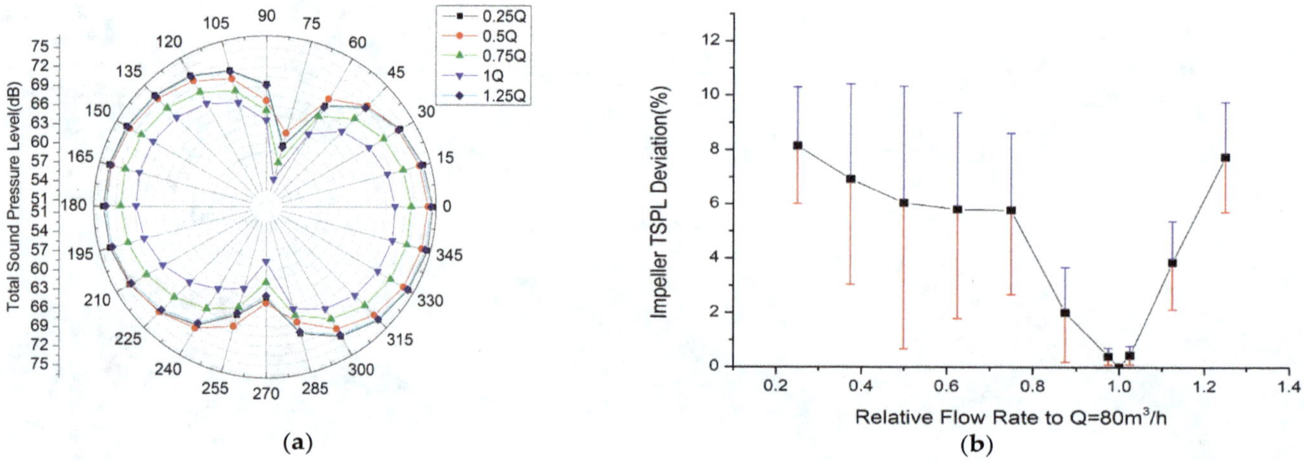

Figure 9. Impeller radiation noise. (**a**) Total sound pressure level (*TSPL*) of directivity field from the impellers; and (**b**) the relationship of *TSPL* deviation vs. the flow rate.

3.3.2. The Volute Radiation Noise

Similarly, the noise generated by the volute is further studied by defining the volute surface as the source of sound. Figure 10a shows the *TSPL* pattern for the volute radiation noise. Interestingly, the radiation field displays very different behavior in comparison with the blade noise discussed above. The *TSPL* profiles exhibit an apparent asymmetric behavior and *TSPL* on the right hand side is much larger than the one on the left. This may be explained by the fact that the volute tongue is the key noise source and the non-uniform distance between the receivers and the tongue leads to the asymmetric distribution. Figure 10b reveals a similar trend shown in Figure 9b. It is interesting to find that the mean *TSPL* levels off in the range of 0.5–0.75 Q. With the increase/decrease of the flow rate vis-a-vis the design value, the *TSPL* deviation increases and the maximum is about 50%. It should be noted that the *TSPL* can reach ~87 dB, which informs us that the noise contribution by the volute is substantial.

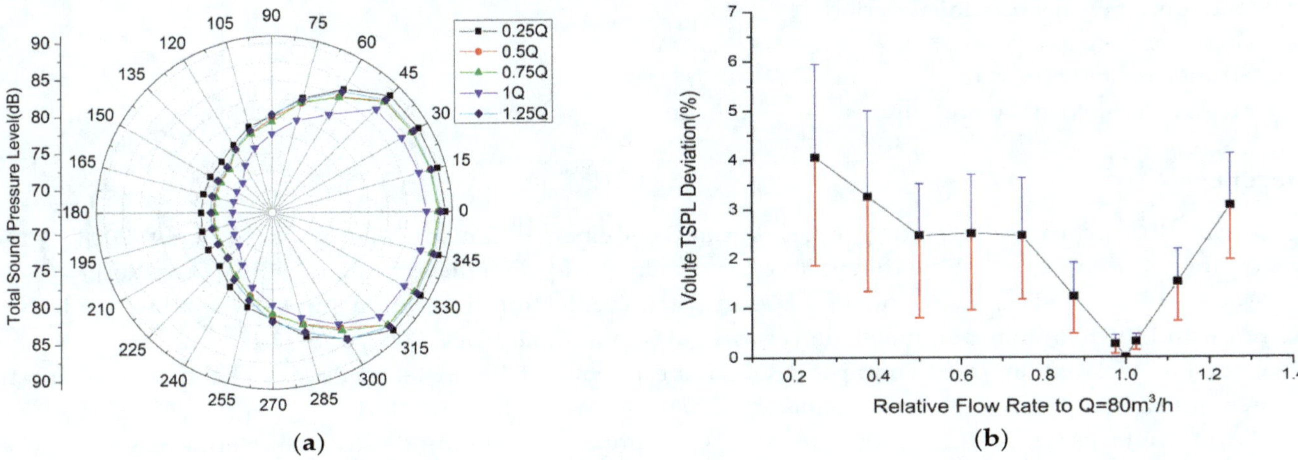

Figure 10. Volute radiation noise. (**a**) Total sound pressure level of directivity field from the volute; and (**b**) the relationship of *TSPL* deviation vs. the flow rate.

4. Conclusions

In terms of a single-stage singe-suction centrifugal pump, a 3D numerical model with the LES closure is developed for elucidation and reliable prediction of the pertinent hydro/aero acoustic behavior. In particular, the FW-H model was employed to predict the noise generation by the blades as well as the volute. Comparisons of numerical predictions with the measured/analytical results reveal that the model can yield good results on the noise and the flow field. The simulations further show that the interaction of the blades with the volute induces high RMS pressure and further leads to additional noise radiation. Moreover, the sources of noise generated by the blade surfaces and the volute surface at different flow rates were studied. It is found that the profiles of *TSPL* of directivity field for the impeller-generated noise demonstrate a typical dipole characteristic behavior, whereas the ones for the volute-generated noise exhibit an apparent asymmetric behavior. In addition, within the flow rate range studied, the design operation (1 Q) generates the smallest *TSPL* than off-design operations. As the flow rates decrease from 1 Q to 0.25 Q, *TSPL* initially increases significantly before 0.75 Q and then levels off afterwards. A similar trend occurs for the cases in the larger flow rates (1–1.25 Q). The *TSPL* deviates with the radiation directivity and the maximum is about 50%. It is also found that *TSPL* by the volute and the blades can reach ~87 dB and ~70 dB at most, respectively.

Acknowledgments: Shenghui Lei would appreciate the financial support by the Industrial Development Agency (IDA) Ireland.

Author Contributions: Ming Gao conceived the initial idea of this research and guided the work; Peixin Dong carried out the simulations and collected and analyzed the data; Shenghui Lei analyzed the data and guided the simulations; Ali Turan contributed valuable scientific discussion; all the authors commented and wrote the paper.

Nomenclature

C_s	Smagorinsky constant
Δ	Size of the grid filter
p'	Far-field acoustic pressure
a_0	Speed of sound
ρ_0	Reference density of fluid
$\delta(f)$	Dirac delta function
$H(f)$	Heaviside function
P_{ref}	Reference sound pressure

K	Step number in a rotational period
n	Rotational speed
$P(t)$	Transient static pressure
$\overline{P}$	Time-averaged static pressure

References

1. Johann, F.G. *Centrifugal Pumps*; Springer: Berlin/Heidelberg, Germany; New York, NY, USA, 2008.
2. Brennen, C.E. A review of the dynamics of cavitating pumps. *J. Fluids Eng.* **2012**, *135*. [CrossRef]
3. Wang, J.; Feng, T.; Liu, K.; Zhou, Q.J. Experimental research on the relationship between the flow-induced noise and the hyaraulic parameters in centrifugal pump. *Fluid Mach.* **2007**, *6*, 1–13.
4. Choi, J.S.; Mclaughlin, D.K.; Thompson, D.E. Experiments on the unsteady flow field and noise generation in a centrifugal pump impeller. *J. Sound Vib.* **2003**, *263*, 493–514. [CrossRef]
5. Parrondo, J.; Pérez, J.; Barrio, R.; González, J. A simple acoustic model to characterize the internal low frequency sound field in centrifugal pumps. *Appl. Acoust.* **2011**, *72*, 59–64. [CrossRef]
6. Liu, H.L.; Dai, H.W.; Ding, J. Numerical and experimental studies of hydraulic noise induced by surface dipole sources in a centrifugal pump. *J. Hydrodyn. Ser. B* **2016**, *28*, 43–51. [CrossRef]
7. Chu, S.; Dong, R.; Katz, J. Relationship between unsteady flow, pressure fluctuations, and noise in a centrifugal pump—Part A: Use of PDV data to compute the pressure field. *J. Fluids Eng.* **1995**, *117*, 24–29. [CrossRef]
8. Chu, S.; Dong, R.; Katz, J. Relationship between unsteady flow, pressure fluctuations, and noise in a centrifugal pump—Part B: Effects of blade-tongue interactions. *J. Fluids Eng.* **1995**, *117*, 30–35. [CrossRef]
9. Langthjem, M.A.; Olhoff, N. A numerical study of flow-induced noise in a two-dimensional centrifugal pump. Part I. Hydrodynamics. *J. Fluids Struct.* **2004**, *19*, 349–368. [CrossRef]
10. Langthjem, M.A.; Olhoff, N. A numerical study of flow-induced noise in a two-dimensional centrifugal pump. Part II. Hydroacoustics. *J. Fluids Struct.* **2004**, *19*, 369–386. [CrossRef]
11. Kato, C.; Yamade, Y.; Wang, H.; Guo, Y.; Miyazawa, M.; Takaishi, T.; Takano, Y. Numerical prediction of sound generated from flows with a low Mach number. *Comput. Fluids* **2007**, *36*, 53–68. [CrossRef]
12. Kato, C.; Kaiho, M.; Manabe, A. An overset finite-element large-eddy simulation method with applications to turbomachinery and aeroacoustics. *Trans. Am. Soc. Mech. Eng. J. Appl. Mech.* **2003**, *70*, 32–43. [CrossRef]
13. Huang, J. Comparison of noise characteristics in centrifugal pumps with different types of impellers. *Acta Acust.* **2010**, *35*, 113–118.
14. Li, K.; Yu, J.; Shi, F.; Huang, A.X. Dimension splitting method for the three dimensional rotating Navier-Stokes equations. *Acta Math. Appl. Sin. Engl. Ser.* **2012**, *28*, 417–442. [CrossRef]
15. Seif, M.S.; Asnaghi, A.; Jahanbakhsh, E. Implementation of PISO algorithm for simulating unsteady cavitating flows. *Ocean Eng.* **2010**, *37*, 1321–1336. [CrossRef]
16. Williams, J.F.; Hawkings, D.L. Sound generated by turbulence and surfaces in arbitrary motion. *Philos. Trans. R. Soc. A Math. Phys. Eng. Sci.* **2010**, *264*, 321–342. [CrossRef]
17. Ghasemian, M.; Nejat, A. Aerodynamic noise prediction of a horizontal axis wind turbine using improved delayed detached eddy simulation and acoustic analogy. *Energy Convers. Manag.* **2015**, *99*, 210–220. [CrossRef]
18. Tao, J.; Lu, X.; Wang, L.; Wang, L.; Li, J. *Methods of Measuring and Evaluating Noise of Pumps*; Chinese Quality Supervision Bureau: Beijing, China, 2013; pp. 4–6.
19. Fluent Inc. *Gambit 2.4 User's Guide*; Fluent Inc.: New York, NY, USA, 2007; pp. 1–17.
20. Choi, H.; Moin, P. Grid-point requirements for large eddy simulation: Chapman's estimates revisited. *Phys. Fluids* **2012**, *24*. [CrossRef]
21. Tsai, C.H.; Fu, L.M.; Tai, C.H.; Huang, Y.L.; Leong, J.C. Computational aero-acoustic analysis of a passenger car with a rear spoiler. *Appl. Math. Model.* **2009**, *33*, 3661–3673. [CrossRef]

6

Source Term Modelling of Vane-Type Vortex Generators under Adverse Pressure Gradient in OpenFOAM

Iñigo Errasti *, Unai Fernández-Gamiz, Pablo Martínez-Filgueira and Jesús María Blanco

Nuclear Engineering and Fluid Mechanics Department, University of the Basque Country, E-01009 Vitoria-Gasteiz, Spain; unai.fernandez@ehu.eus (U.F.-G.); pablo.martinez@ehu.eus (P.M.-F.); jesusmaria.blanco@ehu.eus (J.M.B.)

* Correspondence: inigo.errasti@ehu.eus.

Abstract: An analysis of the generation of vortices and their effects by vane-type vortex generators (VGs) positioned on a three-dimensional flat plate with a backward-facing ramp and adverse gradient pressure is carried out. The effects of a conventional vortex generator and a sub-boundary layer vortex generator are implemented by using a source term in the corresponding Navier-Stokes equations of momentum and energy according to the so-called jBAY Source Term Model. The influence of the vortex generator onto the computational domain flow is modelled through this source term in the Computational Fluid Dynamics (CFD) simulations using the open-source code OpenFOAM. The Source Term Model seems to simulate relatively well the streamwise pressure coefficient distributions all along the flat plate floor as well as certain parameters studied for vortex characterization such as vortex path, decay and size for the two vane-type vortex generators of different heights studied. Consequently, the implementation of the Source Term Model represents an advantage over a fully Mesh-Resolved Vortex Generator Model for certain applications as a result of a meaningful decrease in the cell number of the computational domain which implies saving computational time and resources.

Keywords: flow control; vortex generators; source term; Computational Fluid Dynamics (CFD); OpenFOAM; wind tunnel

---

1. Introduction

The need for bigger capacities and installed power of wind generation systems has allowed the implementation of larger rotor wind turbines. Nowadays, wind power generators of 5–7 MWatt may present rotor blades of 60 m long or even larger.

This meaningful increase in the rotor size and weight of wind turbines has made that the techniques used in the past years to control them have to be improved. Johnson et al. [1] summarized a big part of the most relevant flow control techniques applied in wind turbines to work in an optimal and safe way under diverse atmospheric conditions.

Several flow control devices have been developed in recent decades, Gad-el-Hak [2]. Most of them were firstly designed for aeronautical applications and consequently the initial starting point of research was aeronautics, Taylor [3]. According to Liu et al. [4], flow control devices are regularly also applied in turbo machinery systems. In recent years, researchers have been trying to optimize and install these devices in multi-megawatt wind turbines. Wood [5] and Johnson et al. [1] developed a four layer scheme to conceptually classify these flow control devices.

These control devices can be classified as active or passive ones depending on the control techniques applied as proposed by Aramendia et al. [6]. Passive flow control techniques enhance the turbine's efficiency and load reduction without external energy consumption. Active control techniques need a supplementary energy source to obtain the effect on the air flow and, unlike vortex generators (VGs) and other passive devices, active flow control requires elaborate algorithms to get the most favorable gain (Becker et al. [7]). Johnson et al. [1] carried out an investigation on around 15 control devices for wind turbines. Some of these flow control devices are still checked nowadays on full-scale wind turbines in working conditions.

Vortex generators (VGs) are passive flow control devices that modify the boundary layer fluid motion exchanging momentum from the outer into inner region. Due to this energy transport, the velocity of the inner region increases at the same time as the boundary layer thickness decreases and in turn the flow separation is delayed, Rao et al. [8]. Moreover, Lin et al. [9] proved the effect of drag reduction and lift increase induced by sub-boundary layer vortex generators (VGs).

According to Doerffer et al. [10] and Steijl et al. [11], the streamwise vortices generated by vane-type VGs produce an endless momentum to counter the momentum decrease of the natural boundary layer and the growth of its thickness induced by viscous friction and adverse gradients of pressure. These vane-type VGs are able to reduce or remove flow separation when there is a limited adverse pressure gradient as commented by Velte et al. [12,13]. If the flow separation even appears for cases of large adverse gradients of pressure, the vortices action of mixing will limit the reversed flow region in the shear layer and maintain, to some extent, pressure recovery all along the detached flow.

Fernández-Gamiz et al. [14] and Urkiola et al. [15] analysed the flow effects of a vane-type VG on a flat plate under conditions of negligible streamwise pressure gradient and the primary vortices generated to examine how the physics of the wake past the VGs can be simulated by means of Computational Fluid Dynamics (CFD) computations. In those studies, the CFD simulations were compared with experimental observations. For instance, Gao et al. [16] and Baldacchino et al. [17] focused on the increase of the maximum value of the lift coefficient of a 30% thick DU97-W-300 airfoil designed by the Delft University of Technology by implementing passive VG devices. When increasing the angle of attack, the lift and drag coefficients raised up to values higher than the ones obtained under steady-state conditions.

The so-called micro vortex generators were introduced by Keuthe [18]. These flow control devices were also named as sub-boundary layer vortex generators (SBVGs) according to Holmes et al. [19], submerged vortex generators in Lin et al. [20], low-profile vortex generators by Martinez-Filgueira et al. [21] and micro vortex generators in Lin et al. [22]. Kenning et al. [23] summarized in his study the potential applications of all these vortex generators, including the control of shock-induced separation, smooth surface separation or leading edge separation.

Vortex generators are fixed onto the blades of wind turbines in order to prevent or remove the flow separation as well as to decrease the blade sensitivity. VGs are usually mounted on the suction side of the blade in a spanwise array for instance when a wind turbine does not perform as expected. Accordingly, the implementation of VGs onto the blade could eventually be a valid way to enhance the efficiency of a wind turbine rotor, Schubauer et al. [24] and Bragg et al. [25].

In order to design and optimize the position of the vortex generator on a wind turbine blade, CFD tools can be used as for instance Fernández-Gamiz et al. [26] or Troldborg et al. [27] do. Nonetheless, modelling a full rotor by using a fully-meshed VG computation becomes excessively expensive. As a matter of fact, the size of the vortex generators is generally comparable to the local boundary layer thickness and this implies that a meaningful number of cells around the device in the computational domain is needed in order to accurately model the flow. An alternative way of modelling VGs in the CFD computations is to model the device effect on the boundary layer by using body forces.

Bender et al. [28] established a source term model, the so-called BAY model, based on the Joukowski Lift Theorem and the Thin Airfoil Theory which used body forces. This model was designed for simulating vane-type Vortex Generators in a finite volume Navier-Stokes code and ignores the condition to fully define the geometry in the mesh. In order to calibrate the model, Bender [28] created a test case to compare the results obtained for a source term modelled VG with those obtained for a fully-meshed VG. The test case was based on 24 VGs circumferentially mounted on a pipe in a co-rotating configuration and the results obtained were satisfying. For instance, Dudek [29] successfully evaluated the BAY model for different types of flows in a rectangular duct with a single vane-type VG.

Recently, a better version of the BAY model was designed by Jirasek [30], namely the jBAY model. The improved version was based on the Lift Force Theory based on Bender et al. [28] and brought a more suitable technique for simulating the flow when using VG arrays. Jirasek [30] applied a reduced method for determining the model control points to implement in a effortless way the model and obtain more correct results. The new version of the model was checked by reproducing a single vane-type VG on a flat plate in an S-duct air intake and a high-lift wing configuration. The results showed very good agreement between experimental data and the CFD simulations. Thus, Florentie et al. [31] studied the potential of the jBAY model to reproduce the effects of rectangular VGs positioned on a flat plate and proposed a model optimization.

The main goal of the current study is firstly the implementation of the so-called jBAY Source Term Model into OpenFOAM [32] and secondly to investigate how well the open source CFD simulations are able to mimic the physics of the flow behind modelled rectangular conventional and sub-boundary layer vortex generators (VGs) mounted on a three-dimensional flat plate with a backward-facing ramp under adverse pressure gradient conditions. The satisfactory implementation of the Source Term Model in OpenFOAM would be eventually able to bring new model applications of this open source CFD code. The theoretical background of this study is described in Section 2; Section 3 presents the experimental data; Section 4 describes in detail the computational configuration; the results and their discussion are covered in Section 5 and the conclusions are presented in Section 6.

2. Theoretical Background

2.1. Vortex Generator Setup

The model consists of a single rectangular VG positioned on a flat plate with a backward-facing ramp. The three-dimensional computational domain modelled in OpenFOAM represents the extended test section of the wind tunnel experiment performed by Lin [33] and is shown in the illustration of Figure 1. Note that the single VG is placed upstream the backward-facing ramp. The domain has been designed following a previous simulation study by Konig et al. [34].

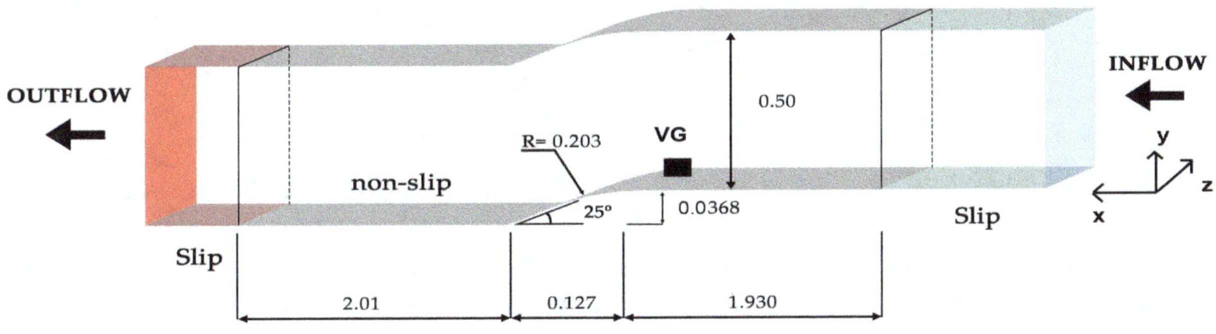

Figure 1. Description of the computational domain representing the extended wind tunnel test section. The domain dimensions are expressed in meters.

The geometry dimensions of the vane-type VG determined by a length L two times the VG height H are depicted in Figure 2a for two different heights $H_1 = 0.8\delta$ and $H_2 = 0.2\delta$ where δ is the local boundary layer thickness just upstream edge of the ramp. The two vortex generators, namely, conventional VG and sub-boundary layer VG of heights $H_1 = 0.8\delta$ and $H_2 = 0.2\delta$ respectively are positioned at distances of 5δ and 2.5δ from the upstream edge of the ramp. Figure 2b illustrates a classical configuration example of a pair of counter-rotating VGs. The rectangular vane has a constant thickness and no sharp edges. The parameter λ represents the spacing between VGs and lines A, B and C are measurement lines along the floor past the VG. The geometry and parameters of the two VG cases are summarized in the table of Section 4. Note that the computational domain has only one vane-type VG oriented to the main flow instead of two vanes in order to reduce the meshing and the computational resources needed. A boundary layer region develops over the flat plate induced by the viscous interaction between the wall and the air flow. The angle of incidence between the main flow direction and the VG vane is $\beta = 15°$ or $\beta = 25°$ depending on the VG case (see Figure 3) and the Shear Stress Transport (SST) turbulence model Menter [35] has been elected because of its capacity to solve whirling flows according to Liu et al. [4].

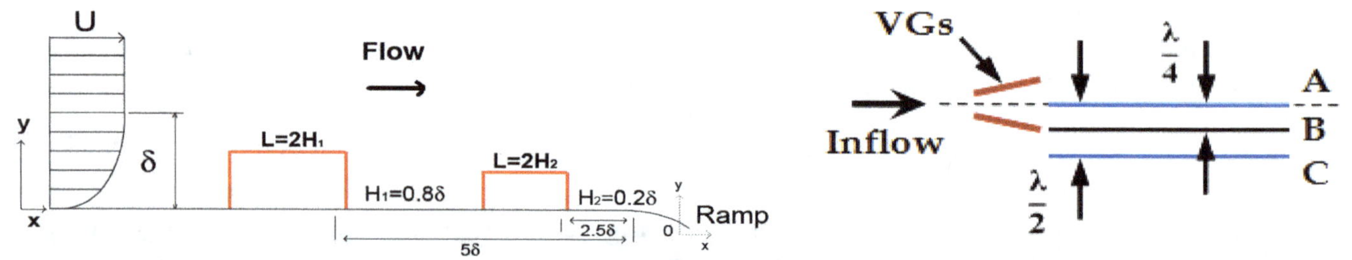

(**a**) The two VGs of heights H_1 and H_2 (in red) positioned on the flat plate. (**b**) Counter-rotating VGs (in red).

Figure 2. (**a**) The two VGs studied of heights $H_1 = 0.8\delta$ and $H_2 = 0.2\delta$ on the flat plate with δ representing the local boundary layer thickness at the upstream edge of the ramp (not at scale); (**b**) Configuration example of a pair of similar counter-rotating VGs with three measurement lines A, B and C past the VG.

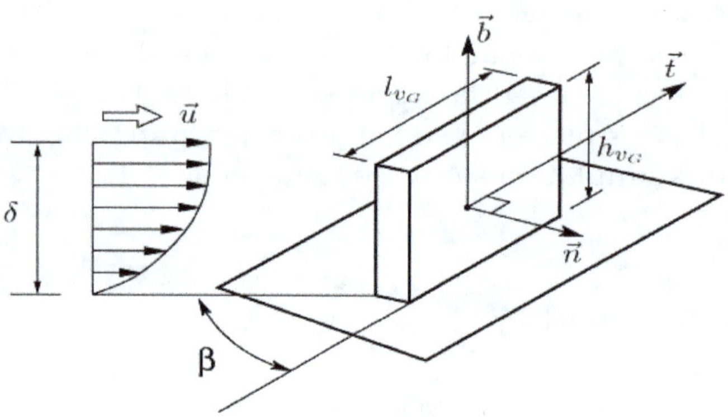

Figure 3. Unit vectors on a rectangular vortex generator (VG) according to the BAY Source Term Model.

The Reynolds number for the CFD simulations is $Re_\theta = 9100$ in order to mimic the experiments explained in the next section and is based on the local boundary local (BL) momentum thickness θ and computed by the following expression:

$$Re_\theta = \frac{U_\infty \cdot \theta}{\nu} \tag{1}$$

where $U_\infty = 40.23\ \mathrm{m \cdot s^{-1}}$ is the free stream velocity, $\nu = 1.9439 \times 10^{-5}\ \mathrm{m^2 \cdot s^{-1}}$ the kinematic viscosity and the momentum thickness θ defined as:

$$\theta = \int_{-\infty}^{\infty} \frac{u}{U_\infty}(1 - \frac{u}{U_\infty})dy \tag{2}$$

where u is the streamwise velocity component and y the vertical coordinate normal to the wall. Thus, the momentum thickness of the local boundary layer δ at the upstream edge of the ramp calculated by the previous equation is $\theta = 4.3967$ mm.

2.2. Source Term Model

Computational Fluid Dynamics (CFD) is a commonly method used to reproduce the air flow downstream the VGs and to estimate the blade efficiency. Furthermore, the combination of this technique of simulation and the corresponding experiments in wind tunnels can be considered as a reliable tool for parametric studies on VG layout. However, these computational methods are time-consuming when designing high-quality meshes and solving the corresponding flow equations.

In the current work, the so-called jBAY Source Term Model presented by Jirasek [30] and based on the widely used BAY Model developed by Bender et al. [28] is implemented. The BAY model was designed for simulating vane-type vortex generators into finite volume Navier-Stokes codes and allows substituting the VG geometry by a subdomain of similar size at the original VG location where a specific body force distribution is then applied. The model is integrated into the CFD code as a so-called source term in the Navier-Stokes momentum and energy equations.

In the model, a force normal is applied to the direction of the local flow and at the same time parallel to the surface. This normal force simulates the side force generated by a vane-type vortex generator. Bender et al. [28] designed this Source Term Model based on the Joukowski Lift Theorem and the Thin Airfoil Theory in order to model the VG influence. Taking into account a rectangular vortex generator (VG), the lift forces on the VG can be computed by the following expression:

$$\overrightarrow{L} = \rho(\vec{u} \times \vec{b}) \cdot \Gamma \cdot h_{VG} \tag{3}$$

where $\overrightarrow{L}$ is the lift force on the VG, ρ the local density, $\vec{u}$ the local velocity, $\vec{b}$ the unit vector defined as $\vec{b} = \vec{n} \times \vec{t}$ with $\vec{n}$ and $\vec{t}$ the unit vectors normal and tangential to the VG respectively (see Figure 3), h_{VG} the VG height and Γ the circulation defined as:

$$\Gamma = \beta \parallel \overrightarrow{u} \parallel l_{VG} \tag{4}$$

where β is the angle of incidence between the flow direction and the VG and l_{VG} the VG length. As expressed in Equation (3), the direction of the lift force $\overrightarrow{L}$ on the VG is obtained by means of the product of the local velocity $\vec{u}$ and the unit vector $\vec{b}$. According to the Joukowski Lift Theorem for 2D airfoils:

$$\overrightarrow{L} = \pi\rho(\vec{u} \times \vec{b})(\vec{u} \cdot \vec{n})S_{VG} \tag{5}$$

where S_{VG} is the vortex generator area ($l_{\vec{VG}} \times h_{\vec{VG}}$) and the lift force on a single cell $\overrightarrow{L}_{cell}$ is computed by the following expression:

$$\overrightarrow{L}_{cell} = \pi\rho(\vec{u} \times \vec{b})(\vec{u} \cdot \vec{n})S_{VG}\frac{V_{cell}}{V_s} \tag{6}$$

where V_{cell} is the volume corresponding to a single cell and V_s the total volume of the grid cells where the model is applied. According to Bender et al. [28] a new term ($\frac{\vec{u}\cdot\vec{t}}{\|\vec{u}\|}$) is introduced in the previous equation in order to obtain the final lift force on a single cell:

$$\overrightarrow{L}_{cell} = C_{VG} \cdot \rho(\vec{u} \times \vec{b})(\vec{u} \cdot \vec{n})\frac{\vec{u} \cdot \vec{t}}{\parallel \vec{u} \parallel}S_{VG}\frac{V_{cell}}{V_s} \tag{7}$$

When applying the BAY Model, a calibration process is required and a Mesh-Resolved VG Model can be used as reference for model calibration. Particularly for this model, the empirical constant C_{VG} of the Equation (7) is a relaxation parameter. This parameter is selected so that the simulated estimations for the lift force distribution fit the results corresponding to the computations of the mesh-resolved VG. Bender et al. [28] proposed that this Source Term Model can operate in two modes, the so-called asymptotic and linear ones. If V_s, defined as the total volume of the grid cells where the force term is applied, differs considerably from the volume of the vane type vortex generator, the model behaves in linear mode and is dependent on the constant C_{VG}. However, the model of the present work has been locally applied, such that the volume V_s is similar to the vortex generator volume. Therefore, the model operates in the asymptotic mode and is non sensitive to C_{VG}. Commonly values for this constant are $C_{VG} \cong 10$, see the study by Jirasek [30]. In the present work, instead of applying forces in all the subdomain cells as the BAY Model specifies, the forces are applied in cells placed just in the outline of the VG geometry, see Figure 4.

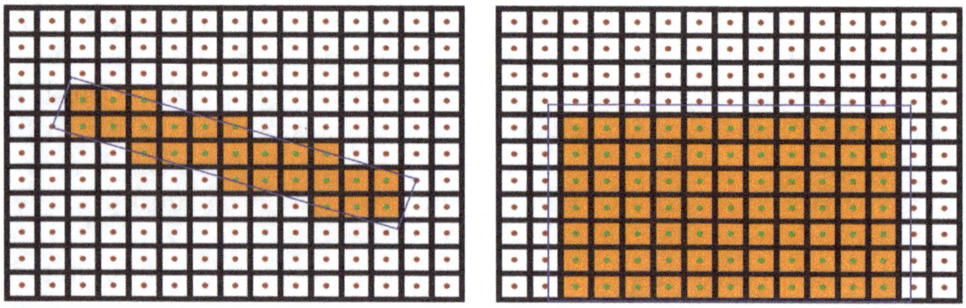

(**a**) Top view. (**b**) Side view.

Figure 4. Cells corresponding to the VG coloured in orange where the body forces are applied. VG dimensions are indicated by the blue line.

Taking into account that the Source Term Model was designed to be user-friendly, any vane size of height H and length L and angle of incidence β can be modelled. When using the OpenFOAM CFD code, only three parameters have to be specified to model the vortex generator (VG): the grid cells where the Source Term Model will be applied, the vortex generator area S_{VG} and the angle of incidence β of the main flow direction with respect to the VG orientation. In Figure 5, two examples of subdomains of the selected mesh cells which represent the two vortex generators (VGs) of heights $H_1 = 0.8\delta$ and $H_2 = 0.2\delta$ in the Source Term Model are shown.

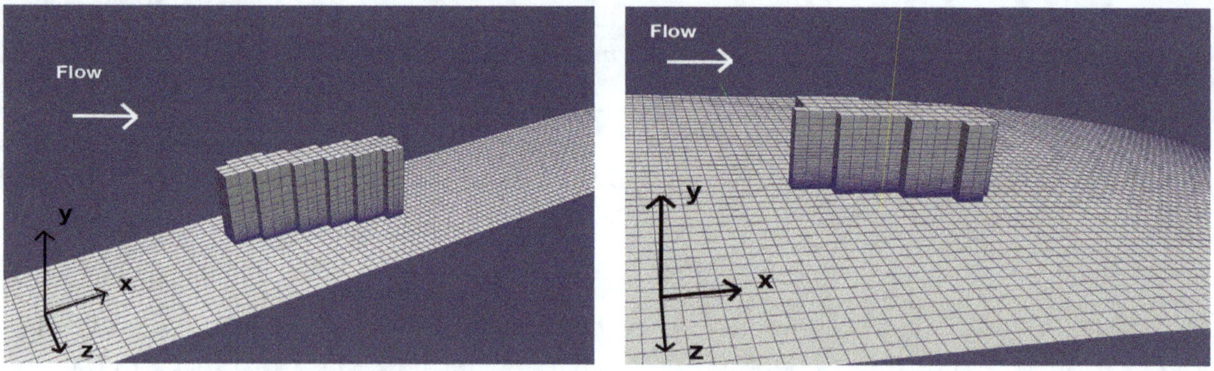

(**a**) Vane-type VG case of height $H_1 = 0.8\delta$. (**b**) Vane-type VG case of height $H_2 = 0.2\delta$.

Figure 5. Selected cells representing the two vortex generators (VGs) of heights H_1 and H_2 according to the Source Term Model and positioned on the flat plate upstream the backward-facing ramp.

The reader should finally note that the Source Term Model can replace the Mesh-Resolved VG Model when modelling VGs in certain cases. In a fully Mesh-Resolved VG Model, a fine mesh in the vicinity of the vortex generator (VG) must be designed in order to capture properly the VG effect on the domain flow. Thus, Fernández-Gamiz et al. [26] designed a fine mesh of approximately eight million cells for simulating the primary vortex induced by a rectangular sub-boundary layer VG on a flat plate in a fully meshed-resolved model. When using the BAY Model, a coarser mesh can be used to capture the VG influence on the flow saving computational time and resources as will be indicated later. However, attention must be paid when defining the mesh because the BAY Model accuracy can eventually be dependent on mesh resolution according to some authors, e.g., Florentie et al. [31].

3. Experimental Data

Experimental data for this study has been taken from the experiments carried out by Lin et al. [33]. The experiments were performed in the NASA Langley 20 by 28-inch Shear-Flow Control Tunnel and free stream velocity of $U_\infty = 40.23$ m·s^{-1}. Baseline flow separation was determined on a 25° sloped backward facing curved ramp of radius $R = 20.32$ cm with a behaviour similar to a classical two-dimensional flow. The ramp spanned the wind tunnel test section width. The local boundary layer thickness δ was approximately 32.51 mm and the Reynolds number was $Re_\theta = 9100$ at the upstream edge of the sloped backward facing curved ramp which was placed approximately at 1.98 m from the wind tunnel test section entrance. In the experiments, different types of passive flow control devices were analysed and between them vortex generators (VGs). Furthermore, sets of vortex generators (VGs) were simulated upstream of the backward-facing ramp. In the present study, only the pressure distribution data corresponding to a counter-rotating conventional vane-type vortex generator (Figure 2) of height $H_1 = 0.8\delta$ and a sub-boundary layer vane-type vortex generator of height $H_2 = 0.2\delta$ where δ represents the local boundary layer thickness at the upstream edge of the ramp will be analysed for comparison with the data obtained from the OpenFOAM simulations. The higher vortex generator of height H_1 was positioned at a distance of 5δ upstream of the ramp and the smaller vortex generator of height H_2 at a distance of 2.5δ.

4. Computational Configuration

Non-commercial open source code OpenFOAM [32] has been used for simulating the vortex and its effects in the present study. The open source CFD code is an object-oriented library package designed in C++ programming language to analyse systems of computational continuum mechanics.

The current computational domain (Figure 6) partially based on a previous simulation study by Konig et al. [34] consists of a flat plate with a backward-facing ramp and adverse pressure gradient where selected cells corresponding to a single rectangular vortex generator according to the Source Term Model namely jBAY previously presented in Jirasek [30] and positioned at a point where the VG height is 80% and 20% the local boundary layer thickness δ at the upstream edge of the ramp for the cases H_1 and H_2 respectively. The incidence angle of the vane with respect to the oncoming flow is $\beta = 15°$ for the $H_1 = 0.8\delta$ vane-type VG case and $\beta = 25°$ for he $H_2 = 0.2\delta$ vane-type VG case as shown in Figure 3. The modelled VGs have an aspect ratio (AR) defined as the relationship between the height H and the length L of AR = 0.5 (Table 1). A representation of the mesh subdomain and selected cells representing the vortex generator is shown in Figure 5. The origin of the computational domain was placed at the ramp center.

Table 1. Vortex generator geometry and parameters corresponding to the vane-type VG cases of height $H_1 = 0.8\delta$ and $H_2 = 0.2\delta$. In the present study: $\delta = 32.51$ mm.

VG Parameter	$H_1 = 0.8\delta$ VG Case	$H_2 = 0.2\delta$ VG Case
Height H	0.8δ	0.2δ
Length L	1.6δ	0.4δ
Incident angle β	15°	25°
Location from the ramp	5δ	2.5δ
Spacing λ	3.2δ	0.8δ

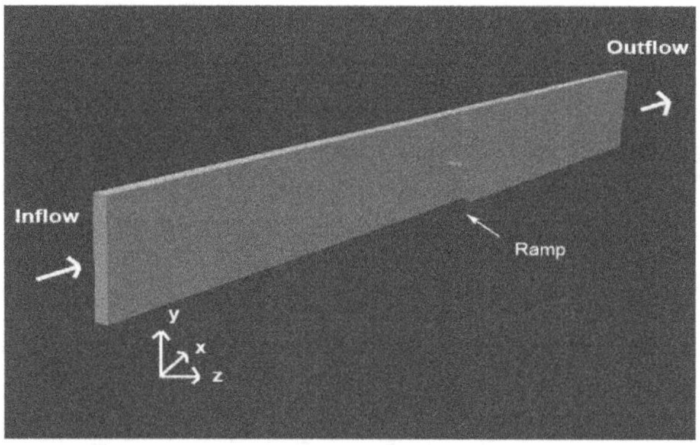

(**a**) Computational domain.

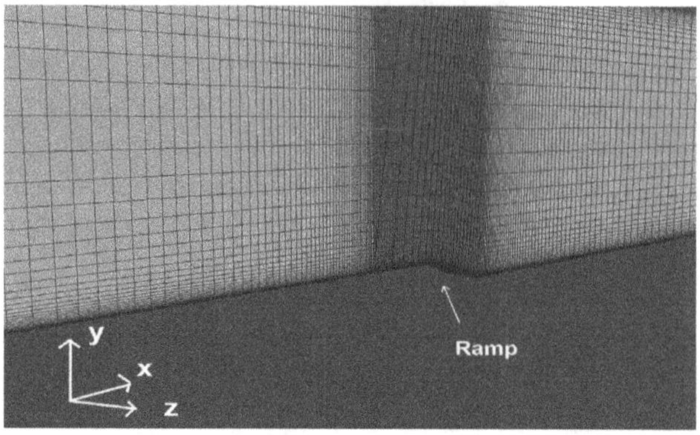

(**b**) Mesh around the ramp region.

Figure 6. Computational domain and mesh detail of the VG positioned on the flat plate upstream the backward-facing ramp.

The variables of particular interest which require boundary conditions are the fluid quantities, velocity and pressure. Non-slip boundary conditions were applied to the flat plate and ramp of the computational domain to allow for boundary layer growth (see Figure 1). The sides of the computational domain have been defined as symmetry condition. The domain consists of only one vane and the symmetry assumption used in the present computations can be justified by previous studies as Gutierrez-Amo et al. [36] and Sørensen et al. [37]. The atmospheric conditions were selected in such a way that a boundary layer thickness of approximately $\delta = 32.51$ mm just upstream edge of the ramp according to Lin [33] was attained in the simulation of the test section of the wind tunnel. As the k-ω Shear Stress Transport (SST) turbulence model designed by Menter [35] was applied, a requirement when setting boundary conditions for the turbulent viscosity ν_t, the turbulent kinetic energy k and the dissipation rate ω was needed. Nearby the flat plate and ramp surfaces, wall functions were not used in the computations. Slip-wall boundary conditions were also implemented at the beginning and end of the flat plate (Figure 1) according to Konig et al. [34].

Two CFD solvers were applied in this three-dimensional simulation. The potentialFoam solver was firstly used to create initial fields to accelerate the convergence process and secondly the simpleFoam solver for the steady-state, incompressible and turbulent flow using the RANS (Reynolds Average Navier-Stokes) equations. This second solver is based on the k-ω SST turbulence model by Menter [35]. This turbulence model combines the Wilcox k-ω model for the close-to-wall regions and the k-ϵ model for the outer regions. The normal forces are applied in the selected cells of the computational domain according to the Source Term Model explained in Section 2 using the topoSet tool.

The computational domain analysed was discretized with a structured-type mesh and hexahedral faces of around 5×10^5 cells for the $H_1 = 0.8\delta$ vane-type VG case and 2×10^6 cells for the $H_2 = 0.2\delta$ vane-type VG case. Full second order linear-upwind schemes for the discretization were implemented for all the simulations. Two quality mesh parameters such as mesh orthogonality and mesh skewness for the two VG cases have been studied in order to analyse the mesh quality and the results are shown in Table 2. Mesh non-orthogonality is an indicator of how orthogonal the pairs of neighboring cells which share a face are and mesh skewness of how optimum the cell shape is in relation with the corner angles. The non-orthogonality parameter should be close to 0 degrees and the maximum skewness should not exceed 0.85 as suggested by Gutierrez-Amo et al. [36] for hexahedral cells to obtain an accurate solution.

Table 2. Quality meshing parameters corresponding to the mesh of the two vane-type VG cases.

VG Case	Non-Orthogonality (Average)	Maximum Skewness (-)
$H_1 = 0.8\delta$	5.44°	0.56
$H_2 = 0.2\delta$	5.32°	0.38

According to Richardson and Gaunt [38] Extrapolation Method, a grid dependency study was performed for three different mesh resolutions: coarse, medium and fine. The method was applied to a vortex parameter, namely the normalized peak vorticity $(\omega_{xmax}\delta)/U_\infty$ analysed in the next section at a plane normal to the streamwise direction located at $x/\delta = -3$ past the vane for the $H_1 = 0.8\delta$ VG case and $x/\delta = -1$ for the $H_2 = 0.2\delta$ VG case. Three parameters were calculated following the Richardson Extrapolation Method: the extrapolated solution RE, the order of accuracy p and the error ratio R. Table 3 exhibits the results obtained in the mesh independency study where a monotonic convergence was fulfilled. The higher resolution of the fine mesh was used in the computations for the two vane-type VG cases analysed. The iterative solution process was carried out in the simulations of the two VG cases until the residual errors dropped below 10^{-4} for the pressure p and 10^{-5} for the velocities U_x, U_y and U_z and the turbulence quantities used, e.g., the turbulent kinetic energy k and the specific dissipation rate ω. The values obtained for the dimensionless distance of the first wall cell for the meshes corresponding to the two VG cases is summarized in Table 4. In both cases the value of y+ is lower than 1 (y+ < 1) as required by the turbulence model adopted.

Table 3. Mesh dependency study based on the Richardson Extrapolation Method performed for the normalized peak vorticity $(\omega_{xmax}\delta)/U_\infty$ on three meshes of Coarse, Medium and Fine resolution.

VG Case	Plane Location	Peak Coarse	Vorticity Medium	$(\omega_{xmax}\delta)/U_\infty$ (-) Fine	RE	p	R
$H_1 = 0.8\delta$	$x/\delta = -3$	74.25	122.78	128.32	127.61	3.13	0.11
$H_2 = 0.2\delta$	$x/\delta = -1$	12.00	27.91	29.54	29.35	3.29	0.10

Table 4. Wall dimensionless distance y+ values for the two VG cases simulated.

VG Case	Minimum	Maximum	Average
$H_1 = 0.8\delta$	0.14	1.29	0.64
$H_2 = 0.2\delta$	0.06	0.71	0.35

5. Results

5.1. Vortex Visualization

Figure 7 shows the vortex visualization based on the velocity field distribution at four planes normal to the streamwise direction and located at normalized distances x/δ respect to the boundary layer thickness δ from the backward-facing ramp where the domain origin is placed. The scale in the y and z axis for the four snapshots in this figure is approximately 13 mm × 13 mm. Thus, the vortex can be clearly visualized for the $H_1 = 0.8\delta$ vane-type VG case as expected. Note that the larger the normalized distance x/δ is, the closer the vortex is to the VG trailing edge (TE). The four planes depicted are located between the selected cells representing the VG and the ramp.

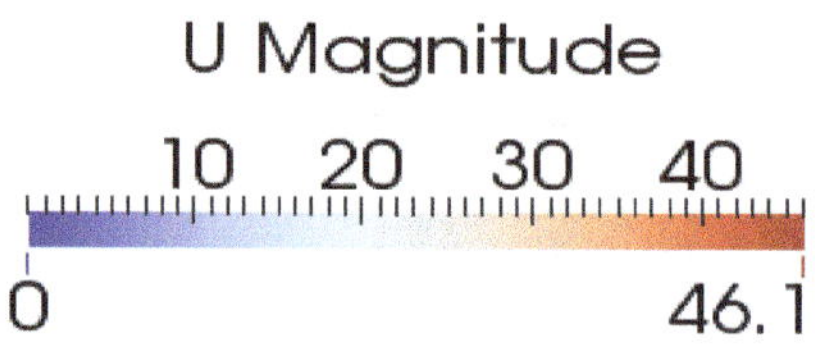

(a)

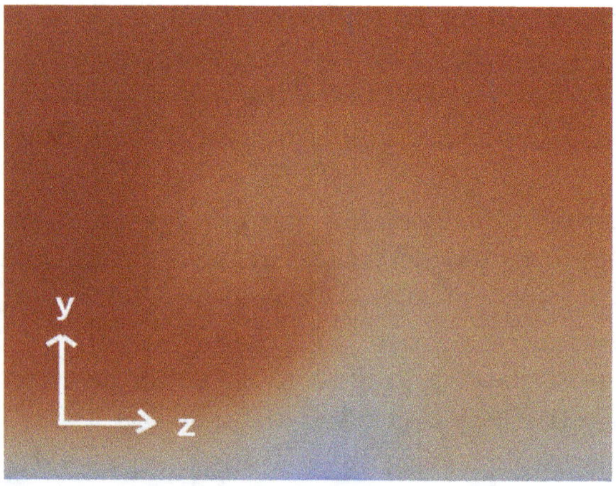

(b) At $x/\delta = -0.45$.

(c) At $x/\delta = -0.35$.

(d) At $x/\delta = 0.0$.

(e) At $x/\delta = +0.35$.

Figure 7. Velocity distribution ($m \cdot s^{-1}$) at planes normal to the streamwise direction and located at normalized distances x/δ from the backward-facing ramp where the origin is placed. $H_1 = 0.8\delta$ vane-type VG case.

Figure 8 shows again the vortex visualization based on the velocity field distribution for the $H_2 = 0.2\delta$ vane-type VG case at four planes normal to the main flow direction. Now, the scale for the four snapshots is approximately 12.2 mm $\times$ 12.2 mm. The formation of the vortex is slightly observed but not as clear as it was observed in the previous $H_1 = 0.8\delta$ vane-type VG case shown in Figure 7. This seems to indicate that the vortex generated by the sub-layer boundary VG has a smaller strength than that corresponding to the conventional and higher VG. For instance, the vortex shape at plane $x/\delta = +0.35$ is not as easy to observe in the figure as the vortex shape is for the $H_1 = 0.8\delta$ vane-type VG case.

(a)

(b) At $x/\delta = -1.50$.

(c) At $x/\delta = -0.90$.

(d) At $x/\delta = -0.30$.

(e) At $x/\delta = +0.35$.

Figure 8. Velocity distribution ($m{\cdot}s^{-1}$) at planes normal to the streamwise direction and located at distances x/δ from the backward-facing ramp. $H_2 = 0.2\delta$ vane-type VG case.

5.2. Pressure Distribution

The effect generated by the $H_1 = 0.8\delta$ vane-type VG placed at $x = -5\delta$ upstream the ramp on the streamwise pressure coefficient distribution c_p along the measurement lines A, B and C (Figure 2b) on the flat plate floor is appreciated in Figure 9. The three curves correspond to the pressure distribution (jBAY model, red rectangles) obtained from the CFD simulation with the jBAY Source Term Model implemented, the distribution data (Baseline, blue circles) obtained from the CFD simulation for a baseline case with no vortex generator positioned and the experimental pressure distribution data (EXP, green triangles) according to Lin et al. [33]. The error between the modelled (jBAY model) and the experimental (EXP) data is also shown in black vertical lines. The reader should note that the center of the backward-facing ramp is placed at the normalized distance $x/\delta = 0$ and the VG is positioned in a negative value of x/δ.

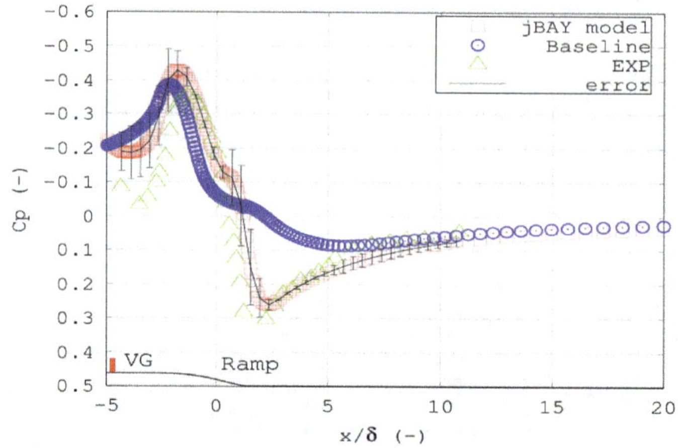

(a) Measurement line A.

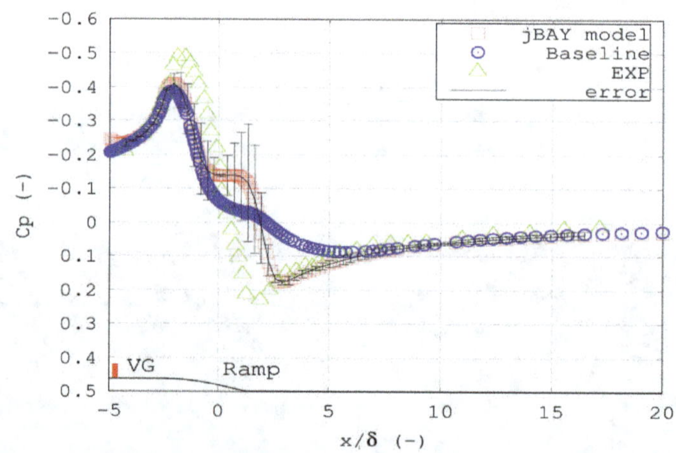

(b) Measurement line B.

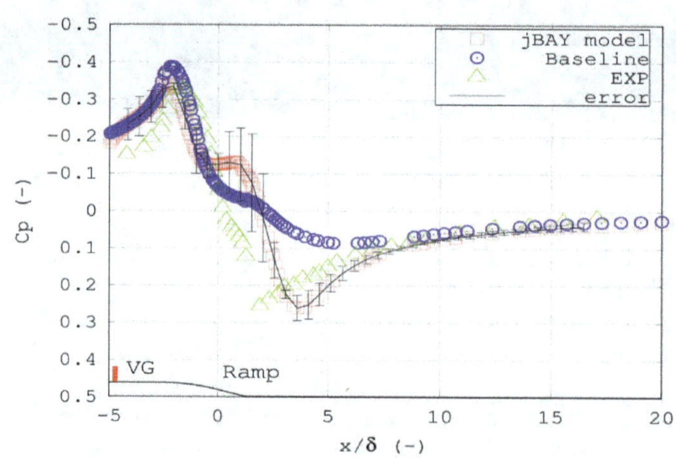

(c) Measurement line C.

Figure 9. Streamwise pressure distribution c_p along the measurement lines A, B and C on the flat plate floor. The red rectangles represent the simulated data (jBAY model), the blue circles the distribution data (Baseline) obtained from the baseline simulation and the green triangles the experimental (EXP) distribution data. Plots corresponding to the $H_1 = 0.8\delta$ vane-type VG case.

According to Lin et al. [33] and Konig et al. [34], the pressure distribution variations and the two peaks observed in the three plots at the beginning and end of the ramp correspond to accelerations and decelerations as expected. The influence of the VGs on the pressure distribution is meaningful when comparing with the pressure distribution for the baseline case.

The results seem to be in agreement with the effects corresponding with those generated by vane-type VGs which considerably increase the pressure gradient between the beginning and the end of the ramp. As expected, the vortex generator is going to bring about variations in the flow structures. When observing the simulated pressure distribution along the measurement line A (Figure 9a), the initial suction peak at the beginning of the ramp is slightly overestimated and underestimated at the end of the ramp.

The gradient of the pressure is quite well matched. Small deviations are appreciated between the experimental (EXP) and the simulated pressure distribution (jBAY model) before and behind the pressure peaks which could be related to the evidence that those regions are placed well within the three-dimensional flow structures induced by the VGs as indicated by Konig et al. [34]. When observing the pressure distribution along the measurement line B (Figure 9b) defined as the floor line starting from the VG trailing edge (TE) and parallel to the main flow (Figure 2b), the initial suction peak at the beginning of the ramp is underestimated although the gradient of the pressure as well as the pressure maximum at the end of the ramp are quite well matched.

Similarly, Figure 9c shows that the simulated pressure distribution (jBAY model) along the measurement line C has a smaller magnitude between the two peaks at the beginning and end of the ramp than the amplitudes obtained along lines A and B. This distribution is shifted towards positive distances downstream the ramp.

Figure 10 depicts the effect generated by the $H_2 = 0.2\delta$ vane-type VG placed at $x = -2.5\delta$ upstream the ramp on the streamwise pressure distribution c_p along the measurement lines A, B and C on the flat plate floor. The error (black vertical lines) between the modelled (jBAY model) and the experimental (EXP) data is again indicated. The amplitude between the peaks is smaller than the amplitude obtained in the previous VG case of height H_1. On the other hand, the simulated pressure distribution (jBAY model, red rectangles) over the ramp as well as at the end of the ramp and for the three plots (Figure 10a–c) is slightly overestimated when comparing with the experimental distribution (EXP, green triangles).

In summary, the comparison of the simulated pressure distribution (jBAY model) with the experiment pressure distribution (EXP) for the $H_1 = 0.8\delta$ and $H_2 = 0.2\delta$ vane-type VG cases depicted in the two previous figures shows a relatively good agreement although relatively small deviations are appreciated, mostly for the $H_2 = 0.2\delta$ vane-type VG case. Thus, the mean absolute percentage error between the modelled and experimental data in the H_1 VG case is 7.57%, 8.10% and 7.10% for the three pressure distributions respectively measured in A, B and C lines. In the H_2 VG case, the mean absolute percentage error is 9.83%, 9.58% and 9.76%. The reason of these little bit larger deviations could be related to the fact that the VG height H_2 is within the sub-buffer zone of the local boundary layer where the shear and viscous stresses are predominant.

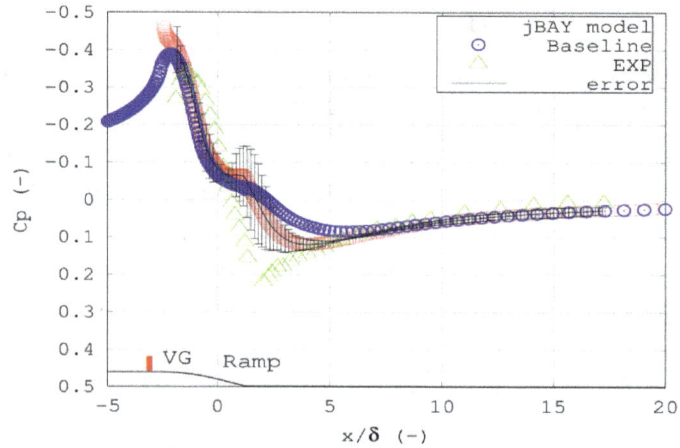

(a) Measurement line A.

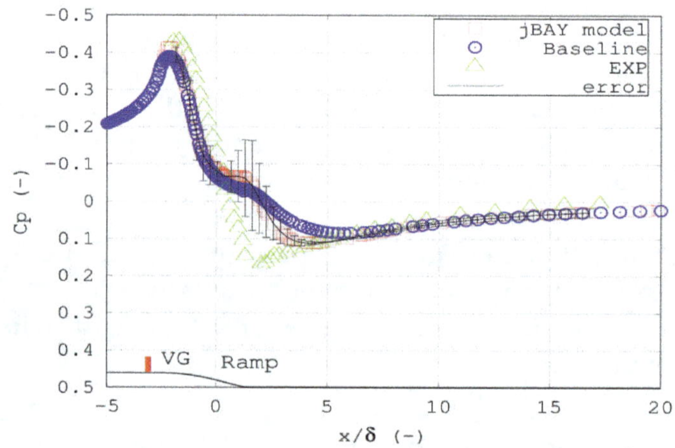

(b) Measurement line B.

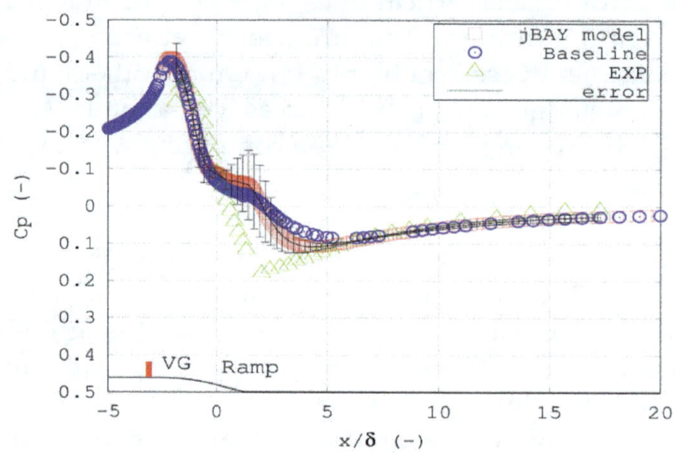

(c) Measurement line C.

Figure 10. Streamwise pressure distribution c_p along the measurement lines A, B and C on the flat plate floor. Plots corresponding to the $H_2 = 0.2\delta$ vane-type VG case.

5.3. Vortex Path

The path or trajectory of the vortex in the vertical (y) and lateral (z) directions can be determined by computing the location of the vortex center generated by the VG all along the downstream axis

x. The vortex center can be analytically defined as the point in a cross-stream plane which has the maximum value of the vorticity, the so-called peak vorticity ω_{max}. In Figure 11, a comparison between the vortex paths corresponding to the $H_1 = 0.8\delta$ and $H_2 = 0.2\delta$ vane-type VG cases is indicated. Streamwise, lateral and vertical coordinates are normalized by the local boundary layer thickness δ to show the effects of VG scaling downstream of the vane.

The vertical path corresponding to the conventional $H_1 = 0.8\delta$ vane-type VG case represented in Figure 11a tends to be parallel to the flat plate and the flow direction far from the backward-facing ramp placed at $x/\delta \cong 0$, but when the vortex approaches the ramp, it starts moving downward probably due to the adverse gradient of pressure created by the ramp. As observed in the same figure for the sub-boundary layer $H_2 = 0.2\delta$ vane-type VG case, the vertical path has a fast downward deviation probably due to the adverse gradient of pressure created by the ramp and its proximity to the ramp. The starting point for both cases is dependent on the position of the trailing edge (TE) of the vane on the flat plate, $x/\delta = -5$ and $x/\delta = -2.5$ respectively. The figure also shows that the vertical trajectories followed by the two vortices are nearly parallel to each other with a quite similar mean negative slope for both cases of approximately 0.009.

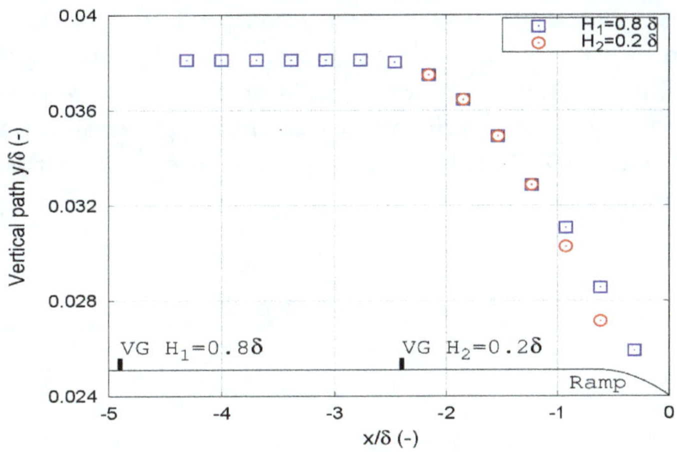

(**a**) Vertical path.

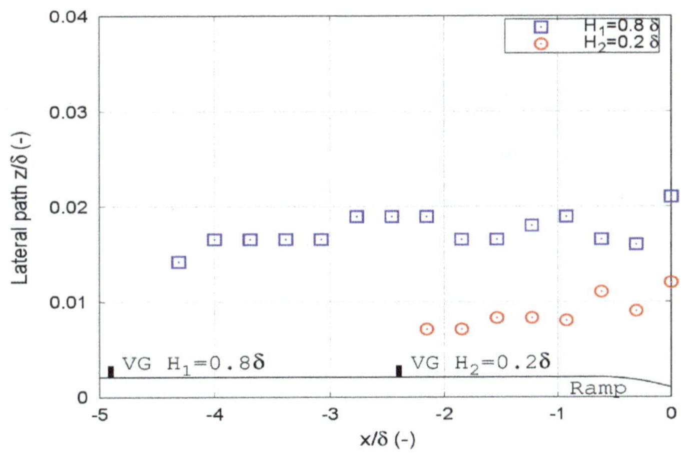

(**b**) Lateral path.

Figure 11. Normalized vertical (**a**) and lateral (**b**) paths function of the normalized distance x/δ from the backward-facing ramp for the $H_1 = 0.8\delta$ (blue rectangles) and $H_2 = 0.2\delta$ (red circles) vane-type VG cases. VGs, flat plate and ramp are also shown (not at scale).

The vortex paths for the $H_1 = 0.8\delta$ and $H_2 = 0.2\delta$ vane-type VG cases in the lateral direction are represented in Figure 11b. Both lateral trajectories seem to show a relatively small increasing trend in the same direction in which the vanes are pointed. As expected, the deviation of the lateral trajectory of the primary vortex described in the spanwise direction (z) by the highest VG is larger than the lateral deviation observed in the lowest VG. The reason for this different trajectory would lay on the fact that the lowest VG is within sub-buffer region of the boundary layer and consequently its effect on the oncoming flow is less significant. The starting point for both cases again depends on the location of the vane on the flat plate. This lateral increasing trend for the two trajectories is almost the same for both cases although for the $H_2 = 0.2\delta$ vane-type VG case the movement occurs in a lower streamwise distance. Thus, the mean positive slopes of the lateral trajectory for the H_1 and H_2 cases are around 0.005 and 0.006 respectively.

5.4. Vortex Decay

In order to illustrate how the vortex decays, the non-dimensional streamwise distribution of the normalized peak vorticity $(\omega_{xmax}\delta)/U_\infty$ can be studied. In Figure 12 the non-dimensional streamwise distribution of the normalized peak vorticity is plotted function of the normalized downstream distance x/δ for the two vane-type cases. The figure depicts the vortex decay in the streamwise direction and ratifies that the peak vorticity ω_{xmax} quite fast weakens past the VG for the two cases analysed. As observed, the maximum of the streamwise peak vorticity is obtained downstream the vane at the position $x/\delta = -2.8$ and $x/\delta = -1.8$ for the $H_1 = 0.8\delta$ and $H_2 = 0.2\delta$ VG cases respectively. After the maximum, the peak vorticity seems to decay in a decreasing exponential way to x/δ which is in concordance with Fernández-Gamiz et al. [39]. As expected, the peak vorticity magnitude decreases when the VG height is reduced.

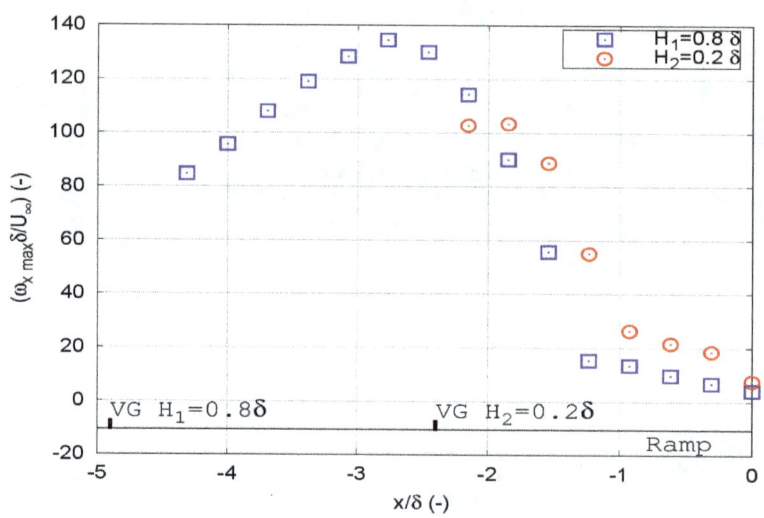

Figure 12. Non-dimensional streamwise peak vorticity $(\omega_{xmax}\delta)/U_\infty$ function of the normalized distance x/δ for the $H_1 = 0.8\delta$ (blue rectangles) and $H_2 = 0.2\delta$ (red circles) vane-type VG cases. VGs, flat plate and ramp also shown (not at scale).

5.5. Vortex Size

The vortex size is a key parameter when modelling vortices. A suitable way to study the vortex size is by means of the half-life radius $R_{0.5}$ defined as the radial distance from the vortex center to the point where the vorticity is half the peak vorticity ω_{max} captured in a cross-stream plane according to Bray [40]. At that point, the measurement errors when calculating the vortex size are negligible and the accuracy degree is high.

Figure 13 shows the vortex size evolution expressed in terms of the normalized half-life radius $R_{0.5}$ for the $H_1 = 0.8\delta$ and $H_2 = 0.2\delta$ vane-type VG cases at different locations x/δ past the VG. This figure shows a increasing trend in the vortex size when moving downstream for both VG cases. This relationship observed between both magnitudes could be formulated by a mathematical expression and this could eventually allow establishing a prediction model between the vortex size and the streamwise distance. As expected, the vortex size generated by the conventional VG of height $H_1 = 0.8\delta$ is larger than the vortex generated by the sub-boundary layer VG of height $H_2 = 0.2\delta$.

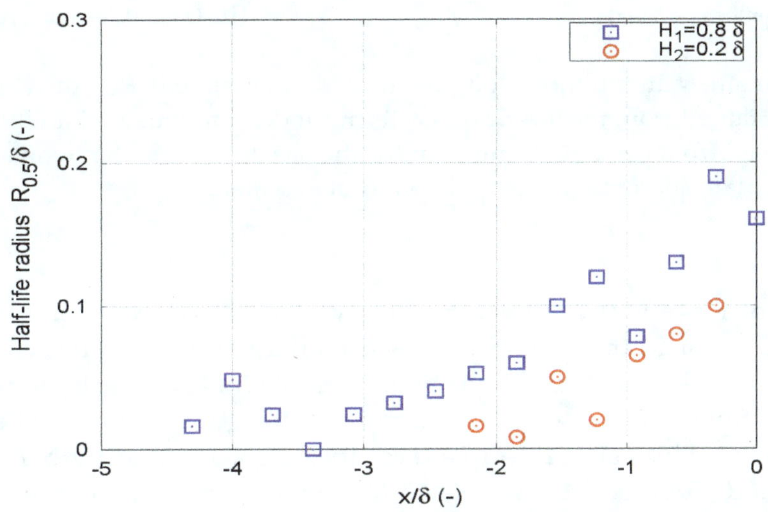

Figure 13. Vortex size evolution expressed in terms of the normalized half-life radius $R_{0.5}$ at difference normalized distances x/δ for the $H_1 = 0.8\delta$ (blue rectangles) and $H_2 = 0.2\delta$ (red circles) vane-type VG cases.

5.6. Wall Shear Stress

Figure 14 shows the wall shear stress distribution along the measurement line B on the flat plate floor for the two vane-type VG cases analysed in the present study. The wall shear stress distribution obtained from the implementation of the jBAY Source Term Model (red rectangles) is compared with that corresponding to the CFD simulation for the baseline case (blue circles). According to Godard and Stanislas [41], the implementation of VGs leads to an increase of the values of the wall shear stress downstream of the vanes in comparison with the cases where passive devices are not implemented. The results show a larger increment between the jBAY and baseline distributions for the $H_1 = 0.8\delta$ vane-type VG case than for the $H_2 = 0.2\delta$ vane-type VG case. This higher increase could be related to the fact that in the case of the conventional VG of height $H_1 = 0.8\delta$ the streamwise vortex induced is stronger and consequently its effects on the wall shear stress bigger than in the case of the sub-boundary layer VG of height $H_2 = 0.2\delta$ whose vortex induced is weaker and so are its effects, for instance, on the wall shear stress. However, these results obtained for the jBAY modelled wall shear stress seem to be again in accordance with those generated by rectangular VGs located on a flat plate which increase the wall shear stress downstream of the VG. According to Godard and Stanislas [41], this increase tends to delay the boundary layer detachment. Two wall shear stress peaks are appreciated at the beginning and end of the ramp which indicate the influence of the ramp on the shear stress. As observed, these two peaks are larger for the $H_1 = 0.8\delta$ vane-type VG case than for the $H_2 = 0.2\delta$ vane-type VG case.

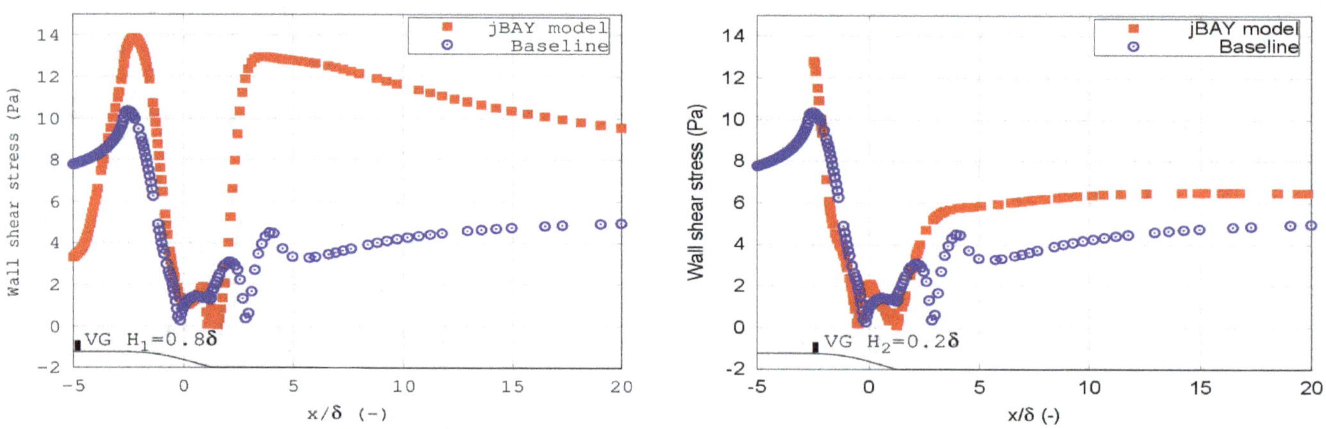

(**a**) $H_1 = 0.8\delta$ vane-type VG case.

(**b**) $H_2 = 0.2\delta$ vane-type VG case.

Figure 14. Wall shear stress distribution along the measurement line B. Red rectangles represent the data (jBAY model) obtained from the jBAY Source Term Model implemented and the blue circles the data (Baseline) obtained from the CFD simulation for the baseline case. The two vane-type VG cases are plotted. VGs, flat plate and ramp are also shown (not at scale).

6. Conclusions

The generation of vortices and their effects by a conventional vortex generator and a sub-boundary layer vortex generator positioned on a three-dimensional flat plate with a backward-facing ramp and adverse gradient pressure has been carried out by means of CFD simulations using the open-source code OpenFOAM. The influence of these two vane-type vortex generators (VGs) on the computational domain flow is implemented by using a source term in the corresponding Navier-Stokes equations according to the so-called jBAY Source Term Model. Steady-state, incompressible and turbulent flow is assumed and Reynolds Average Navier-Stokes (RANS) turbulence modelling is applied in the simulations.

The Source Term Model seems to simulate relatively well the streamwise pressure distributions along the floor of the flat plate for the two vane-type vortex generators studied of heights $H_1 = 80\%$ and $H_2 = 20\%$ the local boundary layer thickness δ at the upstream edge of the ramp. The results obtained for the jBAY modelled pressure distributions are in concordance with the experiments where the influence of these VG devices is measured. However, the streamwise pressure distribution has been slightly underestimated in certain regions far from the ramp and overestimated in other regions by the jBAY Model.

The jBAY modelled pressure distributions for the conventional vortex generator seem to be more accurate than those obtained for the sub-boundary layer vortex generator. The reason of this fact could be related to the height of the sub-boundary layer vortex generator which is within the sub-buffer zone of the local boundary layer where the shear and viscous stresses are predominant. Consequently, the results could potentially indicate that the jBAY Source Term Model presents more difficulties when modelling sub-boundary layer vortex generators.

Other parameters for vortex characterization analysed in the present study such as vortex path, vortex decay and vortex size obtained by means of the jBAY Source Term modelling, are in relatively good agreement with the results expected.

The implementation of the Source Term Model can represent an advantage over a fully Mesh-Resolved Vortex Generator Model for certain application cases due to a meaningful decrease in the cell number of the computational domain with the corresponding saving of computational time and resources. In the present work where the Source Term Model has been applied in the simulations of a conventional and a sub-boundary layer vortex generator of respective heights H_1 and H_2, the total cell number of the mesh could be eventually small when compared with the conventionally used fully Mesh-Resolved Vortex Generator Model. In addition, the jBAY Source Term Model could be quite helpful to reduce the meshing time.

Further research should be done in order to model and characterize vortex generators (VGs) of different dimensions and geometries positioned in surfaces of interest such as high-lift airfoils or ducts by implementing the Source Term Model or an optimized version of the Source Term Model into OpenFOAM or other CFD codes.

Author Contributions: I.E., U.F.-G., P.M.-F. and J.M.B. designed and were responsible for running the numerical simulations, carrying out the post-processing and writing the manuscript.

Acknowledgments: The authors are grateful to the University of the Basque Country UPV/EHU through the EHU 12/26 research program.

References

1. Johnson, S.J.; Baker, J.P.; van Dam, C.P.; Berg, D. An overview of active load control techniques for wind turbines with an emphasis on microtabs. *Wind Energy* **2009**, *13*, 239–253. [CrossRef]
2. Gad-el-Hak, M. *Flow Control: Passive, Active, and Reactive Flow Management*; Cambridge University Press: Cambridge, UK, 2007; ISBN 9780521036719.
3. Taylor, H.D. *Summary Report on Vortex Generators*; Research Department Report No. R-05280-9; Research Department, United Aircraft Corporation: Moscow, Russia, 1947.
4. Liu, Y.; Sun, J.; Tang, Y.; Lu, L. Effect of slot at blade root on compressor cascade performance under different aerodynamic parameters. *Appl. Sci.* **2016**, *6*, 421. [CrossRef]
5. Wood, R.M. Discussion of aerodynamic control effectors concepts (ACEs) for future unmanned air vehicles (UAVs). In Proceedings of the AIAA 1st Technical Conference and Workshop on Unmanned Aerospace Vehicle, Systems, Technologies and Operations, Portsmouth, VA, USA, 20–23 May 2002.
6. Aramendia, I.; Fernández-Gamiz, U.; Ramos-Hernanz, J.A.; Sancho, J.; Lopez-Guede, J.M.; Zulueta, E. Flow Control Devices for Wind Turbines. *Energy Harvest. Energy Effic. Technol. Methods Appl.* **2017**, *37*, 629–655. [CrossRef]
7. Becker, R.; Garwon, M.; Gutknecht, C.; Barwolff, G.; King, R. Robust control of separated shear flows in simulation and experiment. *J. Process Control.* **2005**, *15*, 691–700. [CrossRef]
8. Rao, D.M.; Kariya, T.T. Boundary-layer submerged vortex generators for separation control, an exploratory study. In Proceedings of the AIAA/ASME/SIAM/APS 1st National Fluid Dynamics Congress, Cincinnati, OH, USA, 25–28 July1998.
9. Lin, J.C.; Robinson, S.K.; McGhee, R.J.; Valarezo, W.O. Separation Control on high-lift airfoils via micro vortex generators. *J. Aircr.* **1994**, *31*, 1317–1323. [CrossRef]
10. Doerffer, P.; Barakos, G.N.; Luczak, M.M. Recent Progress in Flow Control for Practical Flows. In *Results of the STADYWICO and IMESCON Projects*; Springer International Publishing AG: Cham, Switzerland, 2017; ISBN 978-3-319-50568-8.
11. Steijl, R.; Barakos, G.; Badcock, K. A framework for CFD analysis of helicopter rotors in hover and forward flight. *Int. J. Numer. Methods Fluids* **2006**, *51*, 819–847. [CrossRef]
12. Velte, C.M.; Hansen, M.O.L. Investigation of flow behind vortex generators by stereo particle image velocimetry on a thick airfoil near stall. *Wind Energy* **2013**, *16*, 775–785. [CrossRef]
13. Velte, C.M.; Hansen, M.O.L.; Okulov, V.L. Helical structure of longitudinal vortices embedded in turbulent wall-bounded flow. *J. Fluid Mech.* **2009**, *619*, 167–177. [CrossRef]
14. Fernández-Gamiz, U.; Velte, C.M.; Réthoré, P.; Sørensen, N.N.; Egusquiza, E. Testing of self-similarity and helical symmetry in vortex generator flow simulations. *Wind Energy* **2016**, *19*, 1043–1052. [CrossRef]
15. Urkiola, A.; Fernández-Gamiz, U.; Errasti, I.; Zulueta, E. Computational characterization of the vortex generated by a vortex generator on a flat plate for different vane angles. *Aerosp. Sci. Technol.* **2017**, *65*, 18–25. [CrossRef]
16. Gao, L.; Zhang, H.; Liu, Y.; Han, S. Effects of vortex generators on a blunt trailing-edge airfoil for wind turbines. *Renew. Energy* **2015**, *76*, 303–311. [CrossRef]
17. Baldacchino, D.; Manolesos, M.; Ferreira, C.; Gonzalez Salcedo, A.; Aparicio, M.; Chaviaropoulos, T.; Diakakis, K.; Florentie, L.; Garcia, N.R.; Papadakis, G.; et al. Experimental benchmark and code validation

for airfoils equipped with passive vortex generators. *J. Phys.* **2016**, *753*, 022002. [CrossRef]
18. Keuthe, A.M. Effect of streamwise vortices on wake properties associated with sound generation. *J. Aircr.* **1972**, *9*, 715–719. [CrossRef]
19. Holmes, A.; Hickey, P.; Murphy, W.; Hilton, D. The application of sub-boundary layer vortex generators to reduce canopy "Mach rumble" interior noise on the Gulfstream III. In Proceedings of the 25th AIAA Aerospace Sciences Meeting, Aerospace Sciences Meetings, Reno, NV, USA, 24–26 March 1987 [CrossRef].
20. Lin, J.C.; Howard, F.G.; Selby, G.V. Small submerged vortex generators for turbulent-flow separation control. *J. Spacecr. Rocket.* **1990**, *27*, 503–507. [CrossRef]
21. Martinez-Filgueira, P.; Fernández-Gamiz, U.; Zulueta, E.; Errasti, I.; Fernández-Gauna, B. Parametric study of low-profile vortex generators. *Int. J. Hydrogen Energy* **2017**, *42*, 17700–17712. [CrossRef]
22. Lin, J. Review of research on low-profile vortex generators to control boundary-layer separation. *Prog. Aerosp. Sci.* **2002**, *38*, 389–420. [CrossRef]
23. Kenning, O.C.; Kaynes, I.W.; Miller, J.V. The potential application of flow control to helicopter rotor blades. In Proceedings of the 30th European Rotorcraft Forum, Marseille, France, 14–16 September 2004; 14p.
24. Schubauer, G.B.; Spangenber, W.G. Forced mixing in boundary layers. *J. Fluid Mech.* **1960**, *8*, 10–32. [CrossRef]
25. Bragg, M.B.; Gregorek, G.M. Experimental study of airfoil performance with vortex generators. *J. Aircr.* **1987**, *24*, 305–309. [CrossRef]
26. Fernández-Gamiz, U.; Errasti, I.; Gutierrez-Amo, R.; Boyano, A.; Barambones, O. Computational modelling of rectangular sub-boundary-layer vortex generators. *Appl. Sci.* **2018**, *8*, 138. [CrossRef]
27. Troldborg, N.; Zahle, F.; Sorensen, N.N. Simulation of a MW rotor equipped with vortex generators using CFD and an actuator shape model. In Proceedings of the 53th AIAA Aerospace Sciences Meeting, Kissimmee, FL, USA, 5–9 January 2015; 10p.
28. Bender, E.E.; Anderson, B.H.; Yagle, P.J. Vortex Generator Modelling for Navier–Stokes Codes. In Proceedings of the 3rd ASME/JSME Joint Fluids Engineering Conference, San Francisco, CA, USA, 18–23 July 1999.
29. Dudek, J.C. Modeling Vortex Generators in a Navier–Stokes Code. *AIAA J.* **2011**, *49*, 748–759. [CrossRef]
30. Jirasek, A. Vortex-Generator Model and Its Application to Flow Control. *J. Aircr.* **2005**, *42*, 1486–1491. [CrossRef]
31. Florentie, L.; Hulshoff, S.J.; van Zuijlen, A.H. Adjoint-based optimization of a source-term representation of vortex generators. *Comput. Fluids* **2018**, *162*, 139–151. [CrossRef]
32. OpenFOAM. Available online: https://openfoam.org (accessed on 1 October 2018).
33. Lin, J.C. Control of Turbulent Boundary-Layer Separation using Micro-Vortex Generators. *Am. Inst. Aeronaut. Astronaut.* **1999**, 1–16. [CrossRef]
34. Konig, B.; Fares, E.; Nolting, S. Fully-Resolved Lattice-Boltzmann Simulation of Vane-Type Vortex Generators. In Proceedings of the 7th AIAA Flow Control Conference, Atlanta, GA, USA, 16–20 June 2014; pp. 2795–3008. [CrossRef]
35. Menter, F.R. Zonal two equation k-w turbulence model for aerodynamic flows. *AIAA J.* **1993**, *93*, 2906. [CrossRef]
36. Gutierrez-Amo, R.; Fernández-Gamiz, U.; Errasti, I.; Zulueta, E. Computational modelling of three different sub-boundary layer vortex generators on a flat plate. *Energies* **2018**, *11*, 3107. [CrossRef]
37. Sørensen, N.; Zahle, F.; Bak, C.; Vronsky, T. Prediction of the Effect of Vortex Generators on Airfoil Performance. *Phys. Conf. Ser.* **2014**, *524*, 012019. [CrossRef]
38. Richardson, L.F.; Gaunt, J.A. The deferred approach to the limit. Part I. Single lattice. Part II. Interpenetrating lattices. *Philos. Trans. R. Soc. Lond. Ser. A* **1927**, *226*, 299–361. [CrossRef]
39. Fernández-Gamiz, U.; Zamorano, G.; Zulueta, E. Computational study of the vortex path variation with the VG height. *J. Phys. Conf. Ser.* **2014**, *524*, 012024. [CrossRef]
40. Bray, T.P. A Parametric Study of Vane and Air-Jet Vortex Generators. Ph.D. Thesis, College of Aeronautics, Cranfield University, Cranfield, UK, 1998.
41. Godard, G.; Stanislas, M. Control of a decelerating boundary layer. Part 1. Optimization of passive vortex generators. *Aerosp. Sci. Technol.* **2006**, *10*, 181–191. [CrossRef]

7

Study on the Effects of Evaporation and Condensation on the Underfloor Space of Japanese Detached Houses using CFD Analysis

Wonseok Oh [1,*] and Shinsuke Kato [2]

1 Graduate School of Engineering, Department of Architecture, The University of Tokyo, 4-6-1 Komaba, Meguro-ku, Tokyo 153-8505, Japan

2 Institute of Industrial Science, The University of Tokyo, 4-6-1 Komaba, Meguro-ku, Tokyo 153-8505, Japan; kato@iis.u-tokyo.ac.jp

* Correspondence: oh-ws@iis.u-tokyo.ac.jp.

Academic Editor: Bjørn Hjertager

Abstract: The purpose of this study is to determine the effects of evaporation and condensation on the underfloor space of Japanese detached houses. In this underfloor space, natural ventilation is applied. A typical Japanese wooden detached house is raised 0.3–0.5 m over an underfloor space made of concrete. The bottom of the underfloor space is usually paved with concrete, and the ceiling which is directly underneath the indoor occupant zone is made of wood. Computational fluid dynamics (CFD) analysis is applied to calculate the rates of the evaporation and condensation generated inside the underfloor under two conditions, namely, a constant (fixed) outdoor environmental condition and a fluctuating environmental condition. In the constant condition, we verified the effects of the outdoor humidity, ventilation rate, and ratio of wetted surface (RWS, ω) on the evaporation and condensation inside the underfloor space. In this condition, the rate of evaporation and condensation was quantified considering the varying outdoor humidity between 0 to 100%, and the RWS ($\omega = 1$ or 0). In addition, the influence of the different ventilation rates at 1.0 m/s for normal and 0.05 m/s for stagnant wind velocities were confirmed. Under fluctuating environmental conditions, the outdoor conditions change for 24 h, so the RWS varies. Therefore, the rate of evaporation and condensation, the amount of the condensed water, and the area of condensation were confirmed. The results were as follows: with a high airflow rate on the underfloor space, the evaporation and condensation phenomenon occurs continuously and is easily affected by outdoor humidity, while under low airflow rate conditions, only the condensation appeared steadily. If the wind velocity is strong, the convective mass transfer on a surface becomes large. In a condition of the outdoor humidity and the airflow rate on underfloor are high, condensation mainly occurs in a corner of the underfloor space due to high evaporation by convection in the mainstream of the airflow. By contrast, when the airflow rate is low, condensation occurs along the air stream. Accordingly, this information could be employed as design considerations for the underfloor space at the architectural design stage.

Keywords: underfloor ventilation; evaporation and condensation; computational fluid dynamics (CFD); ratio of wetted surface (RWS)

1. Introduction

The aim of ventilation of the underfloor space is to prevent moisture generated in the ground from entering the living area and to improve the heating and cooling efficiency of the building, given the large heat capacity of the ground. Moisture is the main problem in wooden buildings, as it causes rotting of the wood and an unpleasant smell inside the building. Even so, the underfloor space is generally ignored by the residents, as it is not only inconspicuous, but also difficult to clean. Generally, moisture originates from the ground (soil), which absorbs water after rain, or from groundwater, and enters the building through the ventilation openings. It is important to block the source of the moisture in order to control the humidity level and to prevent the generation of condensation inside the underfloor space. Therefore, it is crucial not only to prevent ground moisture from entering, but also to eliminate any moisture forming inside from rainwater or high humidity. Since these phenomena influence the health of the occupants and the durability of the buildings, methods, such as paving of the ground or mechanical ventilation are applied occasionally. However, it is useful to predict the climate of the underfloor space in the attempt to control the humidity and to apply an appropriate method during construction, as installing such alternative methods could be costly.

In previous research, the relative humidity, temperature, and pressure inside the underfloor space were measured, the ground moisture evaporation rate was calculated, and subsequently the effects of natural and mechanical ventilation were confirmed [1]. The results showed that the average value of moisture evaporation with unpaved ground was 3.6 g h^{-1} m^{-2} in a naturally ventilated condition and 5.7 g h^{-1} m^{-2} in a mechanically ventilated condition. Research also indicated that the optimum air change rate on the underfloor space was 1–3 air change rate per hour (ACH) throughout the year [2]. Experiments have also been conducted with benzene in the underfloor space to identify the potential health risks associated with toxic soil vapors penetrating into the living spaces [3,4]. These studies employed point measurements. However, the internal humidity was excluded as a factor in the research. Several studies have attempted to confirm the presence of internal humidity by employing computational fluid dynamics (CFD) to determine the distribution of such internal humidity. However, no research has been conducted on the internal condensation of the underfloor space. In the current study, we attempt to determine the effects of internal condensation and evaporation and the probable area where condensation would occur relevant to the natural ventilation. Research on such aspects has not been conducted before.

A Japanese detached house can be simply illustrated as in Figure 1. The height of the underfloor space is 0.3–0.5 m from ground level. As openings are dispersed around the underfloor space for natural ventilation, outdoor moist air flows into the underfloor space, which gives rise to condensation and evaporation. The bottom of the underfloor space is usually paved with concrete and the ceiling, which is directly underneath the indoor occupant zone, is made of wood.

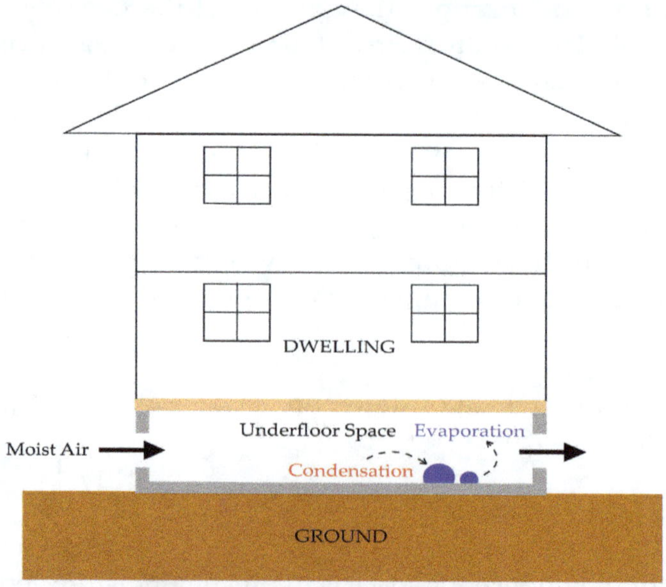

Figure 1. Underfloor ventilation in a Japanese detached house.

There are two main causes of condensation on the underfloor environment, namely, moisture deriving from the outside environment that condenses on the cold surface of the underfloor, and re-condensation that occurs when the condensed water inside the underfloor space evaporates. Once the condensation has been evaporated by ventilation, it becomes another moisture generation source inside the underfloor, raising the indoor humidity and significantly reducing the evaporation effect. The airflow and temperature distribution of the internal environment cause condensation and evaporation locally. Consequently, simple point measurements of the humidity cannot adequately explain the general condensation phenomena on the underfloor space or predict the probable occurrence of condensation. Furthermore, as the inside of the underfloor has a complicated shape, it is important to understand which parts would be vulnerable to condensation and to give due consideration to these vulnerable regions in the design of the house. Generally, in Japan, a detached house is a wooden building and it has an underfloor space, where its bottom touches the ground surface, which is usually paved with concrete to block water seepage from the ground. Natural ventilation is usually employed through underfloor air vents, shown in Figure 2, and mechanical ventilation is generally not introduced.

Figure 2. Underfloor openings for the natural ventilation in Tokyo, Japan.

As many previous studies have conducted research on the conditions of unpaved ground and natural ventilation, we attempted to confirm the internal condensation and evaporation effects on the underfloor space under conditions of paved ground and natural ventilation. It is difficult to determine the distribution of internal humidity by employing experimental methods; therefore, we chose CFD analysis to determine the climate inside the underfloor space. Additionally, the aim of the study is to understand the degree of evaporation and condensation arising because of underfloor ventilation and the distribution of the internal humidity, as well as to identify the area at risk of condensation.

Generally, In CFD, turbulent flows can be predicted through three approaches: direct numerical simulation (DNS), large-eddy simulation (LES), and Reynolds-averaged Navier-Stokes (RANS) equation simulated with turbulence models [5]. DNS gives highly reliable results directly solving Navier-Stokes equations but the required calculation time will be very long. The LES approach is an intermediate modeling technique between DNS and RANS. Turbulent motion can be divided into large eddies and small eddies, where LES is a method that solves the Navier-Stokes equations with filtered large-scale eddies of the turbulent flow except for small-scale eddies. LES can give us more detailed information on turbulence than RANS. However, it still requires a considerable calculation cost. Using RANS, turbulence motion can be quickly predicted and it is the most practical approach.

Computational fluid dynamics (CFD) is also the most effective way to predict the airflow in a building. Predicting airflow patterns in an indoor room by using a CFD analysis was first applied by Nielsen [6]. For the indoor environment prediction, the RANS k–ε turbulence model was used for many years and shows the acceptable results according to many researchers. [7–10]. The standard k–ε turbulence model showed reasonable results to predict the airflows in indoor environment [11]. Moreover, realizable k–ε models usually provide much improved results for swirling flows and flows involving separation compared to the standard k–ε model [5,12,13].

1.1. Weather Data in Tokyo, Japan

The extended Automated Meteorological Data Acquisition System (AMeDAS, 1981–2000) was used as standard-year weather data to estimate the underfloor climate. Tokyo city in Japan was selected as the representative location for our simulation. The detailed weather conditions are described in Figures 3 and 4. Figure 3 shows the correlation of the cumulative relative humidity per month in Tokyo. The wettest months of the year are June to September and a high-humidity climate prevails for more than half of this period, with the relative humidity being more than 75%.

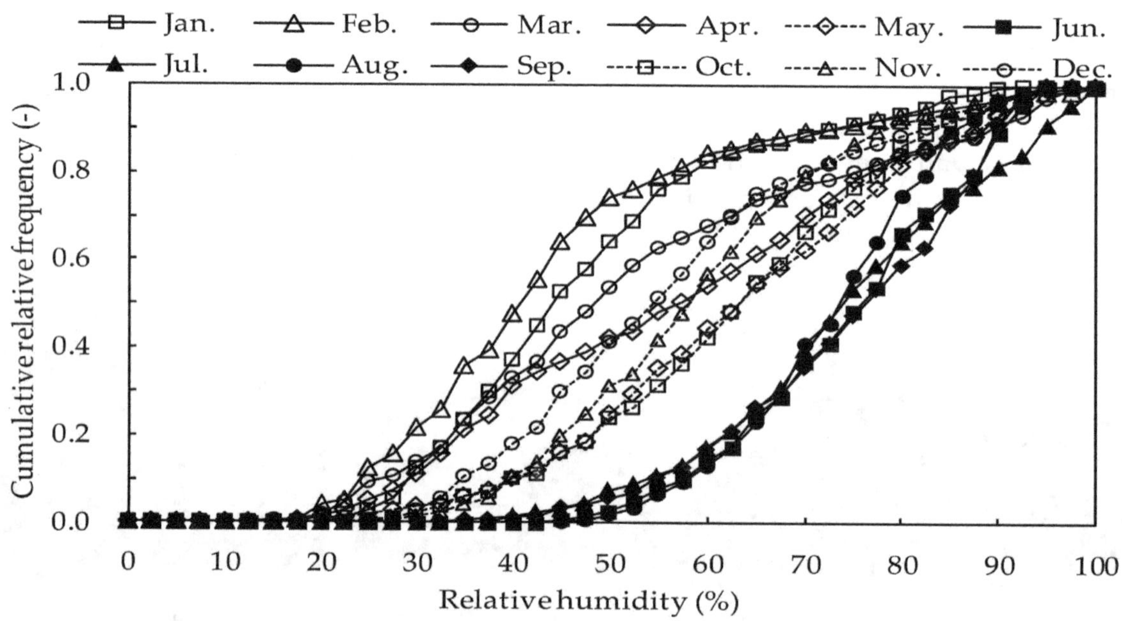

Figure 3. The correlation of the cumulative relative humidity and relative humidity per a month during a year in Tokyo, Japan.

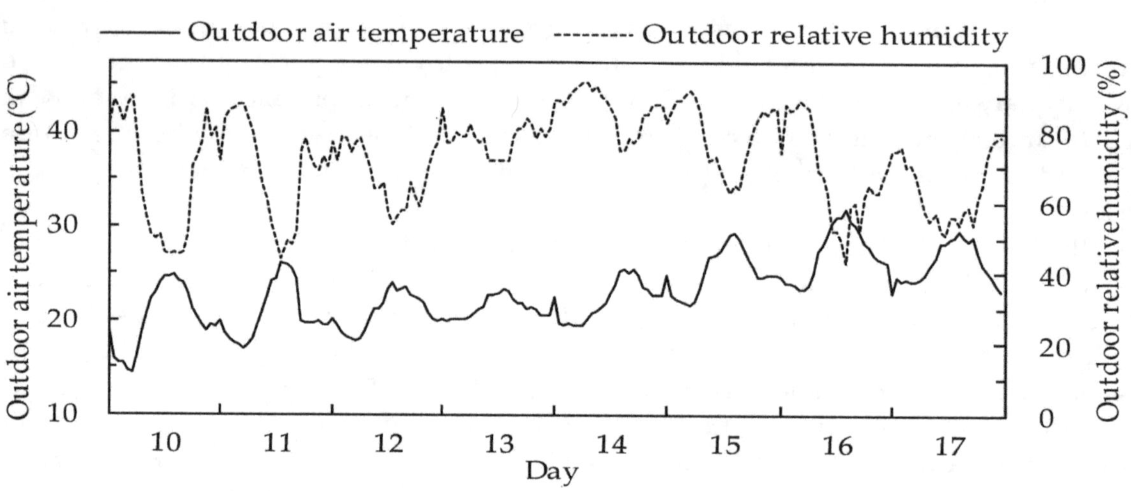

Figure 4. Outdoor air temperature and relative humidity of Tokyo according to the AMeDAS standard weather data (10–17 June).

To consider the worst probable weather situation, the data for 10–17 June were employed in our study, as these days represent the high humidity season in Tokyo. During the rainy season, although the external humidity increases significantly, the temperature of the internal environment remains lower than that of the external environment. This is because the influence of solar radiation is insignificant as the sky is always overcast. Therefore, the risk of internal condensation during this period is expected to increase significantly.

1.2. Research Procedure

In the design of the underfloor space, it is important to consider the interior environmental characteristics and the likely location and time of condensation occurring. However, since the internal airflow is complicated, it is difficult to predict the areas where condensation could occur. Therefore, we employed CFD analyses to determine and confirm the effects of evaporation and condensation relevant to natural ventilation. The underfloor space has dispersed ventilation openings for natural ventilation, with the pattern of the airflow being complex, as this area is divided into several zones by internal partition walls. Evaporation and condensation phenomena arise because of the difference of the partial pressure of water vapor caused by the difference between the surface temperature and the adjacent air temperature right near the surface. It is difficult to confirm this phenomenon inside the whole underfloor space through experiments. This study was conducted relevant to two different circumstances, namely, a constant outdoor environmental condition for 1 h and a fluctuating outdoor environmental condition for 24 h. The rate of evaporation and condensation generated inside the underfloor space was calculated and confirmed relevant to the first condition to determine the influence of outdoor environmental factors. The influence of natural ventilation was determined by quantifying the rate of evaporation and condensation with the internal surface being completely wet ($\omega = 1$) and completely dry ($\omega = 0$), and with the outdoor humidity increasing at a range of 0 to 100%, with a 10–20% interval. Under this condition, the influence of different ventilation rates was also confirmed by verification at a general wind speed condition of 1.0 m/s and a stagnant wind condition at 0.05 m/s. In a fluctuating condition, the ambient air temperature near the internal surface temperature also fluctuated continuously, and evaporation and condensation easily occurred because of the temperature difference. Accordingly, based on these fluctuating conditions, the rate of evaporation and condensation over time and the accumulated volume of condensed water on the surface were determined. The information obtained from our research could be utilized at the architectural design stage to pinpoint the spots vulnerable to and the conditions conducive to the generation of condensation inside the underfloor space.

2. Natural Ventilation on Underfloor Space

Since the geometry of the underfloor space is complex and it has many vents all around, it is difficult to predict the internal airflow patterns. Generally, as the wind blows from the south during a summer season in Japan, it is considered only two conditions of wind speed. During a calculation, the variation of wind direction and wind speed were disregarded in this study. The ventilation rate from inside and outside pressure difference at each opening was calculated.

2.1. Outline of CFD Analysis

The underfloor ventilation rate was calculated by realizable k–ε model with the semi-implicit method for pressure-linked equation (SIMPLE) algorithm and the second-order upwind-convection differencing scheme. The performance of realizable k–ε turbulence model is superior to that of other k–ε turbulence models [13]. Transport equations of airflow are solved for turbulence kinetic energy k (m^2/s^2) and its dissipation ε (m^2/s^3). The turbulent eddy viscosity ν_t is calculated as follows:

$$\nu_t = C_\mu \frac{k^2}{\varepsilon} \tag{1}$$

Transport equation of k and ε is expressed as:

$$\frac{\partial(\rho k)}{\partial t} + \frac{\partial(\rho k u_i)}{\partial x_i} = \frac{\partial}{\partial x_j}\left[\left(\mu + \frac{\mu_t}{\sigma_k}\right)\frac{\partial k}{\partial x_j}\right] + G_k + G_b - \rho\varepsilon - Y_M + S_k \tag{2}$$

and:

$$\frac{\partial(\rho\varepsilon)}{\partial t} + \frac{\partial(\rho\varepsilon u_i)}{\partial x_i} = \frac{\partial}{\partial x_j}\left[\left(\mu + \frac{\mu_t}{\sigma_k}\right)\frac{\partial\varepsilon}{\partial x_j}\right] + \rho C_1 S_\varepsilon - \rho C_2 \frac{\varepsilon^2}{k + \sqrt{\nu\varepsilon}} + C_{1\varepsilon}\frac{\varepsilon}{k} C_{3\varepsilon} G_b + S_\varepsilon \quad (3)$$

where G_k is the generation of turbulence kinetic energy due to the mean velocity gradients, G_b is the generation of turbulence kinetic energy due to buoyancy effect, Y_M is the contribution of the fluctuating dilatation in compressible turbulence to the overall dissipation rate, σ_k, σ_ε are the turbulent Prandtl numbers for k and ε and S_k, S_ε are source terms. The model constants are $C_{1\varepsilon} = 1.44$, $C_2 = 1.9$, $\sigma_k = 1.0$, and $\sigma_\varepsilon = 1.2$. The standard wall functions were applied to calculate the airflow near the wall.

2.2. Simulated Model

The simulation model was selected based on the Institute for Building Environment and Energy Conservation (IBEC) standard model which guides design and evaluation of a typical Japanese house [14]. At the bottom of the building is the underfloor space, which has air vents all around to supply natural ventilation.

The control volume for the calculation was selected properly, i.e., larger than the building model, as shown in Figure 5. Because the wind direction and speed were considered only for the constant southern wind, the south side was set as the inlet and the north side as the outflow (zero gradient condition).

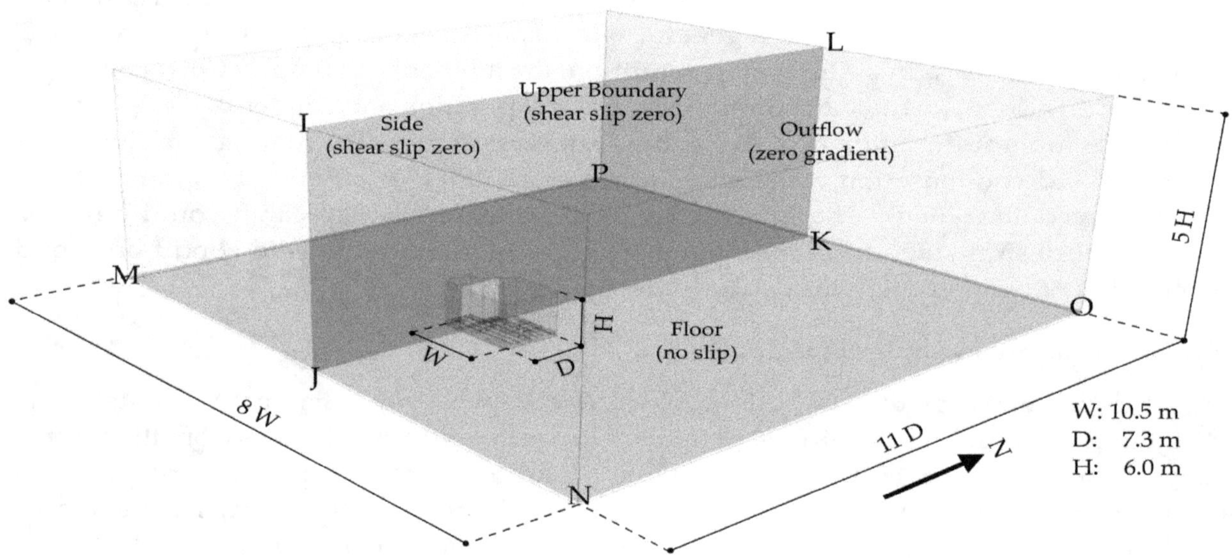

Figure 5. Residential building and control volume for calculating the ventilation rate in sub floor ventilation area.

The inlet wind from the south was created from the power law model to the mean velocity profile in Equation (4):

$$U_z = U_0(z/h)^{0.25} \quad (4)$$

where U_0 represents wind speed, and U_z is the wind speed at a height of z. The specified velocity U_0 takes into account two conditions, namely, a typical 1 m/s condition and a stagnant wind speed of 0.05 m/s, and the reference height h was assumed 1 m. The power law exponent was determined as 0.25 [15]. The logarithm-raw was applied on a surface of the building and the floor as a boundary condition. The side boundary (east, west) and upper boundary conditions were chosen as symmetry conditions (shear slip zero). The calculation was conducted by realizable k–ε turbulence model using a fine grid with 38 million hexahedral cells. Predicting airflow and dispersion around a building, numerical calculation by realizable k–ε turbulence models showed good agreement with the wind tunnel experiment than standard k–ε turbulence model [16]. Plan IJKL is a cross section of the left part of the opening on the southern wall of the building. Plan MNOP refers to the area 0.35 m above ground level.

2.3. Anslysis Results

The results for the two wind speed conditions, namely, 1.0 m/s and 0.05 m/s, are as follows: the wind direction is from the south, with the wind flowing into the underfloor space through two openings in the wall on the southern side, from where it is dispersed into each internal zone and exhausted through the other five openings.

The normalized velocity distribution around the building and inside the underfloor is shown in Figure 6. The airflow rate was calculated by the wind speed of the outdoor and the internal pressure difference of each opening, with the result shown in Table 1. In addition, this result was used as a boundary condition to calculate where the moist air from outside would be drawn into the internal underfloor space.

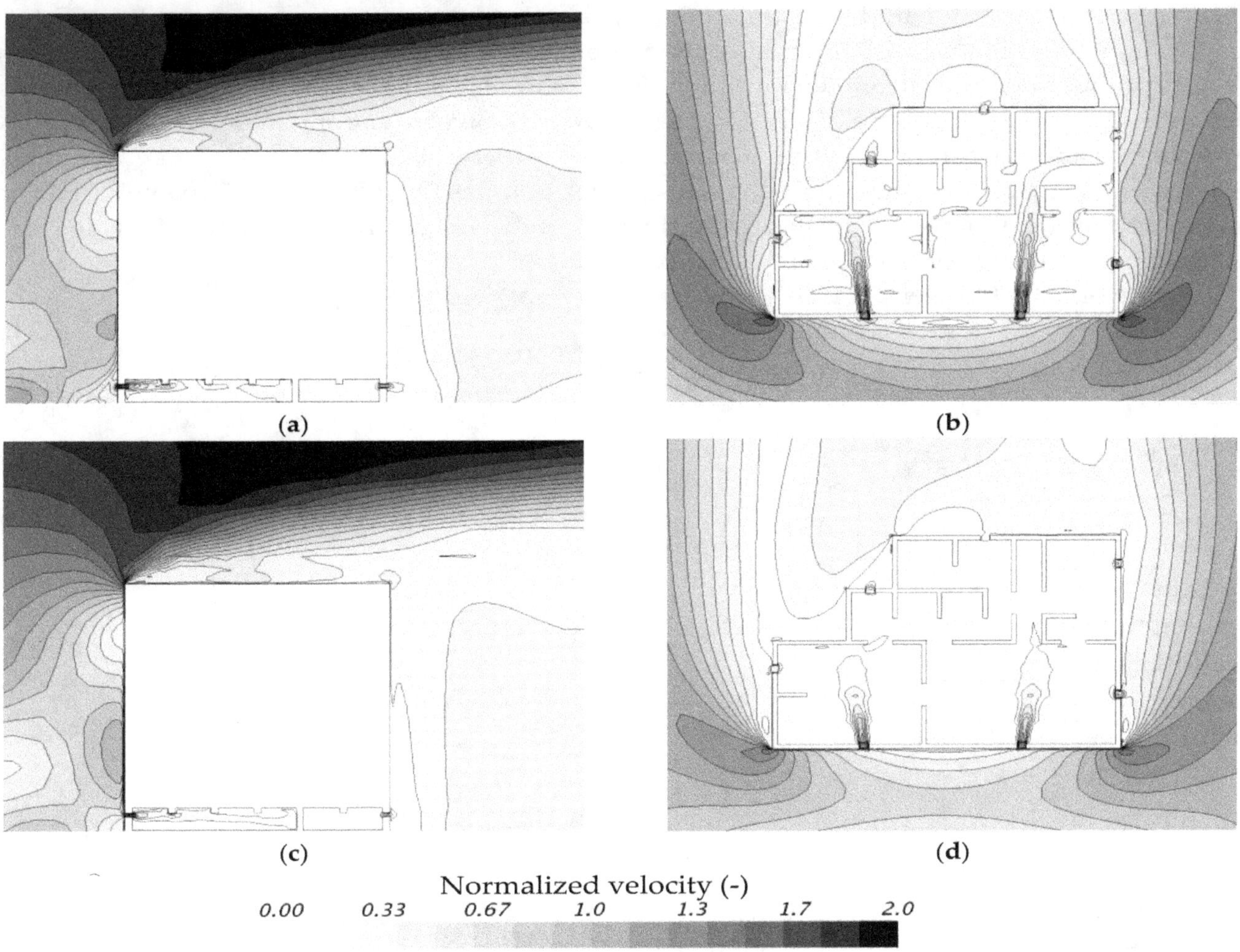

Figure 6. The normalized velocity $\frac{U}{U_o}$ distribution of around the building and inside of the underfloor space in condition of $\frac{U_z}{U_o} = \left(\frac{z}{h}\right)^{1/4}$ velocity profile: (**a**) Scalar velocity distribution in section IJKL, where the velocity condition $U_0(h)$ is 1.0 m/s; (**b**) Scalar velocity distribution in section MNOP, where the velocity condition $U_0(h)$ is 1.0 m/s; (**c**) Section velocity distribution on section IJKL, where the velocity condition $U_0(h)$ is 0.05 m/s; (**d**) Section velocity distribution on MNOP, where the velocity condition $U_0(h)$ is 0.05 m/s.

Table 1. The result of the Airflow rate through the underfloor openings.

Opening	Air Flow Rate ΔQ, m^3/h	
	1 m/s, South	0.05 m/s, South
E1	−75.3 *	−0.3 *
E2	−33.9 *	−1.9 *
W1	−54.5 *	−2.2 *
S1	112.4	4.9
S2	112.9	5.0
N1	−46.9 *	−1.7 *
N2	−33.5 *	−0.8 *

* A negative value means that the airflow is from inside to outside.

3. Evaporation and Condensation Effects on the Underfloor Space

3.1. Outline of Evaporation and Condensation Estimation Method

By modeling the water thickness calculation, the evaporation and condensation phenomena on the surface of the wall could be estimated. This involves the estimation of an additional scalar transport equation for the mass fraction of the water vapor. A source–sink term is considered for the scalar of the condensation and evaporation of the water layer, as well as the latent heat required for the transition. In the instance of a difference between the water vapor content of the water layer on the surface and the cell next to this surface, the model would calculate a rate of evaporation or condensation, depending on the conditions. This model assumes the following [17], namely, the vapor content in the air does not affect the thermal properties of the vapor–air mixture, and the water vapor mass is ignored with respect to the total mass in a cell.

The rate of mass transfer per unit surface, in kg s^{-1} m^{-2}, is given by:

$$\dot{m} = \rho_g \cdot \beta_g \cdot CG \cdot \ln \frac{CG(1 - C_s)}{CS(1 - C_g)} \quad (5)$$

where ρ_g is the density (kg m^{-3})

$$\beta_g = C_{emp} \cdot D_v / L \cdot 0.045 \cdot Re^{0.8} \cdot Sc^{0.43} \ (m/s) \quad (6)$$

C_{emp} is the empirical constant 0.9 (-)
L is a characteristic length (cubic root of cell volume next to fog layer boundary) (m)
D_v is diffusion of vapor in air (m^2/s)
Re is the Reynolds number (-)
Sc is the Schmidt number (-)

$$\begin{aligned} CG &= 0.622 + 0.378C_g \\ CS &= 0.622 + 0.378C_s \end{aligned} \quad (7)$$

C_g is the actual concentration of vapor in air
C_s is the saturation concentration of vapor in air

For a given temperature, the saturation pressure p_s is calculated from the following expression:

$$p_s(T) = 611.85e^{\frac{17.502(T+273.15)}{T-32.25}} \quad (8)$$

The change rate of the water thickness over time can be confirmed from the simulation. In addition, the mass change rate can be obtained from the calculated area of each cell, the water-thickness change rate, and water density. A decrease of the mass change rate indicates that the evaporation phenomenon has occurred and an increase means the condensation phenomenon has occurred. Finally, by determining the mass change rate, the rate of evaporation and condensation of the underfloor space can be calculated.

3.2. Simulated Model

The simulation model was selected based on the IBEC standard, which includes design standards for a Japanese detached house [14]. A two-story residential building is considered in this research, with seven openings on the underfloor space, as shown in Figure 7, which depicts the inside of the underfloor space. The walls and floor are made of concrete, with a basic thickness of 150 mm. The floor of the room corresponding to the ceiling of the underfloor space is made of a wood, supported by a wooden beam section of 105 mm × 105 mm, which is expressed as a dotted line. Figure 8 indicates the air movement passage inside the underfloor space and Figure 9 indicates the ventilation passage exposed to the outside, which is described in Figure 2.

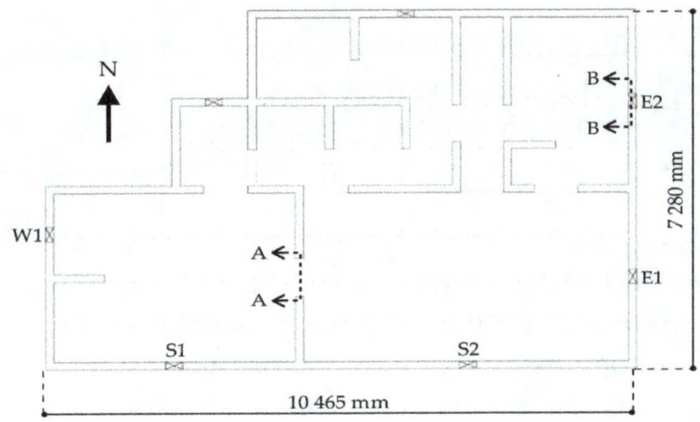

Figure 7. Inside of the underfloor ventilation area.

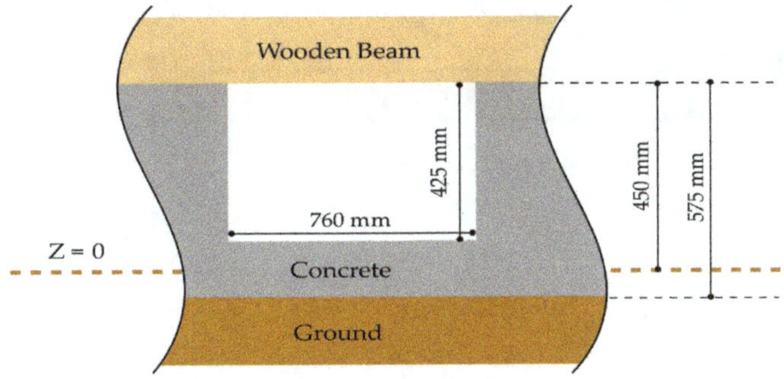

Figure 8. The section view of A-A.

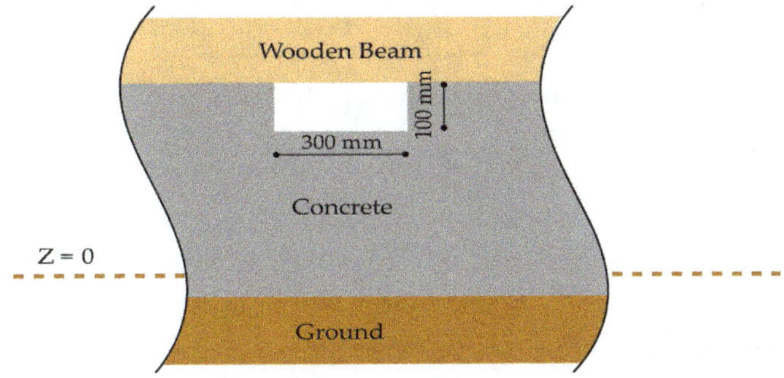

Figure 9. The section view of B-B.

The temperature of underfloor space is low even in the summer season, as it is mostly influenced by the ground temperature, which is not affected by solar radiation and does not have significant temperature changes. Our understanding is that the concrete surface on underfloor space of typical house is maintained low temperature, 20–22 °C during rainy season in Japan. The characteristics of the material should be considered because the surface temperature of the material is time dependent. The material properties for the calculations refer to the American Society of Heating, Refrigerating, and Air-Conditioning Engineers (ASHRAE) data in Table 2.

Table 2. The material properties of underfloor ventilation area and physical conditions.

Material	Density, (kg/m^3)	Specific Heat, (J kg^{-1} K^{-1})	Thermal Conductivity (W m^{-1} K^{-1})	Thickness (m)
Concrete	2240	750	0.53	0.15
Ground (soil)	1500	800	0.33	0.85
Floor (wood)	370	1900	0.11	0.012

The opposite side of the ceiling on underfloor space which represents the floor of the dwelling zone is assumed to remain constant at a temperature of 25 °C, as is the indoor environment. According to the weather data, the temperature of 1 m underground had almost no change in daytime. Therefore, it is assumed that the temperature of the concrete surface contacted with the ground is constant at 18.7 °C during the rainy season. As shown in Figure 10, the room temperature and the ground temperature is constant and the thermal conduction acts purely one-dimensional by the material characteristics of Table 2. The moist air from the outside was set to cause the evaporation and condensation phenomenon on the inner surface of underfloor space. Although some studies have considered the risk of annual condensation considering the absorption-desorption properties of the building materials [18], in this study, only the short-term evaporative condensation due to internal and external environmental conditions was focused on, so that the moisture transfer inside the materials was not considered.

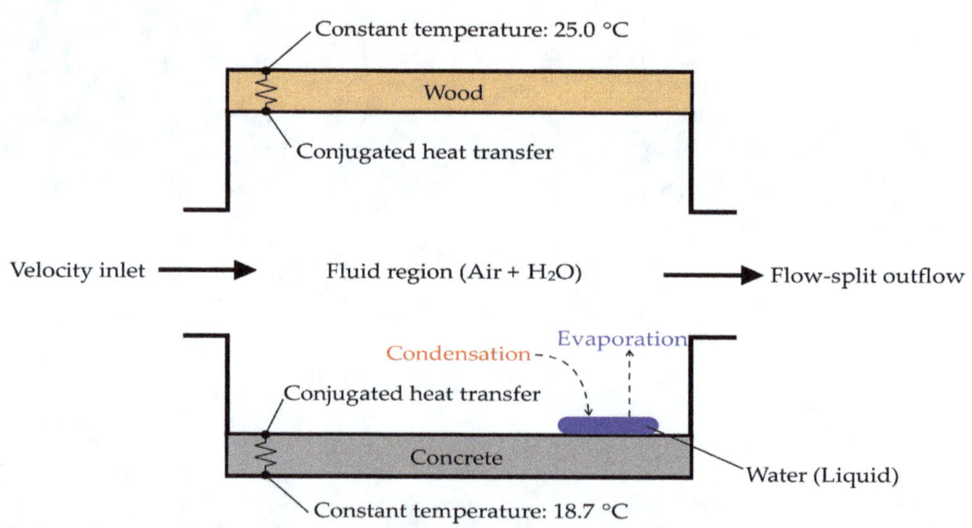

Figure 10. Sketch of the computational domain showing the relevant boundary conditions.

The pressure-velocity coupling for the airflow solution was conducted with the SIMPLE algorithm [19]. The convection and diffusion terms were integrated, using the QUICK difference scheme [20]. The boundary and calculation conditions are shown in Table 3. The parameters of the outdoor environment, which include wind speed (V_S), air temperature (T_S), and humidity (W_S) were applied in openings, the inlet boundary condition. The initial condition of the internal space was set equal to the temperature of the input initial condition. Wind speed conditions of 1.0 m/s and 0.05 m/s were considered, based on the ventilation rates calculated in the previous section. The condition of the inner wall was set to adiabatic and the emissivity was assumed as 0.95.

Table 3. Boundary and calculation conditions.

Turbulence model	Realizable k–ε turbulence model
Algorithm	SIMPLE
Convection	Upwind second-order difference
Numerical schemes	QUICK
Time dependence	Transient calculation
Room size	Described in Figure 9
Inlet	Velocity (V_S), Temperature (T_S), Humidity (W_S)
Outflow	Calculated with mass balance
Concrete-side wall	Standard log-law, emissivity = 0.95
Floor-side wall	Standard log-law, emissivity = 0.95
Initial conditions	Concrete: 20 °C, Floor: 25 °C

The emissivity of inside wall is supposed to be 0.95 and the simulation modeling of underfloor space is illustrated in Figure 11, which is composed of 48,000 hexahedral cells and covered with five prism layers. The value of y+ is lower than 1 on all over the surface of wall. All meshes were calculated by the realized k–ε turbulence model which performs better [13], and enhanced wall treatment, that is to say, two layer all y+ wall treatment which is a hybrid approach that seeks to recover the behaviors of the other two wall treatments in the limit of fine or coarse meshes. When the mesh resolution increases, the result showed better accuracy, while the calculation time has become longer. On the other hand, the results were shown quite similar in case of fine and coarse meshed model [21,22].

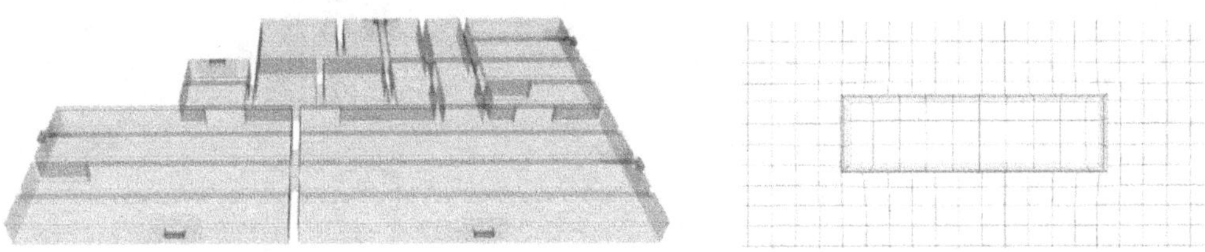

Figure 11. Outline of prism layers and hexahedral meshes prism layers on underfloor space and its opening.

3.3. Wetted Area Ratio

When there is a source of moisture generation inside, the partial pressure of the water vapor in the air is shown to differ, which affects the evaporation and condensation phenomena. Therefore, it is necessary to consider the wetted area ratio on the inner surface:

$$\text{Ratio of wetted surface (RWS, } \omega) = \frac{\text{Wetted surface area (m}^2)}{\text{Concrete surface area (m}^2)} \tag{9}$$

When water is present partially on the surface, the distribution of the internal partial pressure of water vapor differs, depending on its position due to the evaporation. Therefore, even if the ratio of wetted surface is the same, different aspects of evaporation and condensation phenomena are expected to be seen depending on the location of the wet area. However, as it is difficult to generalize the effect of condensation according to the position of the wet area, therefore, only the conditions where the RWS is 1 (all surfaces are completely wet) and the RWS is 0 (all surfaces are completely dry) were considered. Furthermore, a sufficient amount of water was taken into account, so that the ratio of the RWS would not change during the simulation (water thickness = 0.01 m).

If the evaporation phenomenon occurred on all the surfaces under the condition $\omega = 1$, the maximum evaporation effect from underfloor ventilation could be assumed confirmed.

If condensation occurred on all the surfaces in the condition $\omega = 0$, the maximum condensation effect from underfloor ventilation could be assumed confirmed. The wind speed was calculated based on the rate of ventilation at 1.0 m/s and 0.05 m/s. The temperature was set at 25 °C, considering the average temperature during the rainy season in June. In addition, the influence of relative humidity was confirmed by varying it within the range 0–100%, with 20% increasing intervals. A summary is presented in Table 4 of the detail conditions of each study case to confirm the evaporation and condensation effect on the internal underfloor space relevant to the outdoor environmental conditions. The effects of underfloor ventilation relevant to the outdoor environment were investigated by determining the overall average rate of evaporation and condensation inside the underfloor space after a time elapse of 1 h in each study condition. The mass change from evaporation and condensation was calculated for each cell of the surface through the above process.

Table 4. Initial conditions of the film thickness of water on surface of the concrete for the CFD analysis.

Outdoor Condition (25 °C)			Initial Water Thickness on Surface of the Concrete (m)	
Humidity (%)	**Wind Speed (m/s)**	**Wind Direction**	$\omega = 1$	$\omega = 0$
0	1.0	South	0.01	0
10			0.01	0
30			0.01	0
50			0.01	0
70			0.01	0
90			0.01	0
100			0.01	0
0	0.05		0.01	0
10			0.01	0
30			0.01	0
50			0.01	0
70			0.01	0
90			0.01	0
100			0.01	0

3.4. Analysis Results

The results of evaporation and condensation on underfloor space when outdoor environment conditions remained constant for an hour can be identified from the thickness of the water. Accordingly, the results, depending on the differences in the outdoor environmental conditions (wind speed, humidity), were compared by calculating the change per unit time (mg m^{-2} h^{-1}) through the analytical method, as shown in Figure 12.

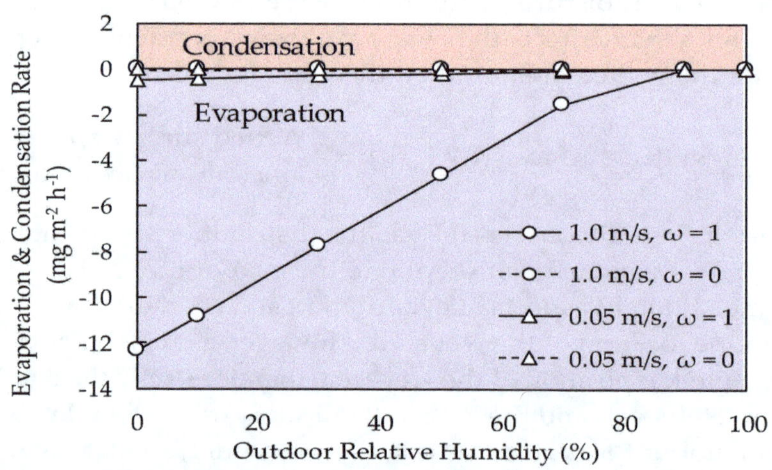

Figure 12. Evaporation and condensation effects relevant to the outdoor environmental conditions and RWS on the surface of the concrete on underfloor space.

Evaporation and Condensation Effect

The drier the air that flowed into the underfloor space and the larger the rate of ventilation, the faster the moisture evaporated from the surface to be exhausted to the outside, leading to the rate of evaporation increasing.

Under conditions of 1.0 m/s wind speed and 0% humidity, and when the entire surface was completely wet (ω = 1), evaporation occurred on all the surfaces, resulting in the largest evaporation rate of 12.3 mg m^{-2} h^{-1}. Considering that the floor area covers 59 m^2, the total evaporation rate of the internal underfloor space was 726 mg/h. When the outdoor humidity was 90%, the evaporation phenomenon reversed and condensation occurred, and, with humidity of 90–100%, the result was 0.4×10^{-3}–0.5×10^{-3} mg m^{-2} h^{-1}. The total internal condensation rate was indicated as 0.24–0.03 mg/h.

Under the condition of ω = 1, the outdoor wind speed decreased 1.0 m/s to 0.05 m/s, and the underfloor space shows a lower evaporation rate. This is because the exhausted volume of water that evaporated from the internal surface was smaller than the 1.0 m/s resulting from ventilation was. The higher the humidity the less the difference between the partial pressure of the water vapor on the water surface and the adjacent air would be, so that the occurrence of evaporation was reduced. Accordingly, with a 0–70% outdoor humidity condition, the evaporation effect was 0.5–0.05 mg m^{-2} h^{-1} per underfloor space on average. On the other hand, with a condition of 90% outdoor humidity, a state of equilibrium was reached, with no evaporation or condensation occurring. When outdoor humidity of 100% was introduced, the evaporation phenomenon reversed and 0.7×10^{-4} mg m^{-2} h_{-1} condensation occurred.

As there was no source of internal water generation when ω = 0, the evaporation phenomenon would not occur under humidity conditions of 0–70%. However, with 90–100% humidity, the moisture introduced from the outside air condensed on the cold surface of the concrete at a condensation rate of 0.03 g/h. The condensation phenomenon was slight, as indicated by the result of 0.4×10^{-2} mg/h. Furthermore, under conditions of 90–100% humidity and ω = 0, the condensation was 0.7×10^{-4} mg m^{-2} h^{-1}, which means all the internal surfaces were dry. As the temperature of the concrete was lower than was the air temperature of the internal underfloor space, the occurrence of condensation had higher significance when all the surfaces were dry (ω = 0) than when they were all wet (ω = 1). This is because the surface temperature of the dry concrete was lower than was the temperature of the water that formed on this surface; therefore, less condensation occurred. As regards the result shown in Table 5, (+) and (−) represent condensation and evaporation, respectively.

Table 5. Water thickness on the surface of the concrete relevant to the outdoor conditions.

Outdoor Condition (25 °C)			Evaporation and Condensation Rate (mg m^{-2} h^{-1})	
Humidity (%)	**Wind Speed (m/s)**	**Wind Direction**	$\omega = 1$	$\omega = 0$
0			−12.3	0
10			−10.8	0
30			−7.8	0
50	1.0		−4.7	0
70			−1.6	0
90			0.4×10^{-3}	0.5×10^{-3}
100		South	0.5×10^{-3}	0.5×10^{-3}
0			−0.5	0
10			−0.4	0
30			−0.3	0
50	0.05		−0.2	0
70			−0.05	0
90			0	0.7×10^{-4}
100			0.7×10^{-4}	0.7×10^{-4}

(+): Positive value corresponds to the condensation. (−): Negative value corresponds to the evaporation.

4. Daily Changes in Fluctuating Environmental Condition

In the second instance, we studied a fluctuation of outdoor environmental conditions over time, as the RWS status varied with the introduction of varying parameter values over time relevant to the outdoor environment. Accordingly, the effect of the ventilation rate of the underfloor space was confirmed by verifying the rate of evaporation and condensation, the amount of the condensed water, and the region where the condensation was likely to occur. Such information could be utilized in the architectural design of residences with underfloor spaces.

When the air conditions changed with time, the indoor conditions of evaporation and condensation fluctuated correspondingly. Furthermore, depending on the airflow, the surface temperature distribution, and moisture distribution, and the position where the condensation can occur would change, so as the RWS. When the condensed water evaporated, it acted as a generation source of water vapor, which affected the occurrence of the evaporation and condensation phenomena. Accordingly, our study confirmed variations over time in the rate of evaporation and condensation, the weight of condensed water, and the area where condensation would likely occur relevant to a rainy day during the rainy season, when the average humidity was high.

4.1. Weather Data

In this part of the study, we considered the outdoor environment during the rainy season, with the highest humidity levels and a high risk of condensation. Figure 13 shows the changes in daily temperature and relative humidity in Tokyo during the rainy season, and absolute humidity is expressed in Figure 14.

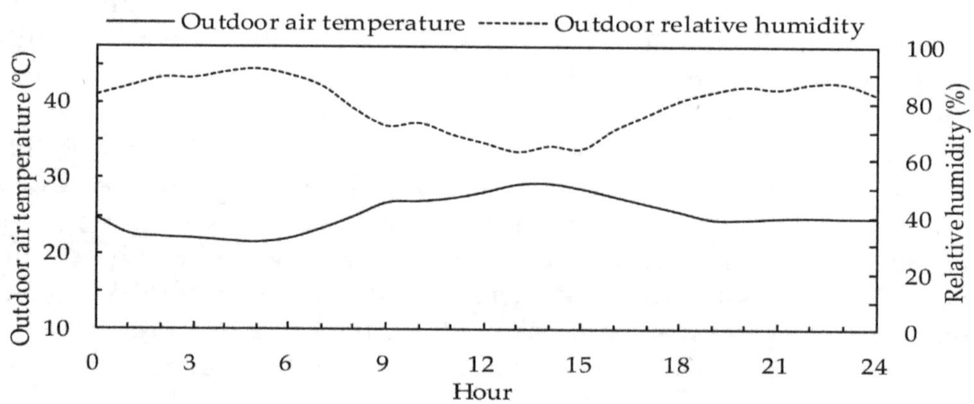

Figure 13. Variations in the daily outdoor air temperature and relative humidity in Tokyo, Japan, according to the AMeDAS standard weather data (15 June).

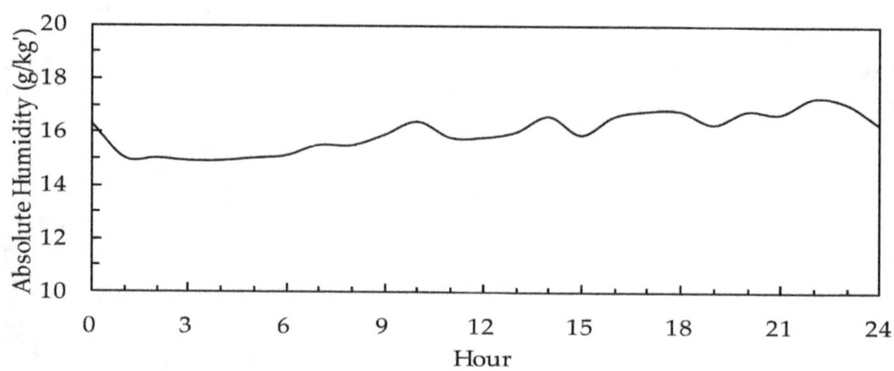

Figure 14. Daily outdoor absolute humidity variations of Tokyo, Japan by AMeDAS standard weather data (June 15).

The average temperature measured was 25 °C, and the highest temperature was measured between 12:00 and 15:00. The relative humidity with the increasing temperature was approximately 65%. The average daily humidity was 74%.

Wind speed conditions of the outdoor air were assigned uniformly by using the calculated ventilation rate of the underfloor, as described in Table 1, assuming a general wind speed and a stagnant wind speed. However, this time, changes in wind direction and wind speed as time elapses were not considered.

4.2. Analysis Method

As shown in Figure 15, the average rate of evaporation and condensation, the amount of condensed water on the surface, and the change in the condensation area over 24 h were confirmed in our study. As changes in wind speed and direction were not taken into consideration, we utilized the values in Table 1, representing the results of ventilation rates at 1.0 m/s and 0.05 m/s of the southern wind. In this instance, unlike the uniform condition of the outdoor environment, RWS changed continuously according to the elapse of time. The simulation calculations were conducted for 24 h, with the method reflected in Table 3. The environmental conditions of the outdoor air complied with the temperature and humidity data of the rainy season in Tokyo, as represented in Figure 13. The initial RWS condition of the internal surface was set at 0, which means that all the surfaces were dry.

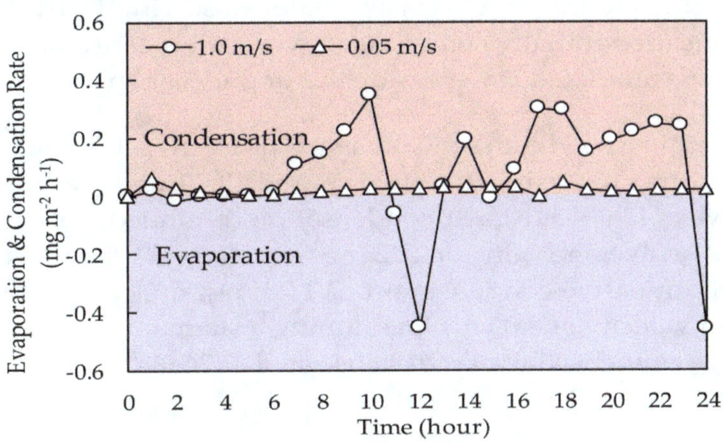

Figure 15. Variation in the average evaporation and condensation rate on the surface of the underfloor space during a day.

4.3. Analysis Results

The evaporation and condensation rates, the amount of the condensed water, and the area of condensation of the underfloor space were observed at hourly intervals over time to identify the effects of the ventilation rate of the underfloor space.

4.3.1. Evaporation and Condensation Effect

The distribution of temperature and pressure inside the underfloor was not uniform because of the influx of outdoor air introduced through the ventilation openings. The value of RWS changed continuously because of the condensation phenomenon. Furthermore, as evaporation and condensation occur differently, depending on the location, the rates of evaporation and condensation was confirmed through the average rate of the entire underfloor space.

When the amount of evaporation was greater than that of condensation, it is expressed as the total average condensation, but when the amount of condensation was greater than that of evaporation, it is expressed as the total average evaporation. The internal humidity of the underfloor would be influenced by the outdoor humidity when the volume of introduced outdoor air increased with an increase in the ventilation rate. In the instance of the outdoor air velocity being 1.0 m/s, the absolute humidity increased with the rising temperature after 6:00 in the morning. The rate of internal condensation continued to increase until 11:00, resulting in a maximum condensation rate of 0.34 mg m^{-2} h^{-1} at 10:00. From 11:00, the condensation rate gradually decreased and the condensed water on the surface evaporated between 11:00 and 12:00, whereas the evaporation rate increased significantly at 12:00 to 0.46 mg m^{-2} h^{-1}. During the rainy season, the humidity remains high throughout the day. Therefore, during the calculated period, the absolute humidity increased with the rise in temperature and the condensation phenomenon appeared again after 13:00 and continued until 23:00. Furthermore, at 24:00, the humidity decreased and evaporation occurred at a largest value of 0.46 mg m^{-2} h^{-1} for the day measured. In a condition of stagnant wind velocity (0.05 m/s), it was not influenced significantly by the outdoor conditions. This means that not only was the outdoor moisture not introduced well but also that the internal moisture was not exhausted well. As time elapsed, the condensation phenomenon continued as the humidity from outside entered the underfloor space; however, the evaporation did not occur for 24 h. Furthermore, the evaporation and condensation phenomena were shown to be in equilibrium at 15:00.

4.3.2. Evaporation and Condensation Effect

We confirmed the overall characteristics of the internal underfloor space over time according to the average value of the amount of condensed water occurring in the entire underfloor space. The occurrence of condensation initially commenced when all the internal surfaces were in the dry state and it subsequently accumulated water inside the underfloor space.

As shown in Figure 16, in a condition of wind speed of 1.0 m/s, the condensation continued until 10:00, with evaporation starting to occur between 11:00 and 12:00, and the amount of condensed water starting to decrease. Between 13:00 and 23:00, condensation continued, with the amount of condensed water being the highest at an average value of 2.22 mg/m^2. At 24:00, the rate of evaporation was high, resulting in an average cumulative condensation of 1.76 mg/m^2 per day. Condensation occurred continuously with a wind velocity of 0.05 m/s, assuming a stagnant environment, and an amount of 0.48 mg/m^2 of cumulative condensed water was observed at 24:00.

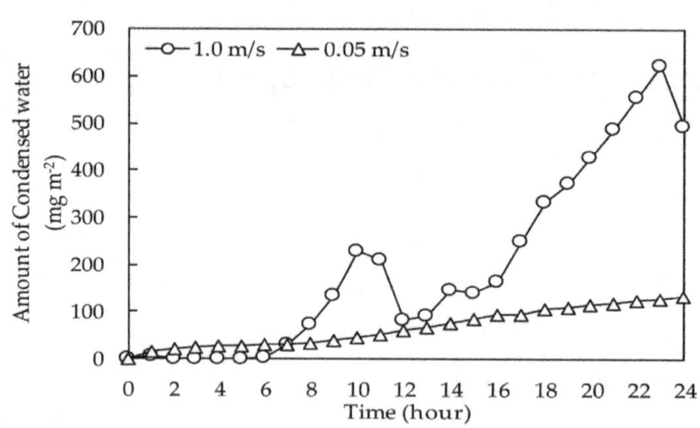

Figure 16. Amount of condensed water of the underfloor space relevant to evaporation and condensation during a day.

Convective mass transfer on surface increases as the air velocity increased. The air temperature increase causes the absolute humidity rise in the moist air due to the change of saturated vapor pressure. These results affect the amount of moist introduced from the outdoor air. As the higher ventilation rate, inflow rate of water vapor into underfloor space increase. This means that the convective mass transfer in a condition of 0.05 m/s is not a dominant comparing to 1.0 m/s. Before 6 o'clock, the evaporation at

1.0 m/s was stronger than the condensation due to water vapor inflow. On the other hand, at 0.5 m/s, the condensation by convection was stronger than the evaporation. In the case of 1.0 m/s, condensation due to water vapor flow is stronger than evaporation due to forced convection as the absolute humidity flowing into the inside increases after 6 o'clock. For this reason, the amount of condensed water in the underfloor space is reversed by 1.0 m/s and 0.05 m/s after 7 o'clock.

4.3.3. Condensed Water Distribution on Surface of Concrete

The area where condensation occurred differed according to the ventilation rates. As shown in Figure 17, at a wind speed of 1.0 m/s, condensation did not occur where the airflow is strong, but did occur nearby the mainstream of the strong airflow, and was likely to occur in the corners of the underfloor space. However, in the northern zone, where the airflow does not reach, condensation hardly occurred at all. The evaporation phenomenon was likely to occur in an area of strong airflow, as mass transfer easily occurred and condensation rarely occurred. In addition, the convective heat transfer from the surface was more pronounced because of the strong airflow when the outdoor air flowed in; therefore, condensation was less likely to occur. In contrast, in a stagnant condition of 0.05 m/s wind speed, condensation occurred mostly near the zone where the airflow was generated. This is because the velocity of the incoming airflow was extremely low and the convective heat transfer on the surface was not large; therefore, the moisture of the incoming outdoor air condensed at the area where the airflow was formed.

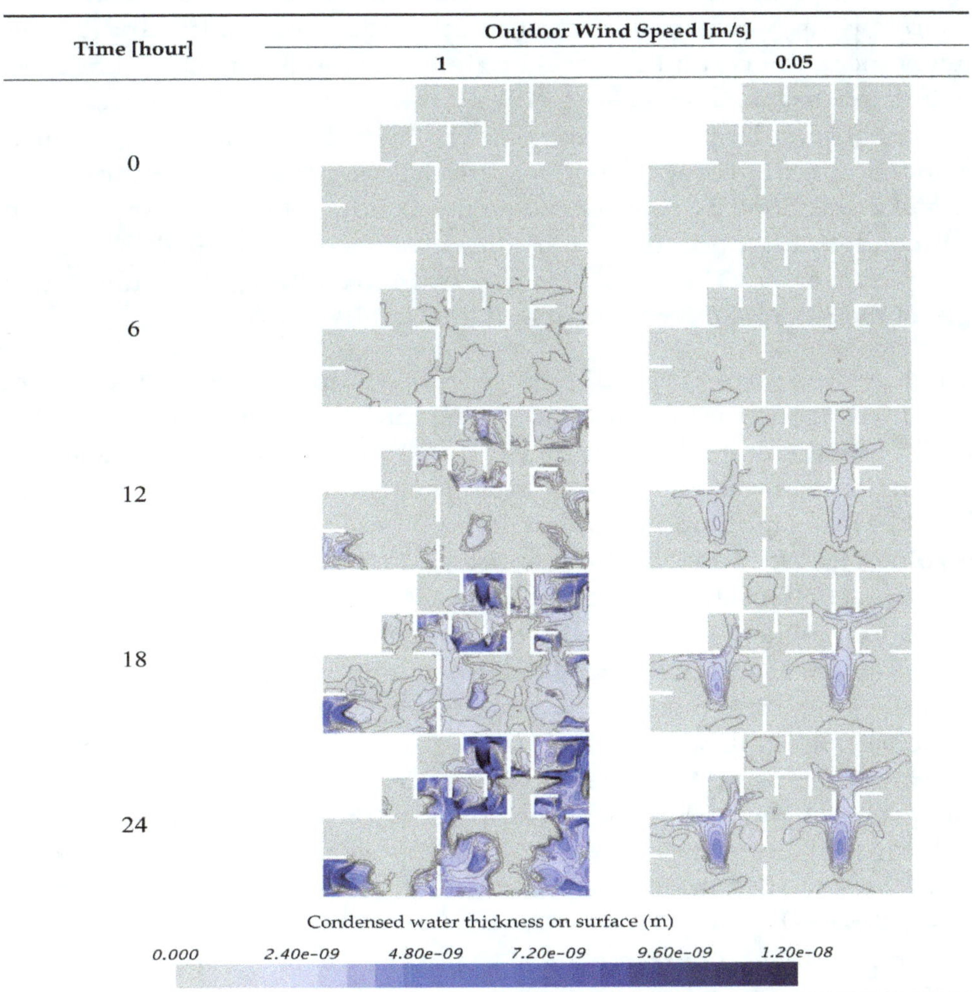

Figure 17. Variation in the condensed water thickness on the surface of the concrete during a day.

5. Discussion

The RWS of the initial internal environment was assumed zero in fluctuating conditions. Under such conditions, dry air was introduced into the underfloor space and the evaporation phenomenon did not occur, even if ventilation were performed. Therefore, the result could not be confirmed and could be different from the actual phenomenon. However, as the relative humidity of the airflow in the initial condition was 80%, the evaporation phenomenon was considered unlikely to occur originally. It is necessary to check the temperature distribution by calculations over a long time in order to confirm an environment similar to the actual environment. However, both the initial surface temperature and the internal temperature of the material were assumed uniform because of the limitation of the calculation load. Therefore, it could differ from the actual environmental conditions.

6. Conclusions

In the case where condensed water exists inside the subfloor, the rate of evaporation was shown to be greater when the airflow rate was high. However, under extreme conditions where the outdoor humidity is 90% or more, condensation occurs instead of evaporation. When the outdoor humidity was less than 90%, the condensation did not occur, regardless of the airflow rate. This means that keeping the inside of the underfloor space free from condensed water and dry is important. In the situation when the outdoor humidity is high, such as the rainy season, internal condensation in the underfloor space easily occurs due to natural ventilation no matter what the airflow rate is. If the airflow rate is high, it is easily affected by outdoor humidity, and the evaporation and condensation phenomena are continuously changing. On the other hand, under low airflow rate conditions, only the condensation appeared steadily. If the wind velocity is strong, the convective mass transfer on a surface becomes large. When the outdoor humidity is high and the airflow rate is high, condensation mainly occurs in a corner of the underfloor space due to high evaporation by convection in the mainstream of the airflow. By contrast, when the airflow rate is low, condensation occurs along the air stream. If condensation occurs once in the corner where the airflow has not reached well, it is difficult to evaporate again. Therefore, it is necessary to control the ventilation rate according to the characteristics of the external environment and internal RWS.

1. We confirmed the influence of the difference in the humidity when the outdoor wind speed was assumed 1.0 m/s and the outdoor air flowing into the room remained constant according to the calculated ventilation rate. In the instance of a RWS value of 1, evaporation occurred actively and the evaporation rate was 12.3 mg m^{-2} h^{-1}. The total volume of the internal evaporation rate was 726 mg/h. The evaporation phenomenon did not occur in a condition of 0–70% humidity, where the value of RWS was 0 and the entire surface was dry. When humidity was high at 90–100%, the total condensation rate of the underfloor was 0.03 g/h.
2. We confirmed the influence of the difference in humidity when the outdoor wind speed was assumed stagnant at 0.05 m/s and the outdoor air flowing into the room remained constant according to the calculated ventilation rate. When the value of RWS was 1, the average evaporation effect per area was determined as 0.5–0.05 mg m^{-2} h^{-1}, and the total volume of the evaporation rate was 29.5–3.0 mg/h. In a condition of 90% humidity of the introduced outdoor air, a state of equilibrium was indicated, with neither evaporation nor condensation occurring. With 100% humidity, the evaporation phenomenon reversed and a condensation phenomenon of 0.7×10^{-4} mg m^{-2} h^{-1} occurred. However, the condensation phenomenon was slight, as indicated by the result of 0.4×10^{-2} mg/h. When the RWS value was zero, with all the internal surfaces dry and humidity of 90–100%, the total condensation rate of the underfloor space was 0.03 g/h.
3. When the outdoor wind speed was 1.0 m/s and the outdoor environmental conditions fluctuated for 24 h, the outdoor humidity remained high and condensation continued. However,

with the temperature rising or humidity decreasing, evaporation occurred between 11:00 and 12:00, and at 24:00. The maximum condensation rate was 0.34 mg m^{-2} h^{-1} at 10:00 and the maximum evaporation rate was 0.46 mg m^{-2} h^{-1} at 12:00 and at 24:00.

4. In a stagnant environment, with an outdoor wind speed of 0.05 m/s, the rate of natural ventilation into the underfloor was low and the effect of the outdoor environment was therefore insignificant. Consequently, condensation continued for 24 h and evaporation did not occur at all.

5. Under a condition of 1.0 m/s wind speed, evaporation and condensation occurred continuously throughout the day and the cumulative rate of condensation of the underfloor space was 1.76 mg/m^2 per day. Under a condition of 0.05 m/s wind speed, the cumulative rate of condensation of the underfloor space from natural ventilation was 0.48 mg/m^2 per day.

6. When the wind speed was 1.0 m/s, condensation did not occur near the strong airflow, but did occur in the vicinity of the airflow, and easily occurred in the corners. Furthermore, as there was less moisture transfer through the airflow, condensation hardly occurred where the airflow could not reach. At a wind speed of 0.05 m/s, assuming a stagnant condition, moisture was transferred to the inside by the airflow and condensation occurred in accordance with the airflow.

Author Contributions: Wonseok Oh performed the simulations, analyzed the simulation results, designed the simulations and wrote the paper. Shinsuke Kato reviewed the paper. All authors have read and approved the final manuscript.

References

1. Kurnitski, J. Ground moisture evaporation in crawl spaces. *Build. Environ.* **2001**, *36*, 359–373. [CrossRef]
2. Kurnitski, J.; Kurnitski, J. Crawl space air change, heat and moisture behaviour. *Energy Build.* **2000**, *32*, 19–39. [CrossRef]
3. Ferguson, C.C.; Krylov, V.V.; McGrath, P.T. Contamination of indoor air by toxic soil vapours: A screening risk assessment model. *Build. Environ.* **1995**, *30*, 375–383. [CrossRef]
4. Krylov, V.V.; Ferguson, C.C. Contamination of indoor air by toxic soil vapours: The effects of subfloor ventilation and other protective measures. *Build. Environ.* **1998**, *33*, 331–347. [CrossRef]
5. Zhai, Z.J.; Zhang, Z.; Zhang, W.; Zhang, Z.; Zhang, W.; Chen, Q.Y. Evaluation of various turbulence models in predicting airflow and turbulence in enclosed environments by CFD: Part 1—Summary of prevalent turbulence models. *HVAC R Res.* **2007**, *13*, 853–870. [CrossRef]
6. Nielsen, P.V. Flow in Air Conditioned Rooms: Model Experiments and Numerical Solution of the Flow Equations. Ph.D. Thesis, Technical University of Denmark, Nordborg, Denmark, 1974.
7. Murakami, S.; Kato, S. Numerical and experimental study on room airflow—3-D predictions using the k-ε turbulence model. *Build. Environ.* **1989**, *24*, 85–97. [CrossRef]
8. Nielsen, P.V. The description of supply openings in numerical models for room air distribution. *Indoor Environ. Tech.* **1991**, *R9251*, 963–971.
9. Nielsen, P.V. Computational fluid dynamics and room air movement. *Indoor Air* **2004**, *7*, 134–143. [CrossRef] [PubMed]
10. Gan, G. Evaluation of room air distribution systems using computational fluid dynamics. *Energy Build.* **1995**, *94*, 665–675. [CrossRef]
11. Zhai, Z.J. Application of computational fluid dynamics in building design: Aspects and trends. *Indoor Built Environ.* **2006**, *15*, 305–313. [CrossRef]
12. Van Maele, K.; Merci, B. Application of two buoyancy-modified k–ε turbulence models to different types of buoyant plumes. *Fire Saf. J.* **2006**, *41*, 122–138. [CrossRef]
13. Shin, T.H.; Liou, W.W.; Shbbir, A.; Yang, Z.; Zhu, J. A new k-ε eddy viscosity model for high Reynolds number turbulent flows. *Comput. Fluids* **1995**, *24*, 227–238.
14. *CASBEE for Home (Detached House) Technical Manual 2007 Edition*; Japan Sustainable Building Consortium & Japan Green Build Council: Tokyo, Japan, 2007; pp. 195–215.

15. Counihan, J. Adiabatic atmospheric boundary layers: A review and analysis of data from the period 1880–1972. *Atmos. Environ.* **1975**, *9*, 871–905. [CrossRef]
16. Tominaga, Y.; Stathopoulos, T. Numerical simulation of dispersion around an isolated cubic building: Comparison of various types of k–ε models. *Atmos. Environ.* **2009**, *43*, 3200–3210. [CrossRef]
17. Sandhu, K.S. *Predicting the Windscreen Demisting Performance Using CAE*; Elsevier: Amsterdam, The Netherlands, 2011; pp. 401–410.
18. Cho, W.; Iwamoto, S.; Kato, S. Condensation risk due to variations in airtightness and thermal insulation of an office building in warm and wet climate. *Energies* **2016**, *9*, 875. [CrossRef]
19. Patankar, S.V.; Spalding, D.B. A calculation procedure for heat, mass and momentum transfer in three-dimensional parabolic flows. *Int. J. Heat Mass Transf.* **1972**, *15*, 1787–1806. [CrossRef]
20. Leonard, B.P. A stable and accurate convective modelling procedure based on quadratic upstream interpolation. *Comput. Methods Appl. Mech. Eng.* **1979**, *19*, 59–98. [CrossRef]
21. Mulvany, N.; Tu, J.Y.; Chen, L. Assessment of two-equation turbulence modelling for high Reynolds number hydrofoil flows. *Int. J. Numer. Methods Fluids* **2004**, *45*. [CrossRef]
22. Salim, S.M.; Cheah, S. Wall Y strategy for dealing with wall-bounded turbulent flows. In Proceedings of the International Multi Conference of Engineers and Computer Scientists, Hong Kong, China, 18–20 March 2009; Volume II.

8

Numerical Analysis on the Effect of the Tunnel Slope on the Plug-Holing Phenomena

Ji Tae Kim, Ki-Bae Hong and Hong Sun Ryou *

School of Mechanical Engineering, Chung-Ang University, Seoul 06974, Korea; sdd322@naver.com (J.T.K.); gbhong@ut.ac.kr (K.-B.H.)

* Correspondence: cfdmec@cau.ac.kr.

Abstract: Preventing the plug-holing phenomena of a natural ventilation system in a shallow underground tunnel is important for improving the ventilation performance, and the tunnel slope has a significant influence on the smoke flow. In this study, the effect of the tunnel slope on plug-holing in a shallow underground tunnel was analyzed by numerical method. The tunnel slope was increased by 0–8 degrees and the fire source was assumed to be 5 MW, which is equivalent to one sedan vehicle. As a result, the possibility of plug-holing decreased as the tunnel slope increased. However, when the tunnel slope is more than 4°, the fresh air from the entrance of the tunnel and smoke are diluted before reaching the shaft, so the flow temperature passing through the shaft is lowered, and the ventilation performance begins to decrease. In particular, plug-holing does not occur at the tunnel slopes of 6 and 8°, but the ventilation performance is expected to decrease because the temperature of the smoke discharged to the shaft is much lower than the general smoke temperature. Therefore, it is necessary to design the natural ventilation system considering the influence of the tunnel slope.

Keywords: plug-holing; tunnel slope; fire; natural ventilation; ventilation performance

---

1. Introduction

Statistics and reports by NFPA (National Fire Protection Association in USA) have shown that smoke inhalation is a major fatal factor in fire accidents [1]. Therefore, it is very important to ensure the good operation of the smoke ventilation system to reduce human casualties.

Generally, the vertical natural ventilation system is widely used in the shallow road tunnel due to low installation and maintenance costs [2]. In order to design an accurate vertical natural ventilation system, it is necessary to understand the tunnel fire phenomena in order to effectively exhaust the smoke, which is generated by the fire.

The plug-holing phenomena is one of the important factors to vertical natural ventilation systems. The plug holing phenomena is defined as: The fresh air beneath the smoke layer is exhausted through the shaft with smoke. Therefore, understanding the plug-holing phenomena, and preventing the plug-holing phenomena, is an important factor in improving the performance of the natural exhaust system.

Some researchers have shown that the horizontal momentum force of the ceiling jet and the buoyancy force of the vertical shaft play an important role in the plug-holing phenomena, and suggests the criterion is the modified Froude number for the effective ventilation system design without plug-holing. Hinkley et al. [3] introduced the plug-holing phenomena and presented a modified Froude number, which is defined as the ratio of the momentum force of the smoke to the buoyancy force in the shaft and the height of the tunnel. J. Ji et al. [4] analyzed the effect of the heat release rate on plug-holing, and proposed a modified Richardson number as the criterion for plug-holing. Additionally, J. Ji et al. [5] shows the influence of the cross-section area and aspect ratio of the shaft.

D. Baek et al. [6,7] presented, through numerical analysis, a modified Froude number with the effect of the hydraulic diameter and tunnel aspect ratio on the plug-holing phenomena.

As reported in previous studies, the effect of geometry is an important factor affecting the ventilation performance of the vertical natural ventilation system. Particularly, it is known that the slope of the tunnel greatly affects the buoyancy force of the smoke flow, the critical velocity, and the back layering length [8–10].

Therefore, the slope of the tunnel relates to the plug-holing phenomena, which reduces the ventilation performance in the vertical natural ventilation system. However, the effect of the tunnel slope on the plug-holing phenomena has not been clearly understood yet.

Therefore, in this study, a numerical study was carried out to analyze the effect of the tunnel slope on the plug-holing phenomena in a shallow underground tunnel.

2. Numerical Details

2.1. Geometry and Boundary Conditions

Figure 1 presents the geometry of the tunnel for analyzing the plug-holing phenomena according to the tunnel slope. Generally, the tunnel slope of the shallow underground tunnel is constructed less than 8 degrees below the urban center to relieve traffic congestion in urban areas. In the shallow underground tunnel, the vertical shaft vents are installed at intervals of about 100 to 180 m [9].

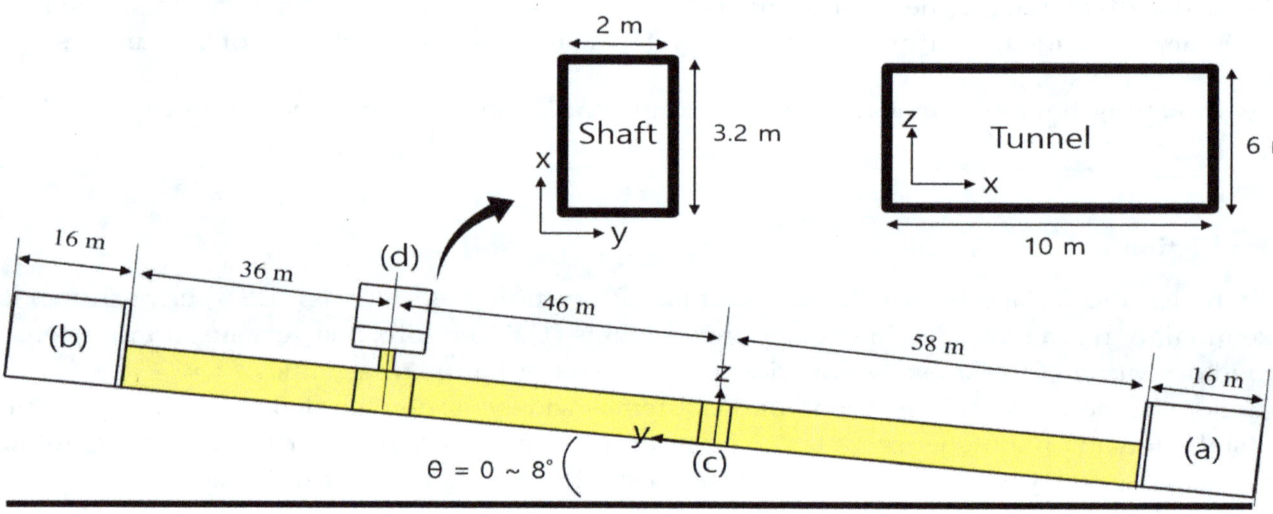

Figure 1. Geometry and boundary conditions of the tunnel.

Therefore, in this study, five tunnel slopes were selected at 0, 2, 4, 6, and 8 degrees. The tunnel is assumed to be a two-lane tunnel, and the width of the tunnel was set to 10 m and the height of the tunnel to 6 m. In addition, the length of the tunnel was set to be installed at intervals of 140 m with repeated shafts. The center of the vertical shaft was located at a distance of 58 m from the entrance of the tunnel, as shown in Figure 1d. The width and depth of the vertical shaft were 3.2 and 2 m, respectively, and the height was 3 m. The walls of the tunnels were assumed to be adiabatic, and zero-gauge pressure was applied to the atmospheric conditions at the exit of both ends of the tunnel and the outlet of the vertical shaft, as shown in Figure 1a,b,d, respectively. The fire source in tunnels is assumed to occur in passenger cars. According to previous studies, the maximum heat release rate (HRR) is known to be about 3–5 MW [11,12]. Thus, in this study the fire source was located at y = 0, as shown in Figure 1a and the maximum HRR was set to 5 MW. The size of the fire was x = 1.2 m, y = 1.2 m, and z = 1 m, and heat release rate applied to the surface was z = 1 m.

2.2. Numerical Methods

The fire dynamics simulator (FDS), developed by the National Institute of Standards and Technology (NIST), was used to analyze the fire-driven fluid flow. NIST performed the verification and validation of the models used in the FDS, and provided validated values such as turbulent Prandtl number, turbulent Schmidt number, convective heat transfer coefficient, and radiation fraction from fire source.

FDS solves the low Mach number Navier—Stokes equation (Ma < 0.3). The large eddy simulation (LES) model was used to calculate turbulence for the fire driven flow. The Deardorff model was used for turbulent viscous model [5]. The gray gas model was used to analyze the radiative heat transfer. Radiation fraction of fire was set to 0.35 and gravity was applied to 9.81 m/s in the $-z$ direction. The atmospheric temperature was set at 25 °C. The grid for numerical analysis was generated at 0.2 m, by grid independency test between 0.1 and 0.4 m in grid size, based on equation (1), which was recommended in the FDS user guide. Where D^* is grid size, $\dot{Q}$ is heat release rate, ρ_∞ is ambient density, c_P is heat capacity of the inflow at constant pressure, T_∞ is ambient temperature, and g is gravity.

Also, near the fire source and the shaft, the grid size was formed at 0.1 m to improve the accuracy of the numerical analysis.

$$D^* = \left(\frac{\dot{Q}}{\rho_\infty c_P T_\infty \sqrt{g}}\right)^{\frac{2}{5}} \tag{1}$$

3. Results and Discussion

Figure 2 shows the temperature contour at x = 0 for the average value about 20 s after the flow due to the quasi-steady state. The tunnel slope was 0 degrees, the high-temperature combustion products generated from the fire source formed a smoke layer along the tunnel ceiling. The smoke layer was formed with a similar temperature field in both directions of the tunnel. The slope of the tunnel was 2 degrees, the flow of the smoke layer was relatively shifted toward the +y direction of the tunnel. Moreover, the tunnel slope was 4 degrees or more, and the flow of the smoke layer formed toward the entrance of the tunnel with a back layering of about 2 m.

Also, the slope of the tunnel was 0 degrees, fresh air flowed in both directions of the tunnel, and a flow field of about 25 °C was formed below the smoke layer. The angle of the tunnel angle was 2 degrees, the temperature of the flow from the fire source to the shaft was about 45 °C near the tunnel floor, as shown in the Figure 2b. The temperature field of 0 and 2 degrees were deeply related with the flow field. The slope of the tunnel was 0 degrees, the flow of the high temperature smoke products spread in both directions from the fire source, and the fresh air flow below the smoke layer was formed from both tunnel ends to the fire source. However, as shown in the Figure 3b, the slope of the tunnel was 2 degrees, and only the flow was formed in the +y direction from the fire source, with some very low velocity flows. Thus, the low velocity flow increased the temperature of the flow near the tunnel bottom.

In addition, as shown in the Figure 2c–e, when the slope of the tunnel was more than 4 degrees, it can be seen that the temperature below the smoke layer was about 30 °C or more, which was higher than the fresh air by more than 5 °C.

As shown in the Figure 3c–e, the flow velocity near the bottom of the tunnel was relatively higher, from the tunnel slope of 4 degrees. Because, the fresh air flow into from the $-$y direction was heated by the fire source and flows only in the +y direction by buoyancy.

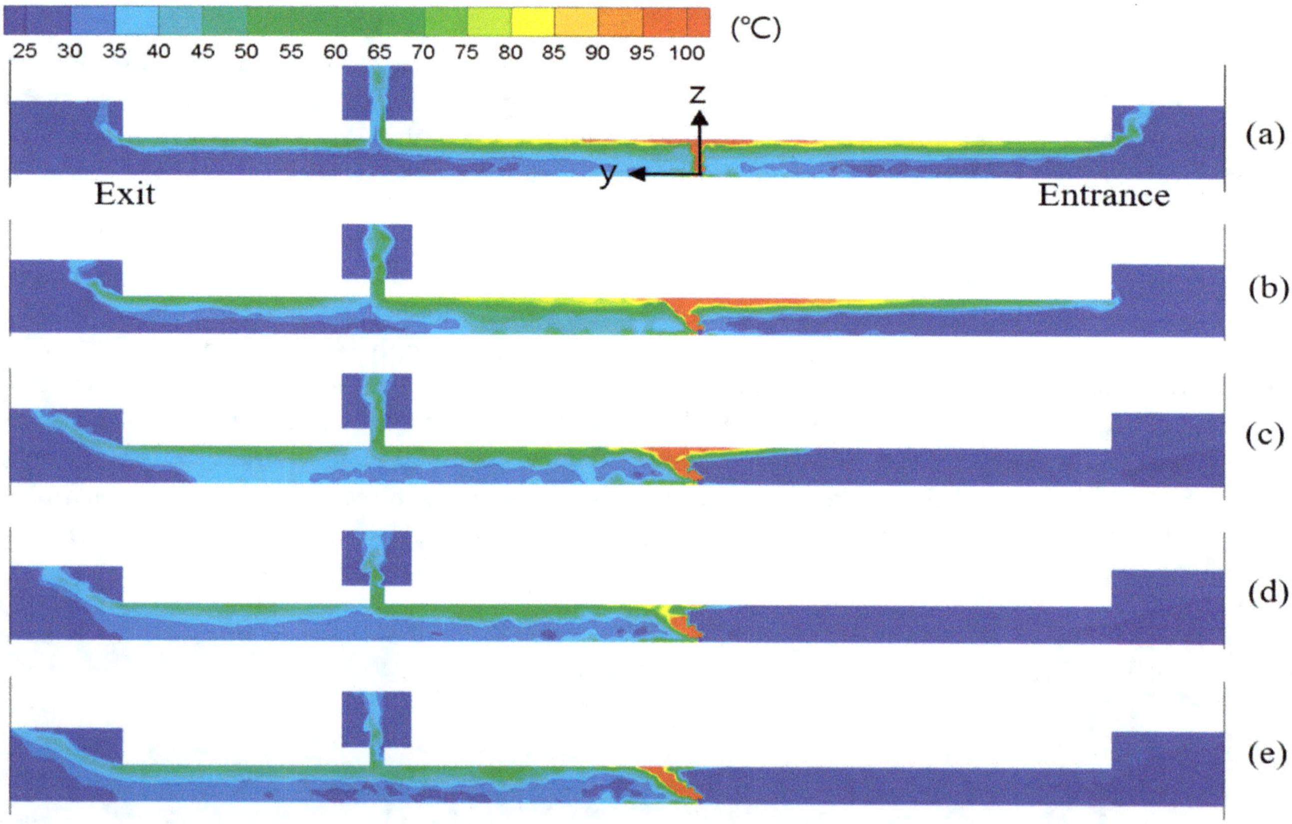

Figure 2. Temperature contour at x = 0: (**a**) 0°, (**b**) 2°, (**c**) 4°, (**d**) 6°, and (**e**) 8°.

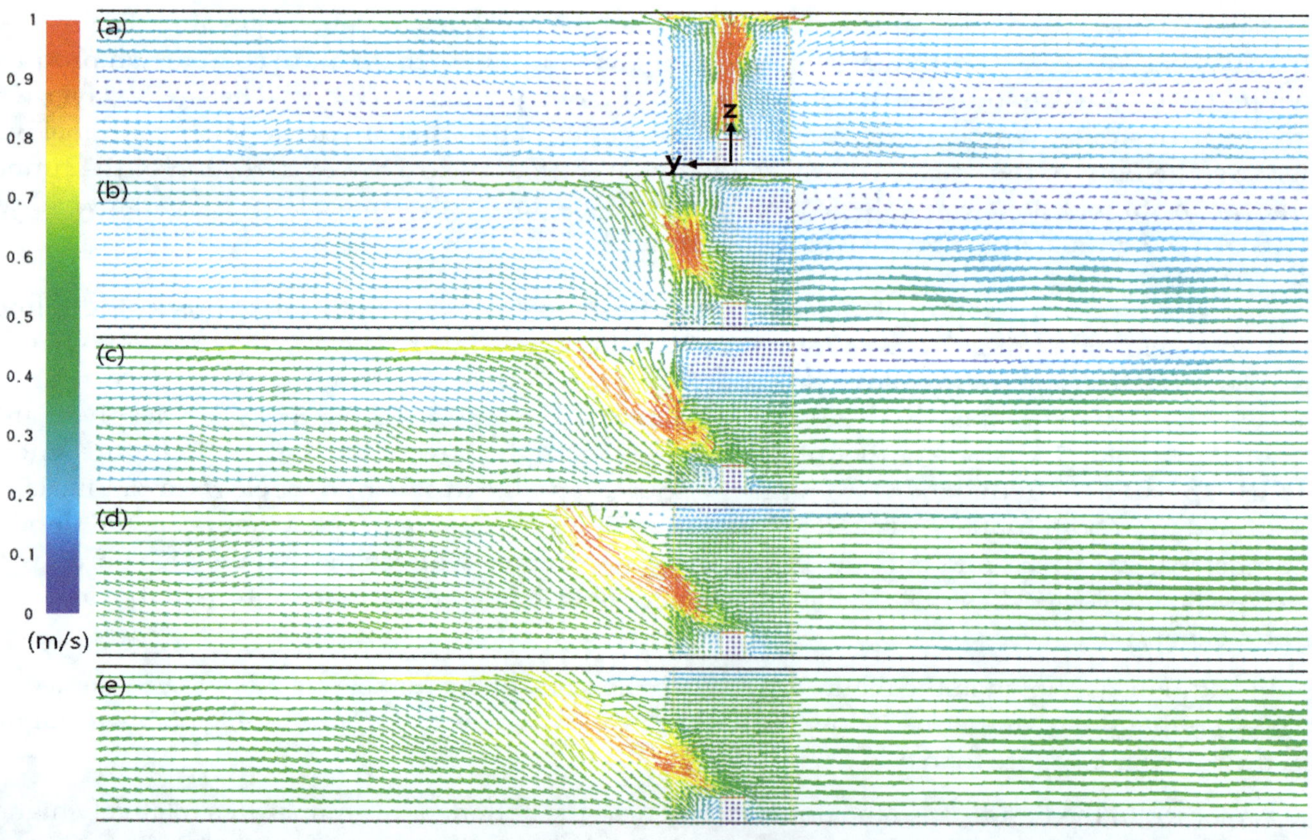

Figure 3. Velocity vector near the fire source at x = 0: (**a**) 0°, (**b**) 2°, (**c**) 4°, (**d**) 6°, (**e**) 8°.

Figure 4 represents the temperature distribution of the flow inside of the shaft. The temperature distribution was obtained from x = 0, y = 0, the dimensionless distance 0 was the starting location of the shaft, and the end location in the y direction was the dimensionless distance 100. The internal temperature of the shaft according to the tunnel slope at the dimensionless distance zero was similar to the smoke layer temperature at the shaft start location.

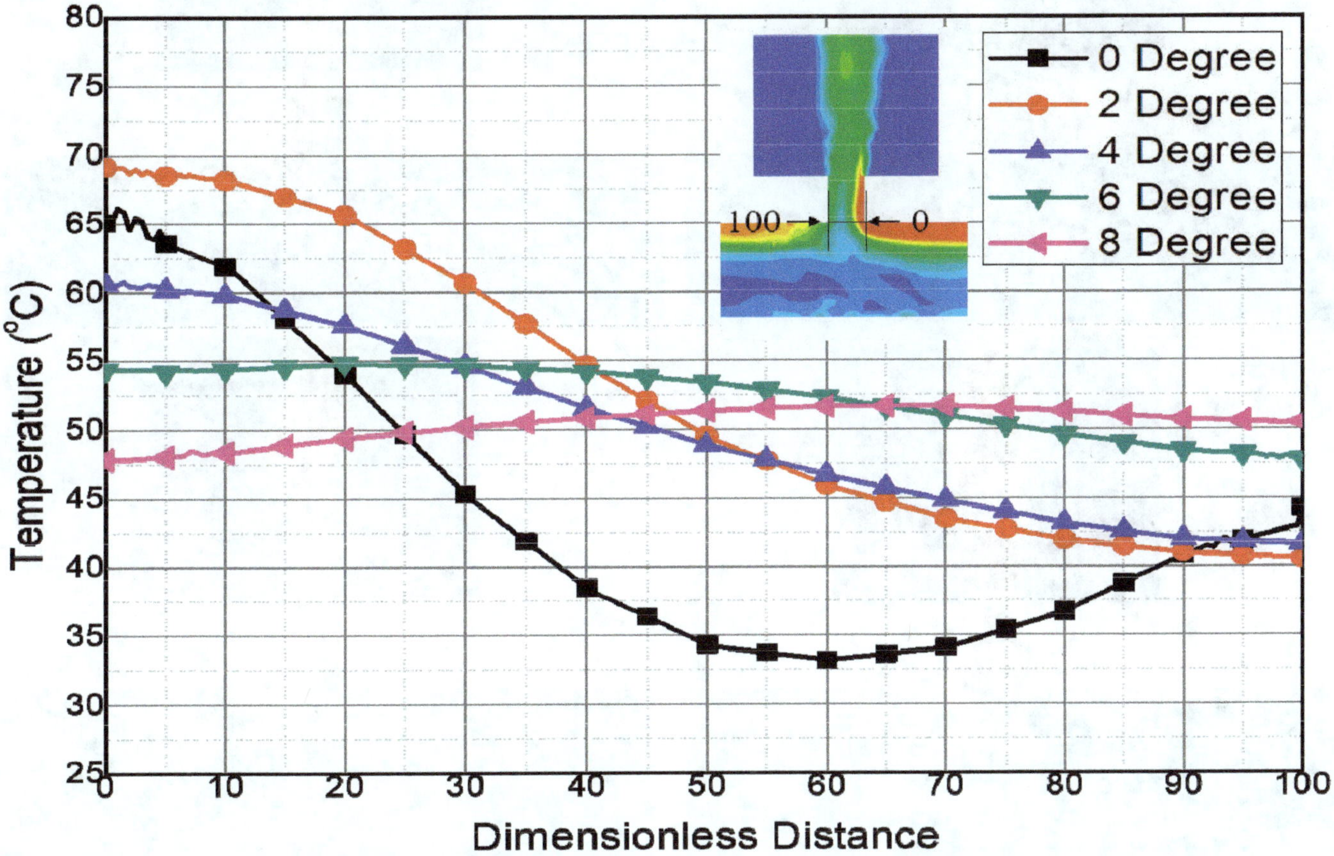

Figure 4. Temperature distributions inside of the shaft.

Generally, when plug-holing occurs, a flow of fresh air that has a lower temperature than the smoke layer flows into the shaft. When the tunnel slope was 0 degrees, the flow temperature inside the shaft was 33.5 °C at the dimensionless distance 60, which means that a flow lower than the temperature of the smoke layer flows into the shaft, which means the possibility of plug-holing is high.

However, the possibility of plug-holing will decrease at the tunnel slope of 2 and 4° because of the temperature at the shaft being higher than the fresh air temperature. In particular, when the tunnel slope exceeds 6 degrees, the maximum temperature deviation inside the shaft is significantly decreased about 8 °C.

These temperature characteristics inside the shaft were deeply related to flow field in the tunnel according to the tunnel slope.

As shown in Figure 5, the smoke layer that flows in the direction of the tunnel exit was exhausted vertically at the shaft. The smoke and the fresh air flows into the shaft and recirculation flow occurs at the region where these two flows collide.

However, as the tunnel slope increases, the velocity of the flow increases with the longitudinal direction of the tunnel. Therefore, as can be seen from the vector near the inlet of the shaft, the flow is oriented longitudinally, to the vertical direction of the shaft, at the tunnel slope of more than 4 degrees.

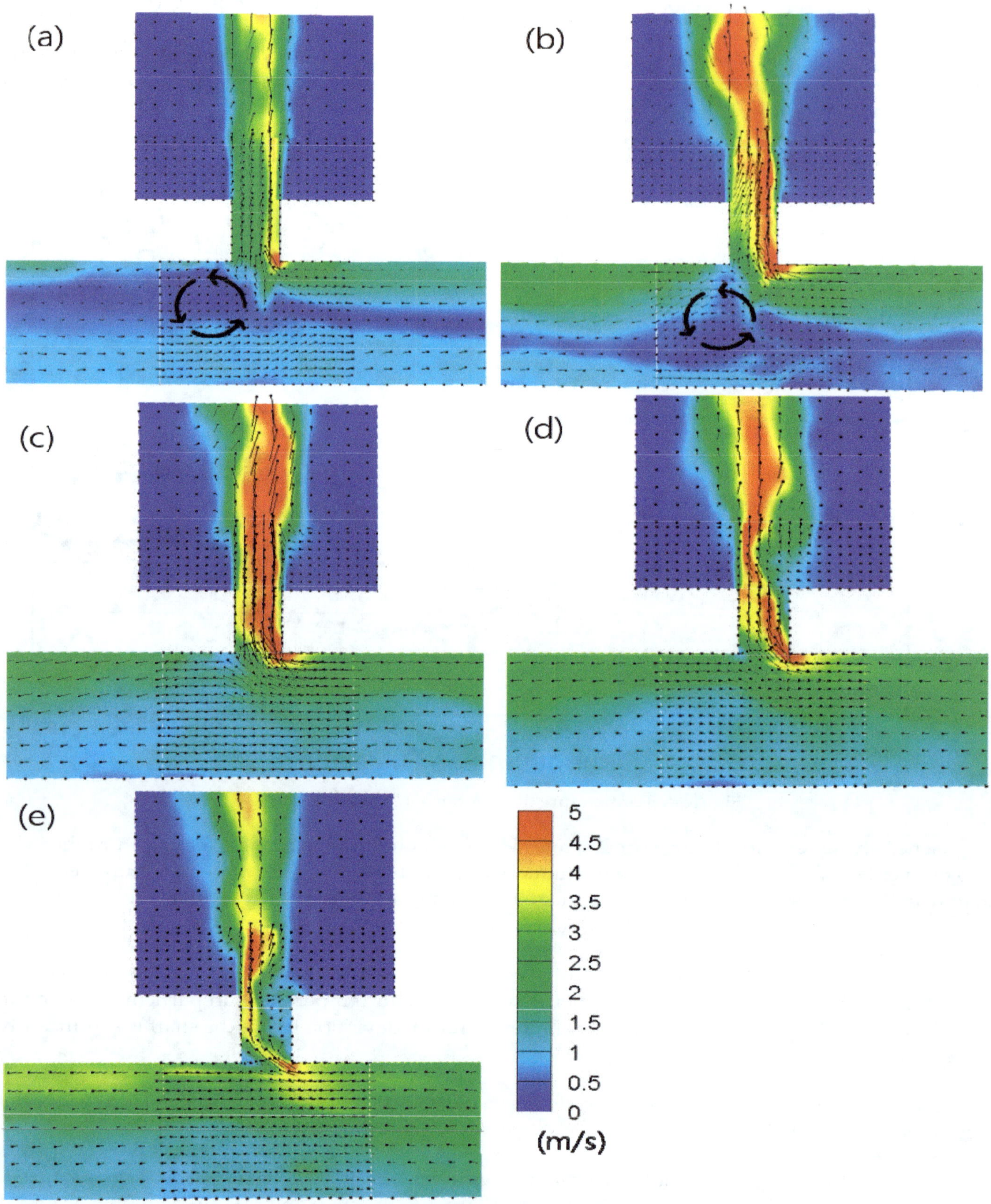

Figure 5. Velocity vector contour (x = 0): (**a**) 0°, (**b**) 2°, (**c**) 4°, (**d**) 6°, and (**e**) 8°.

Figure 6 represents the velocity distribution inside the shaft. The velocity of the shaft is strongly related to buoyancy force, which is influenced by the smoke layer temperature. As shown in Figure 4, the velocity inside of the shaft increases the slope of the tunnel to 4 degrees, and the temperature distribution inside of the shaft decreases by greater than 6 degrees.

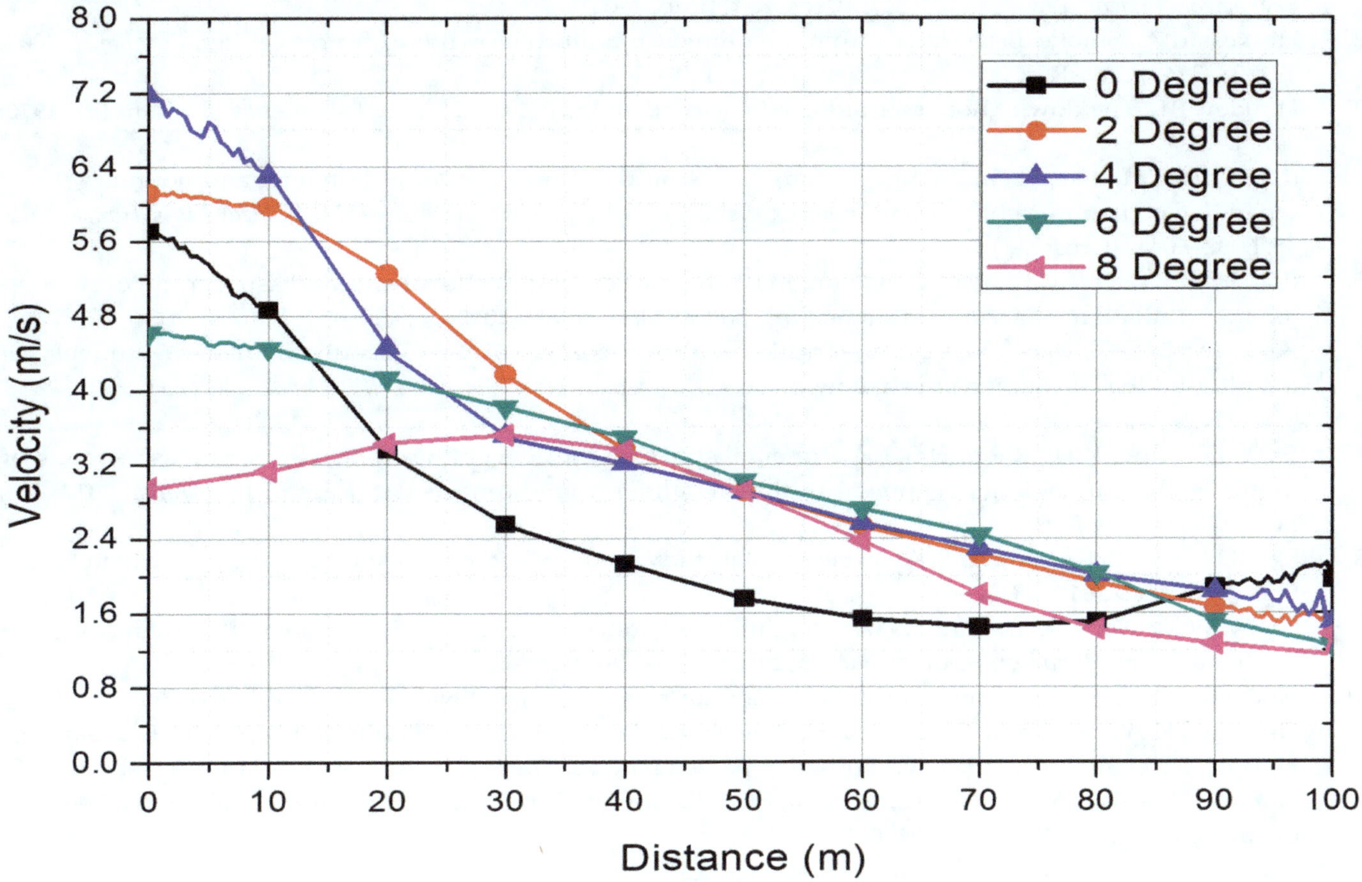

Figure 6. Velocity distributions inside of the shaft.

As shown in Figures 4 and 5, the plug-holing phenomena will not occur even if the slope of the tunnel exceeds 2 degrees. Additionally, the ventilation performance of the shaft is expected to increase because the flow temperature and flow velocity of the shaft are higher than the tunnel slope of 0 degrees, and until the slope of the tunnel reaches 4 degrees. However, above the tunnel slope of 6 degrees the ventilation performance is expected to decrease, because the internal velocity of the shaft is lower than the tunnel slope of 4 degrees.

4. Conclusions

The plug-holing phenomenon and the ventilation performance of the tunnel are affected by the tunnel slope. The possibility of plug-holing decreases as the slope of the tunnel increases. Moreover, when the tunnel slope is 4 degrees, the flow and temperature of the flow exhausted through the shaft are the highest. Therefore, the ventilation performance is the best when the tunnel slope is 4 degrees.

However, when the tunnel slope is greater than 4 degrees, fresh air from the entrance of the tunnel is diluted with smoke before reaching the shaft, thus the possibility of plug-holing is very low above 6 degrees and the ventilation performance begins to decrease. Therefore, it is necessary to design the shaft for the natural ventilation system considering the influence of the tunnel slope.

Author Contributions: Investigation, J.T.K.; Writing-Original Draft Preparation, J.T.K.; Writing-Review & Editing, J.T.K.; Supervision, H.S.R.; Funding Acquisition, K.-B.H.

References

1. Alarie, Y. The toxicity of smoke from polymeric materials during thermal decomposition. *Annu. Rev. Pharmacol. Toxicol.* **1985**, *25*, 325–347. [CrossRef] [PubMed]
2. Heskestad, G. Smoke distributions from fire plumes in uniform downdraft from a ceiling. *Fire Saf. J.* **2004**, *39*, 358–374. [CrossRef]
3. Hinkley, P.L. The flow of hot gases along an enclosed shopping mall a tentative theory. *Fire Saf. Sci.* **1970**, *807*, 1–17.
4. Ji, J.; Gao, Z.H.; Fan, C.G.; Zhong, W.; Sun, J.H. A study of the effect of plug-holing and boundary layer separation on natural ventilation with vertical shaft in urban road tunnel fires. *Int. J. Heat Mass Transf.* **2012**, *55*, 6032–6041. [CrossRef]
5. Ji, J.; Han, J.Y.; Fan, C.G.; Gao, Z.H.; Sun, J.H. Influence of cross-sectional area and aspect ratio of shaft on natural ventilation in urban road tunnel. *Int. J. Heat Mass Transf.* **2013**, *67*, 420–431. [CrossRef]
6. Baek, D.; Bae, S.; Ryou, H.S. A numerical study on the effect of the hydraulic diameter of tunnels on the plug-holing phenomena in shallow underground tunnels. *J. Mech. Sci. Technol.* **2017**, *31*, 2331–2338. [CrossRef]
7. Baek, D.; Sung, K.H.; Ryou, H.S. Experimental study on the effect of heat release rate and aspect ratio of tunnel on the plug-holing phenomena in shallow underground tunnels. *Int. J. Heat Mass Transf.* **2017**, *113*, 1135–1141. [CrossRef]
8. Ko, G.H.; Kim, S.R.; Ryou, H.S. An experimental study on the effect of slope on the critical velocity in tunnel fires. *J. Fire Sci.* **2010**, *28*, 27–47.
9. Chow, W.K.; Wong, K.Y.; Chung, W.Y. Longitudinal ventilation for smoke control in a tilted tunnel by scale modeling. *Tunn. Undergr. Space Technol.* **2010**, *25*, 122–128. [CrossRef]
10. Atkinson, G.T.; Wu, Y. Smoke control in sloping tunnels. *Fire Saf. J.* **1996**, *27*, 335–341. [CrossRef]
11. Okamoto, K.; Watanabe, N.; Hagimoto, Y.; Chigira, T.; Masano, R.; Miura, H.; Ochiai, S.; Satoh, H.; Tamura, Y.; Hayano, K.; et al. Burning behavior of sedan passenger cars. *Fire Saf. J.* **2009**, *44*, 301–310. [CrossRef]
12. Shintani, Y.; Kakae, N.; Harada, K.; Masuda, H.; Takahash, W. Experimental investigation of burning behavior of automobiles. *Fire Saf. Sci.* **2004**, *6*, 1–13.

9

Study of a High-Pressure External Gear Pump with a Computational Fluid Dynamic Modeling Approach

Emma Frosina [1,*], Adolfo Senatore [1] and Manuel Rigosi [2]

1 Department of Industrial Engineering, University of Naples Federico II, Via Claudio, 21-80125 Naples, Italy; senatore@unina.it
2 Casappa S.p.A., Via Balestrieri 1, Lemignano di Collecchio, 43044 Parma, Italy; rigosim@casappa.com
* Correspondence: emma.frosina@unina.it.

Abstract: A study on the internal fluid dynamic of a high-pressure external gear pump is described in this paper. The pump has been analyzed with both numerical and experimental techniques. Starting from a geometry of the pump, a three-dimensional computational fluid dynamics (CFD) model has been built up using the commercial code PumpLinx®. All leakages have been taken into account in order to estimate the volumetric efficiency of the pump. Then the pump has been tested on a test bench of Casappa S.p.A. Model results like the volumetric efficiency, absorbed torque, and outlet pressure ripple have been compared with the experimental data. The model has demonstrated the ability to predict with good accuracy the performance of the real pump. The CFD model has been also used to evaluate the effect on the pump performance of clearances in the meshing area. With the validated model the pressure inside the chambers of both driving and driven gears have been studied underlining cavitation in meshing fluid volume of the pump. For this reason, the model has been implemented in order to predict the cavitation phenomena. The analysis has allowed the detection of cavitating areas, especially at high rotation speeds and delivery pressure. Isosurfaces of the fluid volume have been colored as a function of the total gas fraction to underline where the cavitation occurs.

Keywords: high-pressure external gear pump; numerical modeling; cavitation

---

1. Introduction

As is well known, external gear pumps are widely used in the fluid power field. These pumps are used in both mobile and fixed applications such as in agriculture, construction machines and hydraulic presses. These machines have many advantages like their compactness and low cost, combined with relatively high efficiency and remarkable reliability, a wide range of operating conditions and structural simplicity. Usually, these pumps are designed to work over a wide range of speeds and high delivery pressures as positive displacement units in hydraulic systems.

The working principle of external gear pumps is very simple. The pump consists of two gears. One is connected to the shaft and is called driving gear; the other one is free and is called the driven gear. These gears are meshed with each other. Chambers are created by coupling the gears with sliding bushing blocks, one for each gear. Two plates connect the gears with the pump housing where, usually, ports are located.

Even if the geometry of these pumps is relatively simple, they are object of many studies addressing the improvement of their performance. This research is mainly focused on the study of all the causes of losses inside the components, through internal fluid-dynamics investigations. Several numerical models are available in literature. These studies (analytical, experimental and modeling) are focused on the prediction of the performance of this pump typology.

Thiagarajan et al. [1] demonstrated improvements of the lubrication ability of external gear machines by designing a micro-surface linear wedge added to the lateral surfaces of the gear teeth. The author showed the reduction of torque loss by adopting a computational fluid dynamics (CFD) numerical approach. The results were confirmed by experimentation.

Vacca et al. [2] studied the operation of spur external gear units with a numerical approach. Then, Pellegri et al. [3] applied the new numerical approach described in [2] with a CFD model, whereby CFD solved the film flow and the lumped parameter model evaluated the overall operation of an external gear pump.

Borghi et al. [4,5], predicted the volumetric efficiency of gear pumps using a mathematical model. The model has been validated with experimental data. Mancò et al. [6] have studied an external gear pump using a lumped parameter approach. Also for this research, model results have been compared with experimental data validating the model.

The improvement of the pump performance can be achieved only by correctly designing the internal fluid dynamic of both lateral plates. These grooves designed in the plates can have several functions such as reducing the noise inside the pump, however grooves connecting volumes at different pressures, can significantly affect the volumetric efficiency of gear pumps if not well designed [5]. The effects of grooves inside plates have also been widely studied by Koc et al. [7]. Borghi et al. [8] also did an interesting analysis on the transient pressure in the meshing volume of external gear pumps.

Many literature studies propose examining gear pumps using 2D modeling approaches with deforming mesh and volume remeshing. However, for the study of pumps like external gear ones, even if the 2D modeling approaches give interesting results, they cannot predict the internal flow behavior like 3D models [9,10]. This depends on the fact that during the operation of an external gear pump, the flow is really complex because of the rotation speeds (typically in the range 500–3000 rpm) and high pressure.

These conditions can cause turbulent flows and sometimes phenomena like cavitation. Yoon et al. [10] did an interesting study on external gear pumps using a three-dimensional approach (with an immersed solid method). The model included the decompression slots that cannot be simulated with a 2D approximation. The model has been used to make interesting considerations on the internal pressure peak, local cavitation, and delivery pressure ripple [11]. They have verified the influence of the gear tip and lateral clearances on the pump performance. Those gaps become crucial in the study of these pumps that, as said, work at a very low speed (500 rpm). Castilla et al. [12], have also built up a complete 3D model of an external gear pump developed with an OPENFOAM Toolbox. No simulation models have been found for the study of cavitation.

In this paper, the authors have studied principally the flow behavior in the chambers and in the meshing zone. The implementation in the numerical model of all the groove and leakages is important to achieve the best accuracy in the prediction of working conditions. The present study is focused on the analysis of the internal fluid-dynamic of an external gear pump using a tridimensional numerical approach. As said, in the literature there are few studies done with three-dimensional simulation tools, and as far as we know, no study has been carried out to fully describe an external gear pump under all operating conditions, even cavitation. The purpose of this research is finding an extremely efficient and fast computational methodology able to predict the overall performance of the components and to allow, by means of monitoring certain points, detecting parameters otherwise difficult to experimentally measure. In addition, this analysis describes a methodology that could be widely used to predict when cavitation occurs inside an external gear pump. This point is of extreme interest for engineers involved in the pump design and optimization phases.

The pump under investigation is the Casappa KP30 (Casappa S.p.A., Parma, Italy), shown in Figure 1a. The pump geometry is shown in the exploded vision in Figure 1b. The pump main features are listed in Table 1. An accurate CFD model has been built up using the commercial code PumpLinx® (Simerics Inc., Bellevue, WA, USA). The entire model of the pump has been created and then validated with experimental data collected by the pump manufacturer.

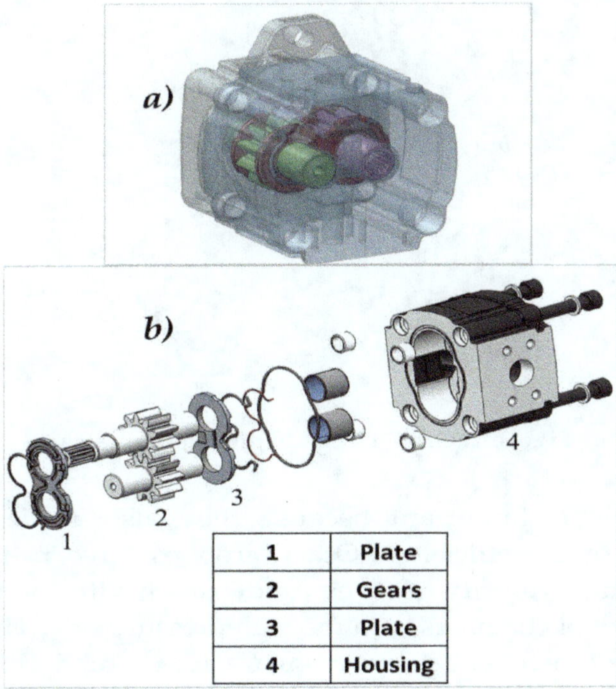

Figure 1. (**a**) Casappa KP30; (**b**) exploded vision of the pump drawing.

Table 1. Pump main features—Casappa KP30.

Pump Main Features	Value
Nominal displacement	44 cm^3
Inlet pressure range	0.7–3 bar
Max. continuous pressure	P1 = 250 bar
Max. intermitted pressure	P2 = 270 bar
Max. peak pressure	P3 = 290 bar
Rotational speed	Min = 350 min^{-1} Max = 3000 min^{-1}

Flow-rate comparisons have been done at three different delivery pressures and oil temperatures while varying the pump speeds. The model demonstrates an accuracy below 2% for all the analyzed working conditions. The validation of the model has been also done on the pressure ripples. Monitoring points have been located in one chamber of each gear. The post-processing of the pressure inside the chambers underlined the areas under low pressure, especially at high pump speeds.

An analysis on the total volume gas fraction has been performed in order to verify if in the chambers of both gears the pressure goes below the saturation value causing cavitation. For this reason, the model has been implemented with an accurate submodel for the prediction of cavitating conditions.

Particular attention has been paid to the connection plates interposed between the ports and gears. In Figure 2, the geometry of one of them is shown. As said, in the grooves drawn inside plates, during the gear rotation, the pressure reduction can cause cavitation. For this reason, the design phase of plates is important and a CFD modeling approach can help optimize those geometries achieving the best solution. The groove created at the delivery side (underlined with the green rectangle in Figure 2, allows the trapping of the fluid in the high-pressure volume, reducing, as a consequence, the pressure spike in the meshing area. On the other hand, the groove instead allows the filling of the chambers from the suction port. In this way, the pressure drop is reduced, limiting the occurrence of cavitation. Those grooves have also the function to reduce the noisiness of the pump during the operation.

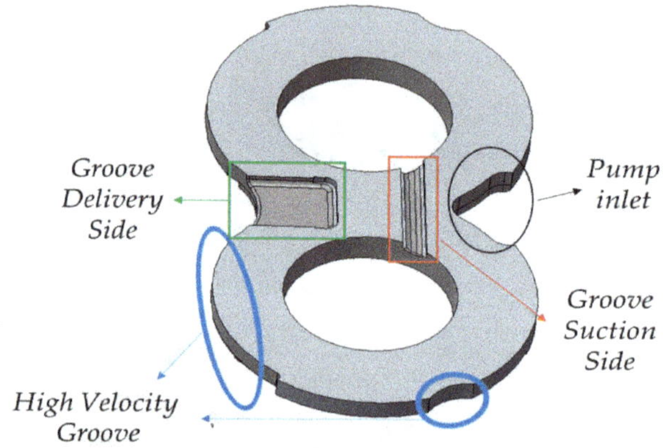

Figure 2. Grooves inside the lateral plates.

The design of grooves is important because they also cause a loss of flow-rate and, as a consequence, of the pump efficiency. Other grooves have been designed on the plates. These volumes are called high velocity grooves (underlined with the ellipse in blue in Figure 2) and connect a defined number of chambers with the high pressure side. This solution affects the forces acting on the gears. In Figure 2 there is also a rectangular vacancy (underlined with the ellipse in black) that allows the connection to the section port of the pump. This fluid volume will be shown better in Figure 3b. In the second section of the paper, the numerical model of the pump is described.

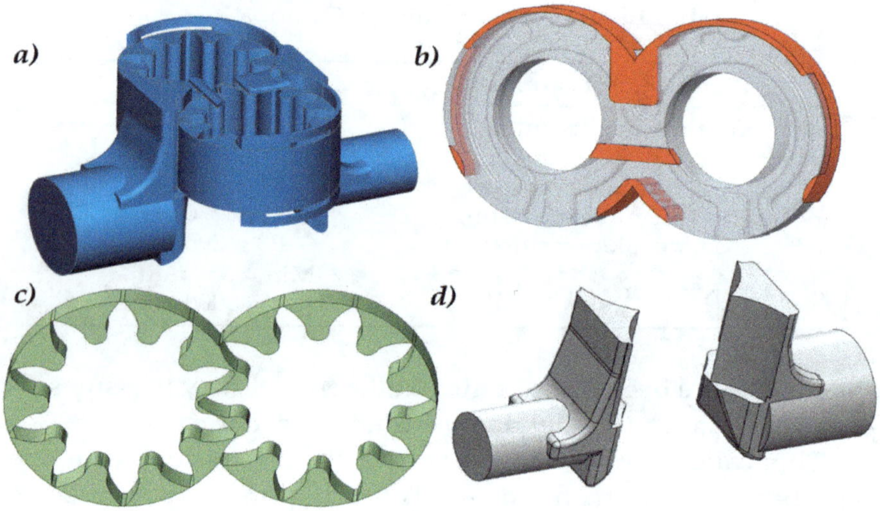

Figure 3. Extracted fluid volume of the external gear pump, (**a**) entire fluid volume; (**b**) fluid volume of the plate; (**c**) fluid volume in rotation; (**d**) fluid volume of the ports.

2. Numerical Model Description

In this section, a numerical model of the studied external gear pump is described. The model has been built up using the commercial code PumpLinx®. Starting from the real CAD geometry of the pump already shown in Figure 1, the fluid volume has been extracted and then meshed. The extracted fluid volume of the entire pump is presented in Figure 3.

The fluid volume has been meshed using a body-fitted binary tree approach. The grid generated with a body-fitted binary tree approach is accurate and efficient because the parent-child tree architecture allows for an expandable data structure with reduced memory storage.

In this architecture of the grid, the binary refinement is optimal for transitioning between different length scales and resolutions within the model. The majority of cells are cubes, which is the optimum cell type in terms of orthogonally, aspect ratio, and skewness, thereby reducing the influence of numerical errors and improving the speed and accuracy. The grid can also tolerate inaccurate CAD surfaces with small gaps and overlaps. Therefore, a body-fitted binary tree approach has demonstrate to be the best methodology to mesh the complex geometry of the pump under analysis. Thanks to the approach adopted, it is possible to create a grid able to take into account the complexity of the geometry, especially in the boundary layer. In fact, also on the surfaces the grid density has been increased without excessively increasing the total cell count. This has been realized using cubic cells, which, as said, are the optimum cell type in terms of orthogonally and improve speed and accuracy. For this reason, in the regions of high curvature and small details the mesh been subdivided and cut to conform it to the surface. From the volume in Figure 3 the mesh has been generated and is shown in Figure 4.

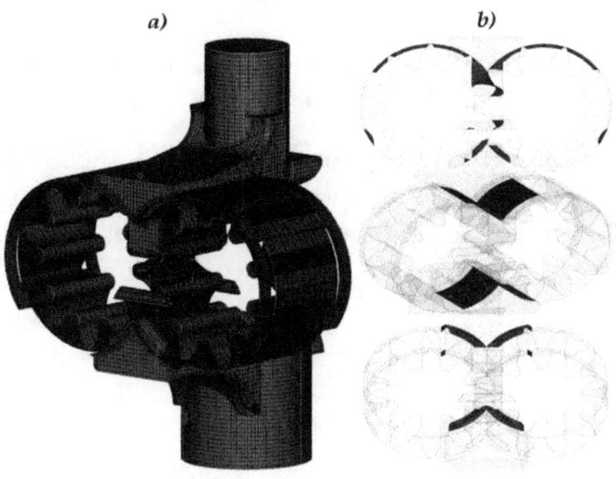

Figure 4. (**a**) Binary tree mesh; (**b**) mismatched grid interface (MGI) between volumes.

A maximum cell size has been chosen for the grid. This parameter defines the maximum cell size in all the fluid volume. In the same way, a minimum cell size has been fixed to limit the minimum dimensions of all cells. Fixing this parameter means that no cell side can be smaller than the minimum cell size. Another important parameter must to be set, which defines the size of cells on surfaces.

Different techniques are available for the treatment of a moving mesh. For positive displacement pumps, it is necessary to use a moving/sliding methodology whereby the stationary and moving volumes are meshed separately. The code, therefore, allows for the simultaneous treatment of moving (gear chambers) and stationary (like suction and delivery ports and grooves in the plate geometry) fluid volumes. Each volume has been connected to the others via an implicit interface called mismatched grid interface (MGI). Each MGI, due to deformations and motion, is updated at each time-step.

2.1. Study of the Gap between Teeth

Before describing the model results obtained it is important to proceed with an analysis on the sizing of the gaps between teeth. This study becomes crucial to correctly predict the performance of an external gear pump. Therefore, in the following a preliminary study done choosing the clearances between teeth of gears is presented.

As shown in Figure 5, the minimum gap between teeth is 5.3 μm. This gap value has been defined after a analyzing the pump geometry and the angles θ and ϕ in Figure 6). The angle θ defines the "driving gear" (Gear 1 in Figure 6) position while the angle ϕ is referred to the "driven gear" (Gear 2 in Figure 6). The angle ϕ is defined as function of the angle θ and the number of teeth z of each gears. This angle ideally is calculated as follows:

$$\phi = \theta + \frac{360}{2z} \tag{1}$$

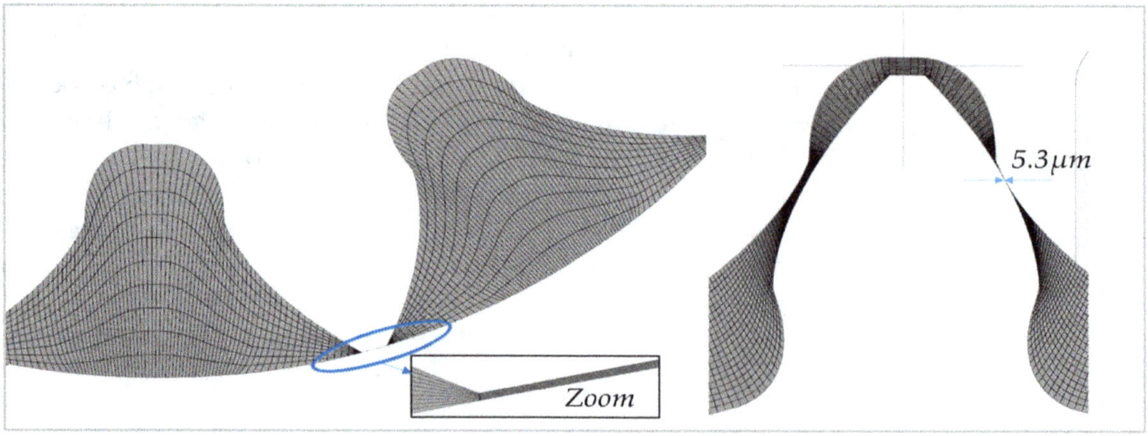

Figure 5. Mesh of the rotating fluid volume and size of the minimum gaps.

Looking at Figure 6, in this configuration gears are perfectly meshing but not in contact. However this meshing is not real, because, as well know, there is no gap between gears in a real working pump. Therefore, in order to correctly reproduce real working conditions, where contact occurs, Equation (1) must be implemented including the shift angle ϕ. Equation (1) becomes:

$$\phi = \theta + \frac{360}{2z} \pm \phi \tag{2}$$

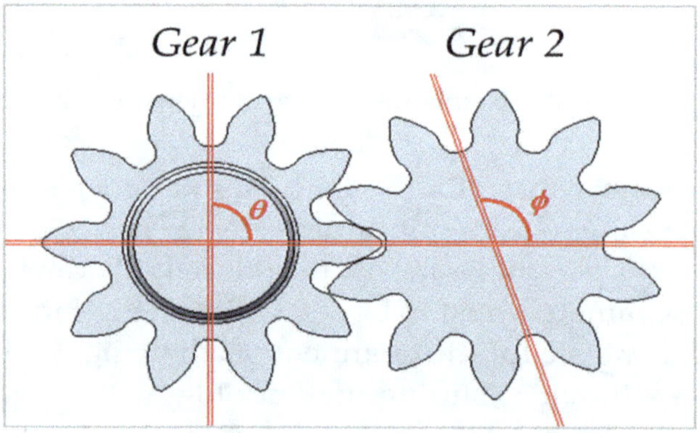

Figure 6. Front view of the gears; angle θ and ϕ.

Therefore, from an ideal gap shown in Figure 6 simulation have been run with three values of ϕ and gap in order to achieve the best accuracy of the model (Figure 7c). The final value of the gap between gears is 5.3 μm.

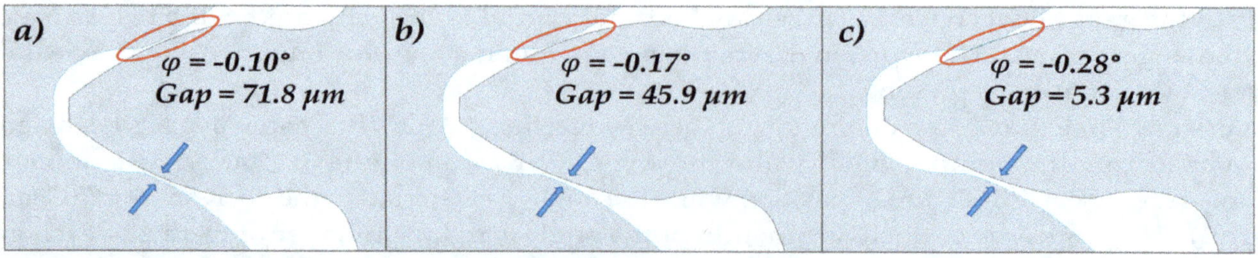

Figure 7. Three different pump configurations: (**a**) ϕ of $-0.10°$ and gap size of 71.8 μm; (**b**) ϕ of $-0.17°$ and gap size of 45.9 μm and (**c**) ϕ of $-0.28°$ and gap size of 5.3 μm.

2.2. Mesh Sensitivity Analysis

A mesh sensitivity analysis has been done on the model by changing the size of the grid. This analysis should be always done in a CFD study to define the best grid for each application and to correctly predict the real working conditions of the pump. Therefore, in this section a mesh sensitivity showing the effects on the model results of variation for example of the clearances between teeth and sliding bushing blocks is presented.

Fluid volumes have been slit in six parts: inlet port, outlet port, outlet high speed grooves, inlet low noise grooves, outlet low noise grooves e rotor. Each volume has been connected to others, as said, via the "MGI" interface (see Figure 4b). Table 2 clarifies the importance of the mesh sensitivity analysis in a CFD study.

A first step has been done analyzing the influence of the grid on the boundary layers. Two different meshes have been created by changing the grid parameters in the boundary layers. These models are called "Mesh 1" and "Mesh 2" where the second one has a finer grid size.

In Table 2, the percentages of error between the suction and delivery flow evaluated by the model are reported. The simulations in Table 2 have been run at 1500 rpm and for two delivery pressures: 50 and 200 bar. As shown in the Table 2, the ΔQ% of both models "Mesh 1" and "Mesh 2", as expected, increases with the delivery pressure.

Table 2. Mesh sensitivity varying the mesh size in the boundary layers, simulations at 1500 rpm.

Delivery Pressure	ΔQ% Mesh 1	ΔQ% Mesh 2
$p_{delivery} = 50$ bar	0.72%	0.24%
$p_{delivery} = 200$ bar	1.78%	1.36%

As shown in the table, improving the quality of the mesh in the boundary layers, the ΔQ% between suction and delivery ports becomes closer to each other. In fact, the reductions from simulations "1" and "2", for both pressure levels, are significant.

After the analysis on the boundary layers, the mesh has been studied improving the quality. Nine models have been created increasing the number of cells. In Figure 8, a comparison of the suction and delivery mass flows is shown. Looking at the graph, it is clear that only with a grid above 200,000 cells does the solution become stable. Each point in Figure 8 corresponds to a model built up with a different grid size.

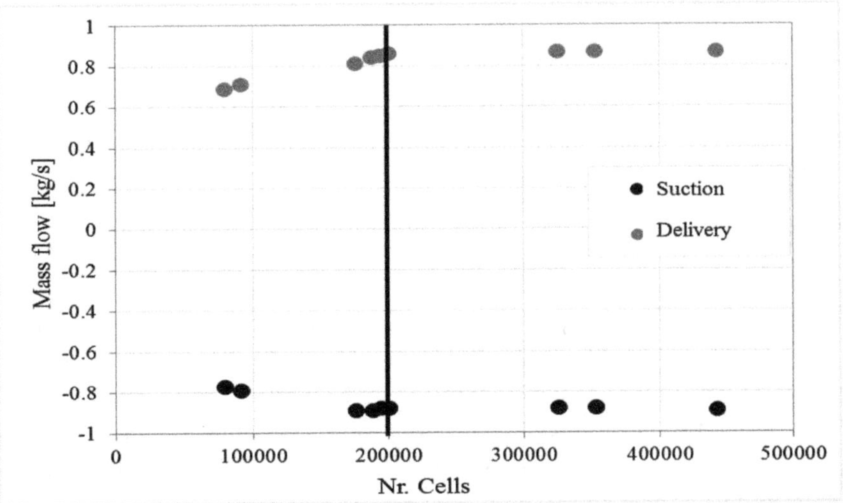

Figure 8. Mesh sensitivity—flow-rate vs cells number at 1500 rpm and a delivery pressure of 50 bar.

Simulations have been run, as already said, has been run on an Intel® Xeon® X5472 CPU @ 3.00 GHz (two processors) with 24 GB RAM. As we know, upon increasing the number of cells the simulation time increases as well. From the analyzed case, the time goes from half an hour for the first point in Figure 8 to 10 h of the last model of almost 500,000 cells. Therefore, in order to achieve the best compromise between accuracy and computational time, the final model consisted of 200,000 cells which is the point underlined in Figure 8.

2.3. Model Setup

The standard k-ε turbulence model has been adopted because for this specific problem it has proven to be accurate. The k-ε model is a robust method demonstrated to provide good engineering results. As well known, in the literature there are other resolution methods more accurate than the k-ε model and RNG k-model (RNG, LES, or DES, etc.). However since the losses due to the viscous stresses are negligible compared to pressure forces the k-ε model can be chosen because it is numerically robust, computationally efficient and it provides good accuracy. In fact, in this application, where the computational time is close to 10 hours per revolution, the adoption of a higher order turbulence model would have increased the computational time with no relevant improvement of the results. The authors of this paper have already used this strategy for many other similar analyses confirming the accuracy of the solutions obtained [13–21].

An analysis on the convergence criteria have been done to achieve the best accuracy results. "R_i" is the residual drop referred to the "i-th volume". It is defined as the difference between the obtained results, for the selected volume, during two subsequent iterations. The correct solution convergence is realized when all the residuals go below the defined R_i.

For all the simulations, the convergence criteria for the pressure, velocity and the vapor mass fraction is 0.01 as shown in Figure 9a. Figure 9a, in fact, presents the trend of the residual drop for each variable in a single time-step. As shown, residuals go below 0.01 (settled $R_{i.}$).

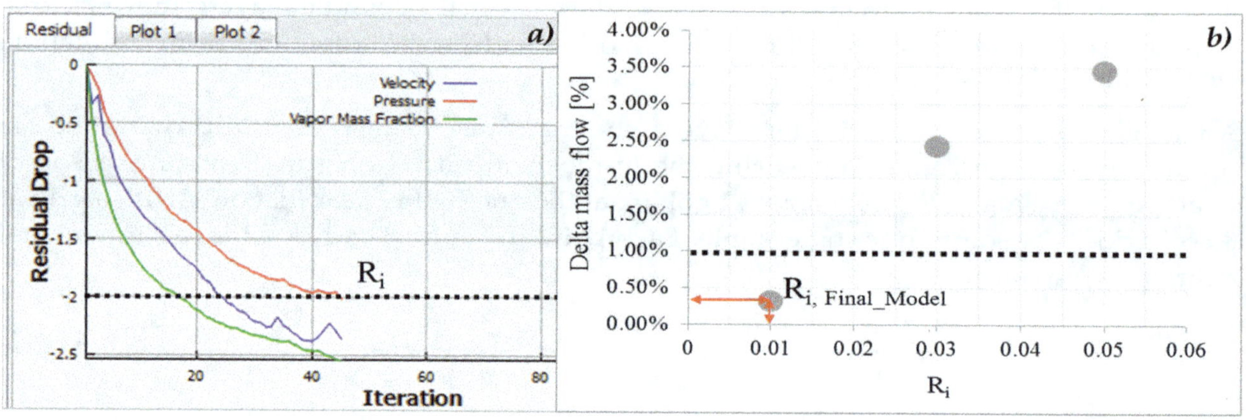

Figure 9. (**a**) Pressure, velocity and vapor mass fraction—residual drop analysis R_i; (**b**) error on the mass flow vs residual drop R_i.

These values have been demonstrated to be suitable for this application. However, a study on the convergence criteria has been also performed varying the R_i. As example, in Figure 9b the R_i is shown as function of the delta mass fraction (%) between the pump inlet and outlet mass flows. As shown, the R_i has been varied from 0.01 to 0.05 with steps of 0.02.

The maximum error on the delta mass fraction between the pump inlet and outlet mass flows has been fixed at 1%. It has been demonstrated that the best solution can been achieved with a R_i of 0.01 (Figure 9b). Other analyses have been run reducing the R_i for the to lower value without improving the delta mass flow percentage.

Before running simulations, it is important to underline that in this study the model has been built up including an accurate submodel able to predict cavitation. In the following subsection the study of cavitation inside an external gear pump has been done. In fact, one of the targets of this research was to detect cavitation, in order to predict the worst working conditions for the pump.

Since one of the aims of the research is the investigation on the ability of the pump to avoid cavitation, the model has been equipped with a robust submodel, available in the code, able to predict with high accuracy cavitating conditions. The chosen cavitation module accounts for all three important real liquid properties: cavitation, aeration, and liquid compressibility. This model builds on the work of Singhal et al. [22]. A module is used to predict the aeration and cavitation in a liquid system, therefore allowing for the calculation of cavitation effects, when the pressure in a specified zone of the fluid domain falls below the saturation pressure and vapor bubbles form and then collapse as the pressure rises again [22]. Many physical models for the formation and transport of vapor bubbles in liquids are available in literature, but only few computational codes offer robust cavitation models. This is due to the difficulty on handling gas/liquid mixtures with very different densities. Even small pressure variations may cause numerical instability if they are not optimally treated [22].The original cavitation model proposed by Singhal et al. describes the vapor distribution using the following formulation [13–16,22]:

$$\frac{\partial}{\partial t}\int_{\Omega(t)}^{0} \rho f \mathrm{d}\Omega + \int_{\sigma}^{0} \rho((v - v_{\sigma})n) f \mathrm{d}\sigma = \int_{\sigma}^{0}\left(D_f + \frac{\mu_t}{\sigma_f}\right)(\nabla f n)\mathrm{d}\sigma + \int_{\Omega}^{0}(R_e - R_c)\mathrm{d}\Omega \tag{3}$$

where D_f is the diffusivity of the vapor mass fraction and σ_f is the turbulent Schmidt number. In the present study, these two numbers are set equal to the mixture viscosity and unity, respectively. The vapor generation term, R_e, and the condensation rate, R_c, are modeled as [13,21]:

$$R_e = C_e \frac{\sqrt{k}}{\sigma_l} \rho_l \rho_v \left[\frac{2}{3}\frac{(p - p_v)}{\rho_l}\right]^{\frac{1}{2}} (1 - f_v - f_g) \tag{4}$$

$$R_c = C_c \frac{\sqrt{k}}{\sigma_l} \rho_l \rho_v \left[\frac{2}{3}\frac{(p - p_v)}{\rho_l}\right]^{\frac{1}{2}} f_v \tag{5}$$

in which the model constants are $C_e = 0.02$ and $C_c = 0.01$.

The final density calculation for the mixture is done by [13–16,22]:

$$\frac{1}{\rho} = \frac{f_v}{\rho_v} + \frac{f_g}{\rho_g} + \frac{(1 - f_v - f_g)}{\rho_l} \tag{6}$$

The model by Singhal et al. has been extended to include non-condensable gases, finite rate and equilibrium dissolved gas. The fluid model, in addiction, accounts for liquid compressibility. This is critical to accurately model pressure wave propagation in liquids. The liquid compressibility is found to be very important for a high-pressure system and the systems in which water hammer effects are relevant.

3. Model Validation

The pump has been tested on a dedicated test bench at Casappa S.p.A. The test bench is shown in Figure 10. Simulations have been run under the same conditions tested in the lab with pump speeds of 1000, 1500 and 2500 rpm, oil temperatures of 50 and 80 °C and delivery pressures of 50, 150 and 250 bar.

The test bench layout is shown in Figure 10b. There are two strain gauge sensors P1 and P2 (both from WIKA®, Lawrenceville, GA, USA, scale: 0–40 bar and 0.25% FS accuracy) and P2 at the delivery side (by WIKA®, with a scale of 0–400 bar and an accuracy of 0.25% FS). The pressure sensor P4 is piezoelectric (made by KISTLER®, Winterthur, Switzerland, scale: 0–1000 bar,

140 kHz natural frequency, 0.8% FS accuracy) while the traducer P5 is piezoresistive (ENTRAN®, Strainsense Ltd, Milton Keynes, UK, scale: 0–350 bar, 450 kHz natural frequency, compensated temperature, 0.5% FSO accuracy).

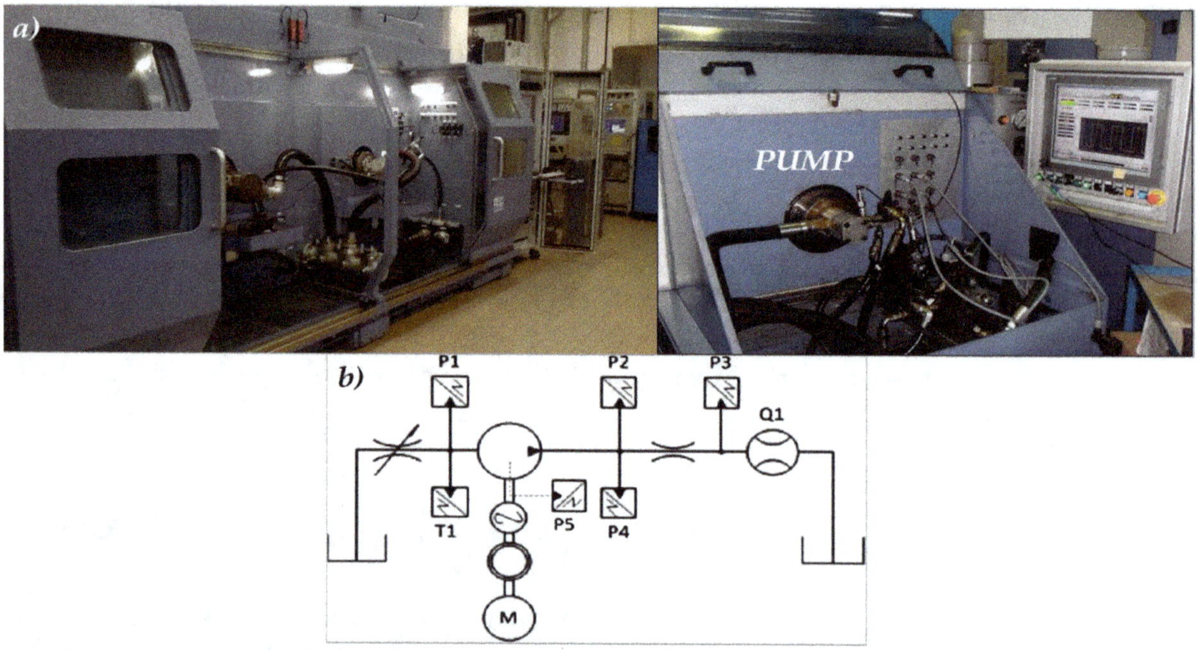

Figure 10. (**a**) Test bench of Casappa S.p.A.; (**b**) test bench layout.

The flow-rate mater Q1 is a VSE® VS1 (VSE.flow, Neuenrade, Germany), scale 0.05–80 l/min, 0.3% measured value accuracy. There is an HEIDENHAIN®, ERN120 θ encoder (HEIDENHAIN, Traunreut, Germany), 3600 r/min, a limit velocity 4000 r/min and a period accuracy of 1/20.

In Figure 11a–c, a first comparison between model and experimental data is shown for three different pump speeds: 1000, 1500 and 2500 rpm. In the graph, the continuous line are the experimental data while the dashed ones are the simulation results.

Simulations and test, as said, have been done for three delivery pressure conditions: 50, 150 and 250 bar. The comparisons have been done on the torque requested on the driving gear (gear 1 in Figure 6).

By analyzing Figure 11, it is possible to appreciate that the model results are really close to the experimental data. However, there is a gap between model and experimental data that depends also on the fact that the model does not consider torque due to mechanical friction.

The experimental data in thick lines represents the absorbed torque by the pump measured by a torque meter. This means that experimental data takes into account, of course, the "hydraulic" component of torque (volumetric efficiency) even of all the friction (mechanical efficiency). This is the reason why the simulation results are lower than the experimental data. In Figure 12 the delta torque between experimental and model results is presented. This ΔC represent the toque lost due to friction.

Model validation has been done also comparing at the delivery flow-rate varying the delivery pressure. This comparison is shown in Figure 13 at the oil temperatures of 50 and 80 °C and at pump speeds of 1500 and 2500 rpm.

It can be noticed that delivery flow rate decreases by increasing the delivery pressure for both the model and experimental data; the error percentage between model and experimental data is always below 2%.

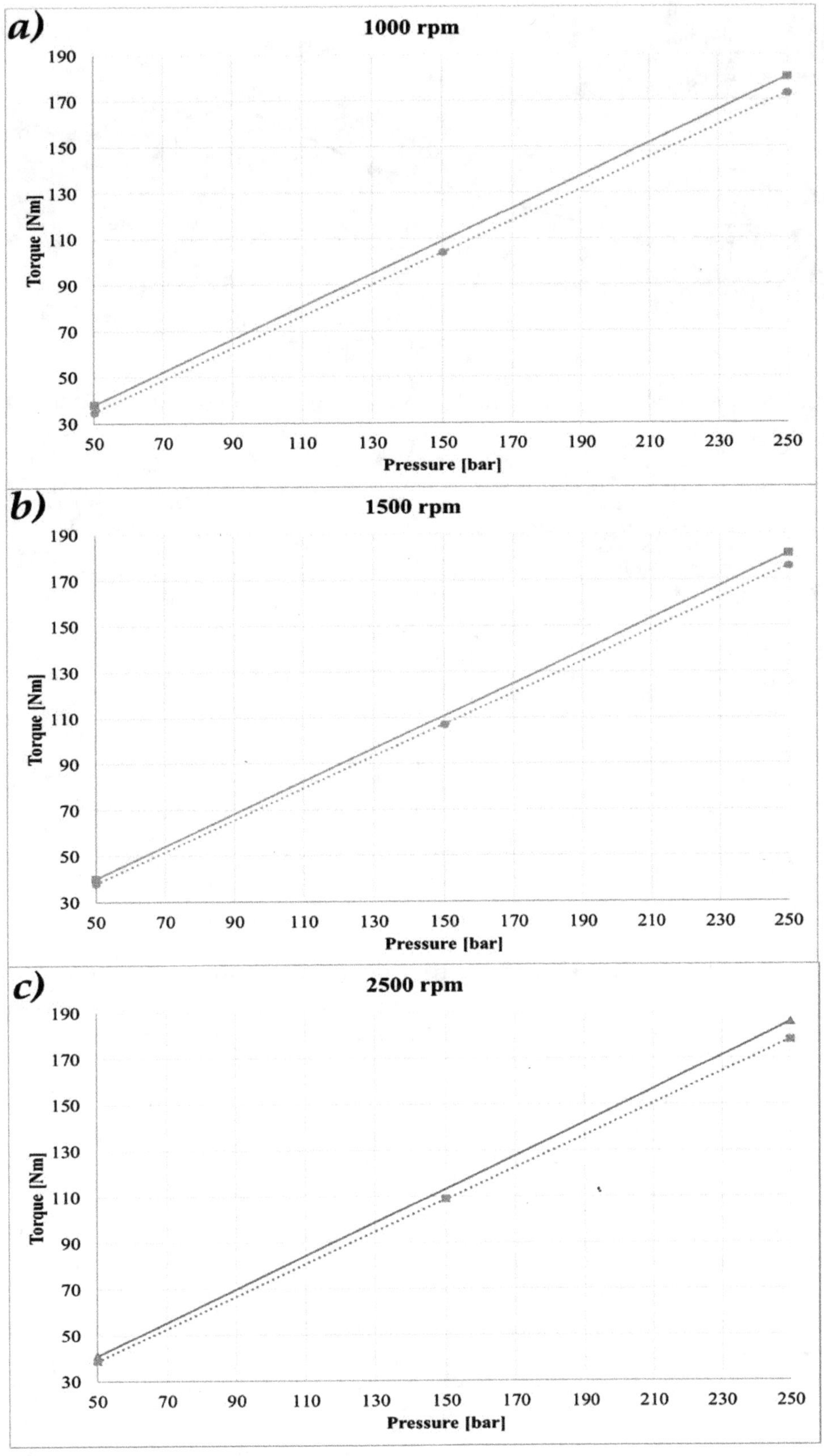

Figure 11. Absorbed torque—comparison between model and experimental data. (**a**) 1000 rpm; (**b**) 1500 rpm and (**c**) 2500 rpm.

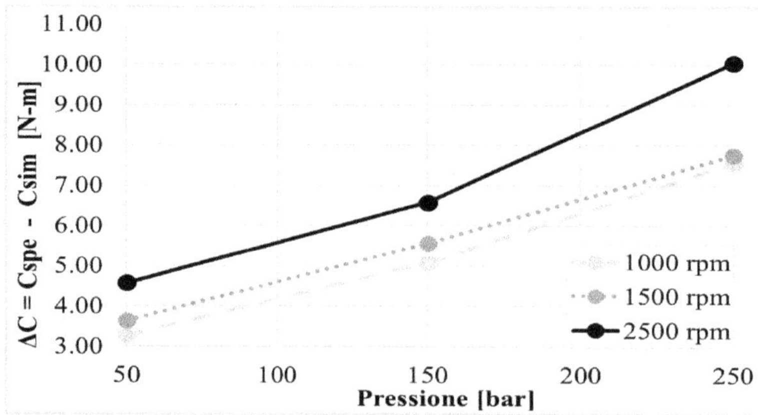

Figure 12. Delta torque between experimental data and simulation data at 1000, 1500 and 2500 rpm.

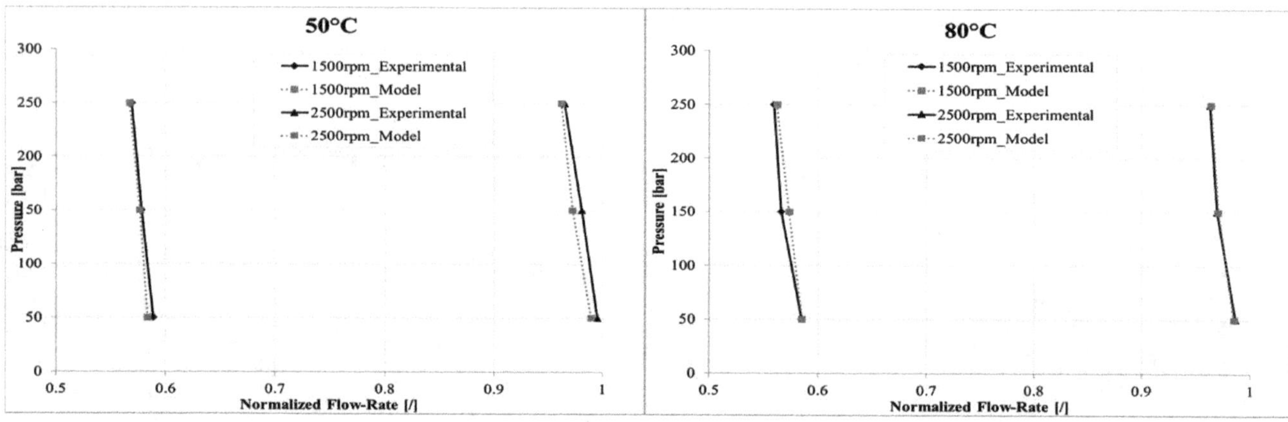

Figure 13. Pressure vs. delivery flow-rate comparison at 50 and 80 °C.

4. Model Results

The validated model has been analyzed in order to study the pump in depth. A three-dimensional CFD modeling approach, as said, has many advantages. In fact, considering the real geometry of the pump, it is useful to visualize parameters that are impossible or onerous to obtain experimentally such as the velocity magnitude of the fluid in leachates.

Figure 14 shows a first result of the numerical model with the velocity magnitude countering through the leakages between gears and the static body. The area under investigation in the pictures is shown in Figure 14c. In fact, figures refer to the leakages between the first pressurized chamber and the last chamber at low pressure.

Figure 14a,b, in particular, show the results in the gaps by increasing the pressure from 50 to 250 bar for oil temperatures of 50 and 80 °C, respectively, at the same pump speed. The velocity magnitude scale in all figures is always in the 0–13 m/s range.

In Figure 14a, by comparing graphs at the same oil temperature it is possible to distinguish the flow velocity magnitude profile through gaps. In fact, as consequence of the higher pressure drops between the two following chambers (from 50 to 250 bar), the flow velocity through the gap between the tooth and the pump housing increases and, as shown in pictures in Figure 13, flow becomes a jet.

Looking for example at Figure 14a, the flow velocity through the gap with a delta pressure of 50 bar is below 5 m/s. It increases in correspondence of the connection between the gap and the chamber at the lower pressure. This velocity profile is, of course, quite similar with a higher pressure drop, however the values increase. For a Δp of 150 bar the maximum velocity value registered in the analyzed fluid volume is of almost 9 m/s. It becomes 13 m/s for the delta pressure of 250 bar.

It is interesting to underline the effect of the T_{oil} on the flow velocity and as consequence, on the pump volumetric efficiency. For all the analyzed Δp, the velocity at 50 °C is lower than for an oil temperature of 80 °C. This depends by modification of the fluid properties with the temperature.

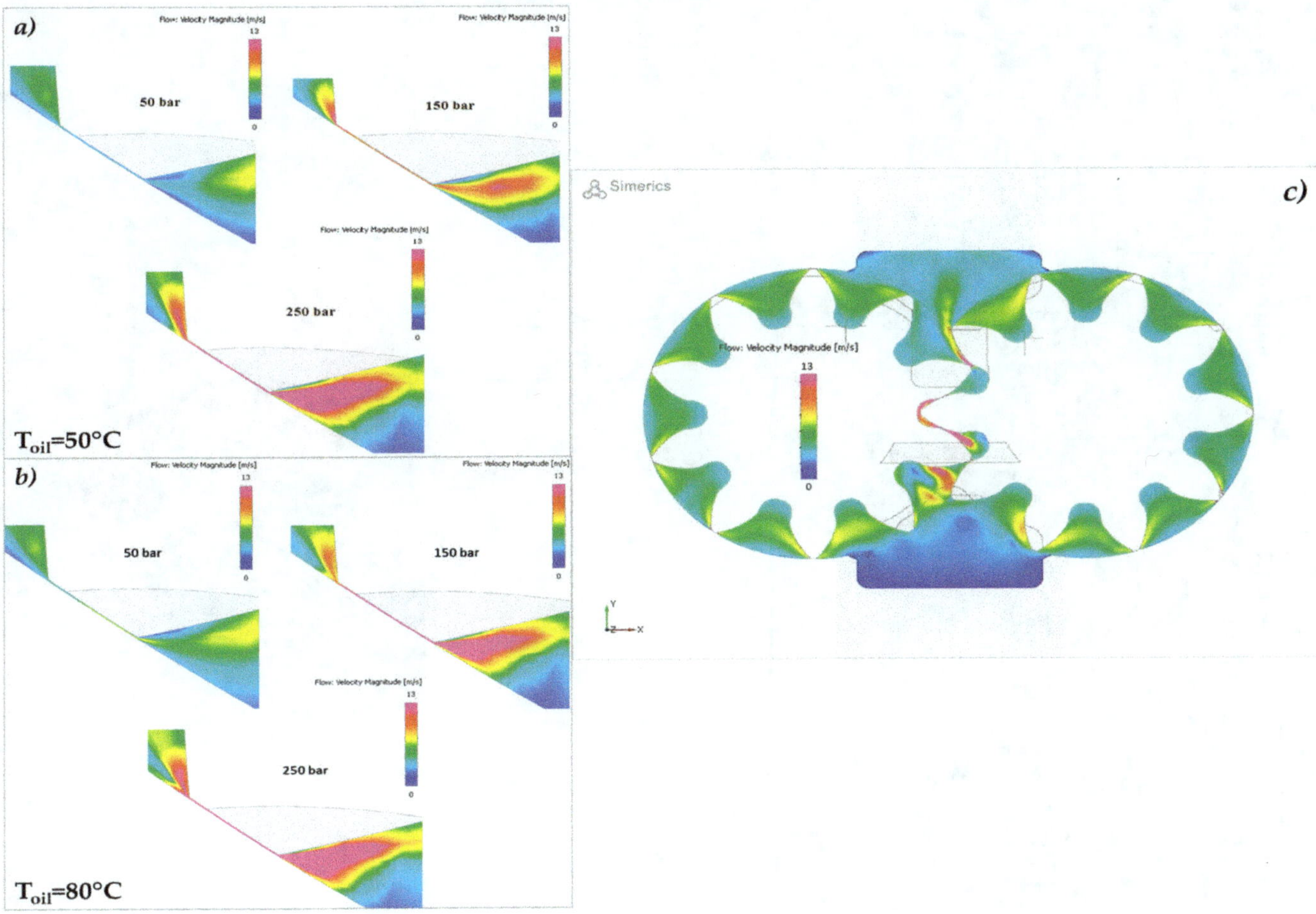

Figure 14. Velocity magnitude at (**a**) 50 °C; (**b**) 80 °C and (**c**) zoomed area.

In Figure 15, the pressure evolutions inside the chambers are diagrammed for a shaft revolution at the operating condition listed below:

- Oil temperature of 50 °C
- Pressure drop of 50 and 250 bar
- Pump speed of 1500 and 2500 rpm.

As is well known, it is also possible to obtain experimentally the results in Figure 15, however it is not easy. Therefore, the best solution is to obtain these important data with modeling techniques.

Data shown in graphs of Figure 15 allow evaluating the pressure evolution inside a chamber in rotation for each gear. In particular, the lines in red refer to the chambers of the gear 1, while the blue ones correspond to gear 2.

The graphs can be interpreted as follows. From the suction port the flow fills the generic chamber (obtained by coupling the two gears) and is compressed. Looking at Figure 15, it is clear that each chamber has a pressure evolution during the 360 degrees but the low-pressure zone has a lower angular extension. It is, in fact, more or less 90 degrees and depends by the pump geometry.

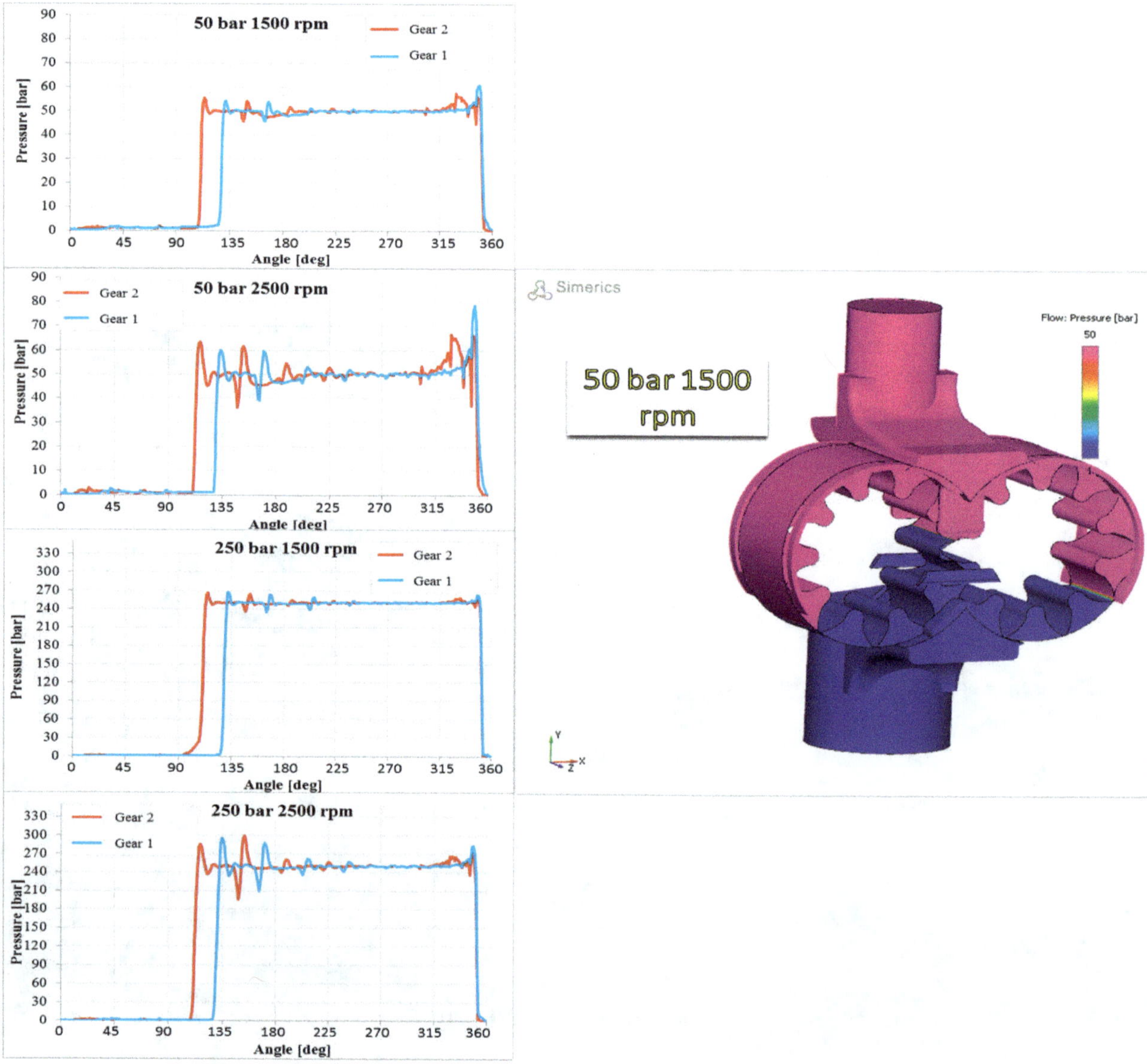

Figure 15. Pressure evolutions for two pressure levels at T_{oil} of 50 °C.

As expected, when the chamber is connected to the delivery port, it has an oscillating pressure around the delivery value (in this case 50 and 250 bar). The pressure is smoother around the mean value with a lower pump speed (1500 rpm). At 2500 rpm instead the oscillations around the mean value are high for both delivery pressures. The pressure spikes inside the chambers, at the higher delivery pressure level, are clear for both gears' fluid volumes. Those investigations, especially the evaluation of the spikes values, are really important for engineers to design or optimize a pump.

The results in Figure 15 are also of extreme interest by visualizing the pressure-axis in the range 0–3.5 bar and reducing the angle-axis to the only the degrees of connection with the pump suction side. In this way, it is possible to appreciate the behavior of the pressure when the chamber is connected to the suction port. These results are shown in Figure 16, providing a preliminary investigation of the cavitation phenomena. Cavitation, however, it analyzed in depth in Section 5.

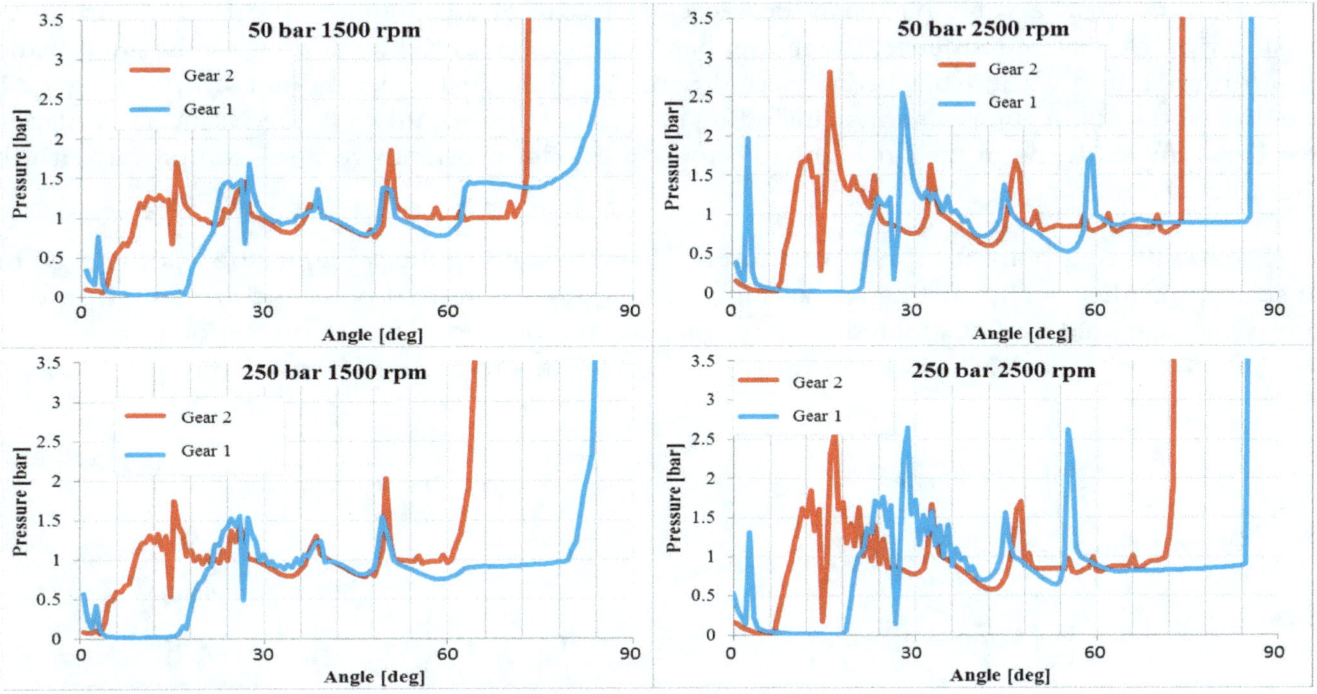

Figure 16. Pressure evolutions for two pressure levels in the range 0–3.5 bar, at T_{oil} of 50 °C.

However before studying the cavitation phenomena a study of the pressure ripples at the delivery side of the pump has been performed. For this reason the model has been implemented adding a supplementary volume in accordance with the experimental tests. This duct has the same geometry of the real one, installed on the test bench in order to correctly follow the experimental delivery pressure ripples. The geometry of the duct is shown in Figure 17a while in Figure 17b the final model is presented. In the model a monitoring point located at the same position of the pressure transducer has been added and a section set like the real lamination valve of the test bench has been added too. This section therefore simulates the lamination valve, and for this reason the entire model simulated with high accuracy the pressure waves inside the delivery duct.

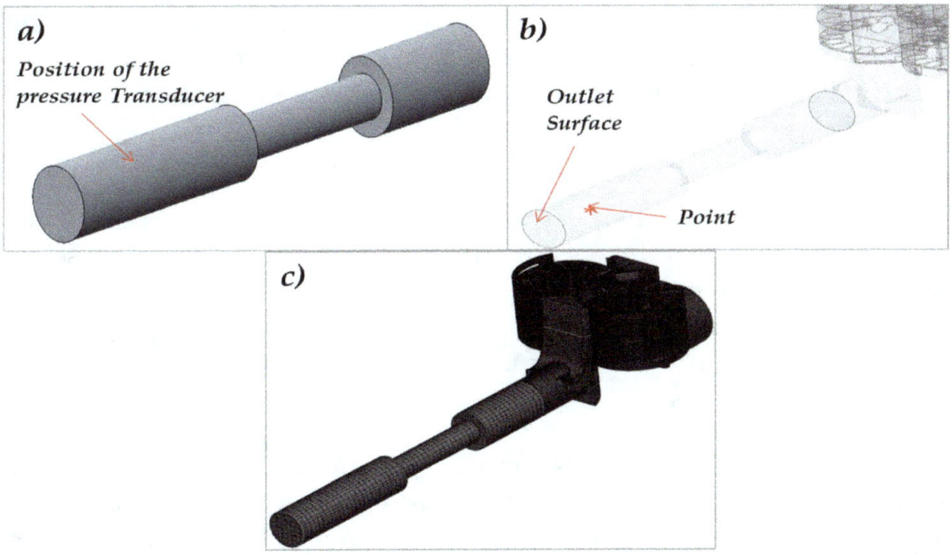

Figure 17. (**a**) Calibrated orifice geometry, (**b**) pressure ripples monitoring point; (**c**) mesh of the entire model.

During the tests a valve has been located close to the pump delivery in order to amplify the pressure ripples. As consequence, the model has been implemented by adding a calibrated orifice. The additional fluid volume has been inserted to replicate the real experimental setup. In Figure 17, the entire model of the pump is shown, including the fluid volume of the calibrated orifice. Simulations have been run with the model in Figure 17 and the model results have been compared with the experimental data.

The monitoring point for the simulation has been located in the same position of the pressure transducer installed on the test bench, as shown in Figure 17b; in that point the pressure ripple has been both measured and calculated. Comparisons have been done at two rotational speeds and three pressure levels at 50 °C (see Figure 17). The diagrams in Figure 18 have been normalized to a reference pressure.

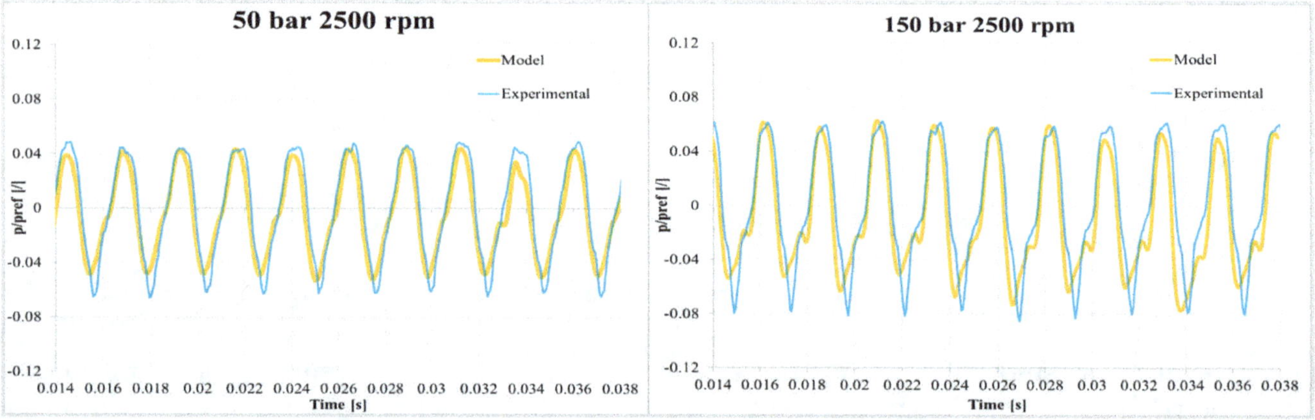

Figure 18. Pressure ripple comparison at 50 bar and 150 bar, 2500 rpm and 50 °C.

As said, the model results in Figure 18 have been obtained changing the delivery pressure at 2500 rpm. Comparison between the model and experimental data shows a good agreement in all the analyzed conditions, confirming the accuracy of the numerical model. The results, in fact, show that the timing and ripples have the same amplitude as the experimental data.

Pressure ripples show that the amplitude of the oscillation increases with the delivery pressure. In the graphs the y-axis has been fixed, therefore, looking at the ripples, it is clear that by increasing the delivery pressure the amplitude increases as well. This happens for both pump speeds.

5. Cavitation

In this section, the model results have been analyzed verifying if cavitation occurs under particular pump-operating conditions. The model, as already said in the corresponding section, includes the capability of prediction of the occurrence of cavitation. Thanks to the three-dimensional visualization of the total gas volume inside the pump, it is possible also to find when the cavitation occurs and where bubbles are located. The study of this phenomenon becomes important for engineers in the pump design and optimization phases.

In Figure 19, a front view of the pump is shown. The volume is colored as a function of the total gas fraction that varies in the range 0–1. The gas volume fraction is the volume fraction of the free non-condensable gas (NCG) in a liquid for a selected volume. In Figure 19, there are presented four sections relative to gears rotations of $\theta = 9°$, $18°$, $27°$ and $36°$. The simulation has been done under fairly critical conditions for the cavitation: 2500 rpm and 250 bar.

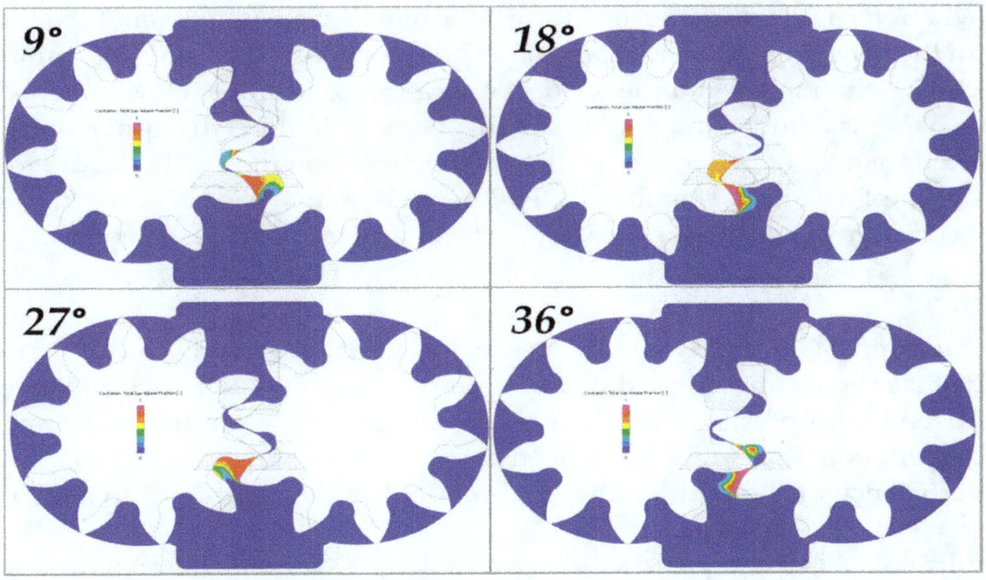

Figure 19. Distribution of the total gas fraction in a section view of the pump at 250 bar, 2500 rpm and 50 °C for a shaft rotation of $\theta = 9°$, 18°, 27° and 36°.

Areas in blue correspond to a concentration of free gas equal to zero while areas colored in magenta are at a percentage of gas of 100%. As shown, the areas in magenta are located in areas where the rotating chamber, from the condition of minimum, starts to increase its volume. Therefore, in correspondence with the meshing area, an intense and sudden cavitation occur. As well known, this phenomenon causes noise and damage to parts and therefore must be avoided. Figure 20 shows two diagrams of the total gas volume fraction inside the chambers of gears. The total gas volume fraction, as said, varies in the range 0–100%.

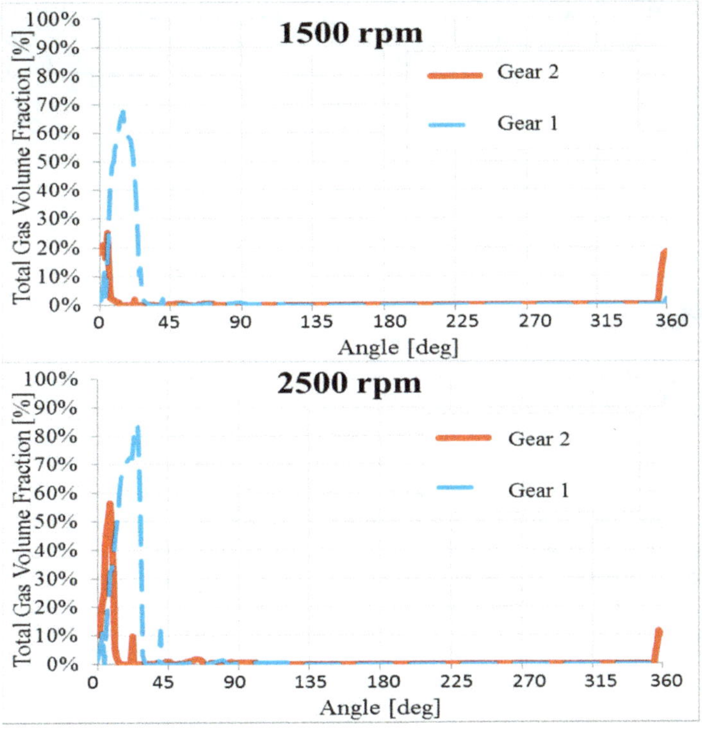

Figure 20. Total gas fraction behaviors at 1500 and 2500 rpm.

Signs of cavitation could be appreciable for values higher than 80%. Even if the total gas fraction in the chamber of the driving gear at 2500 rpm has a spike between 0° and 45° this pump demonstrates to not be subject to cavitation, even under critical operating conditions. However, the methodology described in this paper is demonstrated to be suitable because it offers the opportunity to fully study the complex fluid dynamics of external gear pumps. Results showing cavitation are really important for the optimization of a pump. Therefore, a CFD modeling analysis becomes fundamental in the design phase of components to achieve the best performance for all working conditions.

6. Conclusions

A three-dimensional CFD study a high-pressure external gear pump has been descried in this paper. The model has been realized based on a geometry of a pump manufactured by Casappa S.p.A. The model has been built up using a commercial code including accurate tools to predict the flow turbulence and cavitation. Leakages have been taken into account in order to correctly estimate the volumetric efficiency of the pump. The final model has demonstrated to achieve an accuracy close to 1%.

Model results have been compared with experimental data obtained on a dedicated test bench by the pump manufacturer. Thanks to the good agreement of the data the model has been run under some conditions (such as at high pressure value) which are very challenging from a modeling and numerical point of view. Simulations have found that at the higher pressure and speeds the pump has a tendency to cavitate. For this reason, monitoring points have been located inside the chamber volumes of both gears.

The pressure behavior underlined that the pressure, especially inside the driving gear chamber, goes below the oil saturation pressure. This has been confirmed by visualizing the total gas fraction distribution in sequential images in a pump cross-section. This study summarized a methodology able to completely describe the operation of pumps such as external gear pumps. All the working parameters have been analyzed and, where possible, compared with experimental data. Research has confirmed the importance of CFD, which can be a valuable instrument for engineers to study some working anomalies like cavitation.

Acknowledgments: This research is a result of a research collaboration between the Fluid Power Research Group (FPRG) of the Department of Industrial Engineering of the University of Naples "Federico II" and Casappa S.p.A. We appreciate the technical contribution from Federico Monterosso and Micaela Olivetti of OMIQ s.r.l.

Author Contributions: Emma Frosina and Manule Rigosi built up the model. Adolfo Senatore supervised the study. Emma Frosina wrote the paper.

Nomenclature

$c_{eq,k}$	Coefficient
CFD	Computational fluid dynamic
C_c	Cavitation model constant
C_e	Cavitation model constant
D_f	Diffusivity of vapor mass fraction
dS	Infinitesimal surface, m^2
f_v	Vapor mass fraction
f_g	Non-condensable gas mass fraction
G_t	Turbulent generation term
MGI	Mismatched grid interface
N_{driving}	Number of the teeth of the Gear 1
n	Pump velocity rotation, rpm
$p_{u,k}$	k-th inlet pressure, bar
$p_{d,k}$	k-th outlet pressure, bar
p_s	Oil supply pressure, Pa

Q_{teo}	Theoretical flow-rate, l/min
Q_{real}	Real flow-rate, l/min
R_i	Residual drop
R_e	Vapor generation rate
R_c	Vapor condensation rate
rpm	Pump speed
S'_{ij}	Strain tensor
T_{oil}	Oil temperature, °C
V	Pump displacement, cm^3/rev
Vi	i-th volume, m^3
Greek Letters	
β	Bulk modulus, bar
γ	Angle step, deg
ε	Turbulence dissipation
η_{vol}	Volumetric efficiency
Θ	Angle of rotation driving gear, deg
μ	Fluid viscosity, Pa·s
μ_t	Turbulent viscosity, Pa·s
v	Velocity, m/s
$\rho(p_k)$	Fluid density at pressure k, kg/m^3
ρ_g	Gas density, kg/m^3
ρ_l	Liquid density, kg/m^3
ρ_v	Vapor density, kg/m^3
τ	Stress tensor
θ	Angle of rotation driven gear, deg
ϕ	Shift angle, deg
Ω	Volume, m^3

References

1. Thiagarajan, D.; Dhar, S.; Vacca, A. Improvement of lubrication performance in external gear machines through micro-surface wedged gears. *Tribol. Trans.* **2017**, *60*, 337–348. [CrossRef]
2. Vacca, A.; Guidetti, M. Modelling and experimental validation of external spur gear machines for fluid power applications. *Simul. Model. Pract. Theory* **2011**, *19*, 2007–2031. [CrossRef]
3. Pellegri, M.; Vacca, M. A CFD-radial motion coupled model for the evaluation of the features of journal bearing in external gear machines. In Proceedings of the Symposium on Fluid Power and Motion Control FPMC2015, Chicago, IL, USA, 12–14 October 2015.
4. Borghi, M.; Milani, M.; Paltrinieri, F.; Zardin, B. Studying the axial balance of external gear pumps. In Proceedings of the SAE Commercial Vehicle Engineering Congress, Chicago, IL, USA, 1–3 November 2005.
5. Borghi, M.; Zardin, B.; Specchia, E. External gear pump volumetric efficiency: Numerical and experimental analysis. In Proceedings of the SAE 2014 World Congress and Exhibition, Detroit, MI, USA, 8–10 April 2009.
6. Mancò, S.; Nervegna, N. Simulation of an external gear pump and experimental verification. In Proceedings of the JHPS International Symposium on Fluid Power, Tokyo, Japan, 13–16 March 1989.
7. Koc, E.; Ng, K.; Hooke, C.J. An analysis of the lubrication mechanisms of the bush-type bearings in high pressure pumps. *Tribol. Int.* **1997**, *30*, 553–560. [CrossRef]
8. Borghi, M.; Milani, M.; Paltrinieri, F.; Zardin, B. Pressure transients in external gear pumps and motors meshing volumes. In Proceedings of the SAE Commercial Vehicle Engineering Congress, Chicago, IL, USA, 1–3 November 2005.
9. Strasser, W. CFD investigation of gear pump mixing using deforming/agglomerating mesh. *J. Fluids Eng.* **2007**, *129*, 476–484. [CrossRef]
10. Castilla, R.; Gamez-Montero, P.J.; Ertuürk, N.; Vernet, A.; Coussirat, M.; Codina, E. Numerical simulation of the turbulent flow in the suction chamber of a gear pump using deforming mesh and mesh replacement. *Int. J. Mech. Sci.* **2010**, *52*, 1334–1342. [CrossRef]

11. Yoon, Y.; Park, B.H.; Shim, J.; Han, Y.O.; Hong, B.J.; Yun, S.H. Numerical simulation of three-dimensional external gear pump using immersed solid method. *Appl. Therm. Eng.* **2017**, *118*, 539–550. [CrossRef]
12. Castilla, R.; Gamez-Montero, P.J.; del Campo, D.; Raush, G.; Garcia-Vilchez, M.; Codina, E. Three-dimensional numerical simulation of an external gear pump with decompression slot and meshing contact point. *ASME J. Fluids Eng.* **2015**, *137*, 041105. [CrossRef]
13. Senatore, A.; Buono, D.; Frosina, E.; Santato, L. Analysis and simulation of an oil lubrication pump for the internal combustion engine. In Proceedings of the ASME International Mechanical Engineering Congress and Exposition, San Diego, CA, USA, 13–21 November 2013.
14. Frosina, E.; Senatore, A.; Buono, D.; Pavanetto, M.; Olivetti, M.; Costin, I. Improving the performance of a two way flow control valve, using a 3D CFD modeling. In Proceedings of the ASME International Mechanical Engineering Congress and Exposition, Montreal, QC, Canada, 14–20 November 2014.
15. Frosina, E.; Senatore, A.; Buono, D.; Stelson, K.A.; Wang, F.; Mohanty, B.; Gust, M.J. Vane pump power split transmission: Three dimensional computational fluid dynamic modeling. In Proceedings of the ASME/BATH 2015 Symposium on Fluid Power and Motion Control, Chicago, IL, USA, 12–14 October 2015.
16. Frosina, E.; Buono, D.; Senatore, A.; Stelson, K.A. A modeling approach to study the fluid dynamic forces acting on the spool of a flow control valve. *J. Fluids Eng.* **2016**, *139*, 011103. [CrossRef]
17. Pellegri, M.; Vacca, A.; Frosina, E.; Senatore, A.; Buono, D. Numerical analysis and experimental validation of gerotor pumps: A comparison between a lumped parameter and a computational fluid dynamics-based approach. *Proc. Inst. Mech. Eng. Part C J. Mech. Eng. Sci.* **2016**, *1989–1996*, 203–210. [CrossRef]
18. Schleihs, C.; Viennet, E.; Deeken, M.; Ding, H.; Xia, X.; Lowry, S.; Murrenhoff, H. 3D-CFD simulation of an axial piston displacement unit. In Proceedings of the 9th International Fluid Power Conference, Aachen, Germany, 24–26 March 2014; pp. 332–343.
19. Vacca, A.; Franzoni, G.; Casoli, P. On the analysis of experimental data for external gear machines and their comparison with simulation results. In Proceedings of the IMECE2007 ASME International Mechanical Engineering Congress and Exposition, Design, Analysis, Control and Diagnosis of Fluid Power Systems, Seattle, WA, USA, 11–15 November 2007.
20. Borghi, M.; Zardin, B. Axial balance of external gear pumps and motors: Modelling and discussing the influence of elastohydrodynamic lubrication in the axial gap. In Proceedings of the IMECE2015 ASME International Mechanical Engineering Congress and Exposition, Advances in Multidisciplinary Engineering, Houston, TX, USA, 13–19 November 2015.
21. Zhou, J.; Vacca, A.; Casoli, P. A novel approach for predicting the operation of external gear pumps under cavitating conditions. *Simul. Model. Pract. Theory* **2014**, *45*, 35–49. [CrossRef]
22. Singhal, A.K.; Athavale, M.M.; Li, H.Y.; Jiang, Y. Mathematical basis and validation of the full cavitation model. *J. Fluids Eng.* **2002**, *124*, 617–624. [CrossRef]

10

Effects of the Second-Stage of Rotor with Single Abnormal Blade Angle on Rotating Stall of a Two-Stage Variable Pitch Axial Fan

Lei Zhang, Liang Zhang, Qian Zhang *, Kuan Jiang, Yuan Tie and Songling Wang

School of Energy, Power and Mechanical Engineering, North China Electric Power University, Baoding 071003, China; zhang_lei@ncepu.edu.cn (L.Z.); zhangliang@ncepu.edu.cn (L.Z.); jiangkuan@ncepu.edu.cn (K.J.); tieyuan@ncepu.edu.cn (Y.T.); wsl@ncepu.edu.cn (S.W.)

* Correspondence: zhangqian@ncepu.edu.cn.

Abstract: It is of great value to study the impact of abnormal blade installation angle on the inducement mechanism of rotating stall to achieve the active control of rotating stall in an axial fan. Based on throttle value function and SST k-ω turbulence model, numerical simulations of the unsteady flow process in stall condition of an axial flow fan with adjustable vanes were carried out, and the influence mechanism of abnormal stagger angle of a single blade in the second stage rotor on induced position and type of stall inception and evolution process of rotating stall were analyzed. The results show that compared with synchronous adjustment of blade angle, the blade with abnormal stagger angle will cause the increase of flow rate at the beginning of stall and make the fan fall into an unstable condition in advance. The existence of blade with abnormal angle does not cause the change of the induced position and type of stall inception and the inducement mechanism of rotating stall, which are the same as the axial fan with normal blade angle. Moreover, the single blade with abnormal deviation angle has important impacts on the 3D unsteady evolution process from stall inception to stall cell formation in two rotors.

Keywords: two-stage axial fan; numerical simulation; abnormal blade installation angle; rotating stall

---

1. Introduction

With the wide application of the two-stage variable pitch axial fan, its safety and reliability has become increasingly vital in engineering applications. Rotating stall is the instability phenomenon that occurs when the fan operates at flow rates below the design conditions, which can not only deteriorate the flow field, generate noise [1], and threaten the service life of rotor blades owing to the alternating stress, but even cause fan blades to break and endanger the operation of the fan seriously. Hence, it is necessary to study the inducement mechanism of rotating stall, which is significant for improving the performance and safety of the axial fan and realizing the active control of stall [2].

Initially, Emmons et al. [3] theoretically studied the mechanism of instability and believed the separation of airflows caused by the imbalance resulting from manufacturing and installation of blades and any other reasons lead to the occurrence of rotational stall. In fact, the system instability is the result of flow collapse, and the internal flow of impellers is the internal cause and mechanism of instability phenomenon, such as rotating stalls. At present, great progresses have been made in researches of the mechanism of stall experimentally [4]. The mechanism of rotating stall in a centrifugal fan has been experimentally studied by Hou [5], who found the intermittent phenomenon of energy exchange when the fan works under weak rotating stall and the stall on the pressure side of the G4-73 fan blades; he further pointed out that stall at the pressure side of blades requires two conditions, i.e., the actual inlet flow of fan being greater than that at designed point and the inlet prewhirl being caused by deflector adjustment, separately. However, experimental researches still have limitations. With the development of computer technology and computational methods, it was found that numerical simulations can be used to analyze the flow inside flow passages in more details, and are expected to break through the constraints of experimental researches. Zhang et al. [6] conducted a numerical study on the mechanism and dynamics of rotating stalls in a centrifugal fan, revealing that the relative positions in circumferential directions and the propagation velocity of stall cells are important factors affecting the total pressure fluctuation and the characteristic frequency in the fan.

Researches on the inducement mechanism, patterns, and impact factors of stall inceptions are the premise of active control of stall. Researches on stall inceptions are a hot spot at present. Many scholars have conducted a lot of experiments and numerical simulations to attain a deeper understanding of the patterns, mechanism, and impact factors of stall inceptions in turbo machinery [7–9]. Based on the M-G model, Greitzer and Moore predicted that stall inceptions should appear before rotating stall occurs, considering the stability of the compression system [10]. There are two stall inception patterns in turbo machinery that are generally accepted currently, namely the spike type stall inception and model type stall inception. Day [11] first discovered the spike wave experimentally which is characterized by short term in time and small scale in space, and which is frequently originated in the tip area and has a high rotation speed. In the meantime, its propagation speed and range rapidly increased. Through the measurement of pulsation of the axial velocity in a single-stage axial compressor, Mcdougall [12] first discovered the model type stall inception characterized by large-scale disturbance, smaller rotation speed, and lack of specific time and origin of space. In recent years, relevant researches at home and abroad have shown that the adjustment of structural and operating parameters can change complex flow fields in flow passages, affecting the forms of stall inception [13,14]. Bianchi S et al. [15] conducted an experimental study on stall inception patterns under different installation angles of rotor blades in a low-speed axial fan. It was found that owing to the interactions of inflow, the blade tip leakage flow and the back flow from end walls, the stall inception patterns belong to the spike-type stall at the designed angle and the modal wave type inception at an angle less than the designed one. Gaetani et al. [16] found experimentally that the intensity of separation vortexes in turbo machinery is related to the axial spacing, which affects the wake disturbance in stages and the mixing intensity of mainstream. It was found that leakage vortexes in the tip can produce significant unsteady fluctuations under certain flow conditions and affect the total performance of compressors by looking through the relevant literature [17]. Tong et al. [18] studied the relationship between the unsteadiness of leakage vortexes and rotating stall and discovered that the places that experience the unsteady activity of leakage vortexes are the source of the spike-type stall inception as it has been experimentally proved. Vo et al. [19] combined experimental data and multichannel numerical simulations to deeply investigate the influence of the tip clearance flow on the spike-type stall inception and the inducing conditions. The tip leakage flow directly affected by the size of the tip clearance is thought to be the unsteady flow in the top region caused by the pressure difference between the two sides of the blade, and the main factor that induces the spike-type stall inception. A large number of experimental and numerical studies have confirmed the close relationship between the tip clearance flows and stall inception from various perspectives [20–22]. Choi [23] conducted a numerical study on stall in the same fan at different speeds and found that the development of stall is affected by speed changing and stall inception patterns are closely related to the speed. Domestic researches on stall inception patterns and the influence factors mainly focus on the tip clearance, the axial spacing, the rotating speed [24] and so on now. It has been showed in the literature that an abnormal regulation of the stagger angle may have an important impact on the fan performance and the stability control. However, the influence on the rotating stall inception patterns and the evolution process of stall has often been taken not seriously.

During the actual operation of two-stage variable pitch axial fan, the nonsynchronous adjustment of one or several rotating blades under variable working conditions occurs sometimes, due to the block caused by the ash accumulation of the petiole, the installation errors after the blade inspection and other reasons. Researches on the abnormal deviation from the design blade angle mainly focus on the aerodynamic and noise performance at present [25]. Li et al. [26] studied the aerodynamic performance of axial fan when the abnormal deviation of the blade angle occurred. By analyzing the simulation results, they found that the performance of fan is significantly deteriorated and the noise is increased once the deviation of the stagger angle occurs, and the main source of noise is the tip and leading edge of the suction surface of blades. Through researches based on numerical simulations on characteristics of the internal flow and performance curves of a single blade in an OB-84 axial fan under normal and abnormal deviations, Ye et al. [27] found that with the increase of installation angle, the loss of the trailing edge on abnormal blades increases and the gradient on the pressure side decreases from the leading edge to the trailing edge in the presence of abnormal deviation of the blade angle; the abnormal installation angle has a significant influence on the radial total pressure distribution in intermediate blade regions, and the fan efficiency is significantly decreased. However, the impact of the abnormal deviation on stall inceptions has not been reported. The local flow field and interaction between two stages will be changed with a stagger angle anomaly of one or several rotating blades in the two-stage variable pitch axial fan, and the collapse of the local flow field may induce the fan to experience stall in advance.

In the research of this paper, the influence of blade with an abnormal stagger angle on stall inception patterns, starting positions and induced mechanism are studied through simulations in the presence of an abnormal deviation of a second-stage blade angle in a two-stage axial fan.

This paper is organized as follows. The numerical calculation method, including the physical model, computational method, mesh generation, and the grid independence verification, is described in Section 2. The static pressure distribution at the exit, the relative speed at monitoring points and the dynamic characteristics in flow passages under different deviations of the stagger angle of blade are discussed in Section 3. Finally, the conclusions of this work are summarized in Section 4.

2. Computational Method

2.1. Geometric Model

The axial fan model consists of four components: a bell mouth, two stage rotors, two stage guide vanes, and a diffuser, as drawn in Figure 1, which shows the geometric model of a two-stage variable pitch axial fan. The main structural parameters of the fan are listed in Table 1.

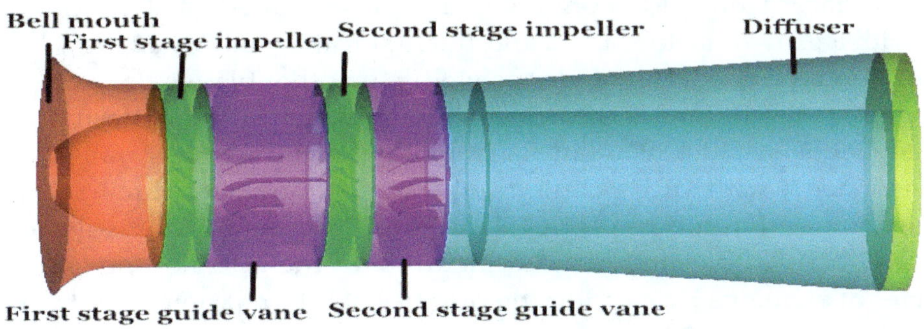

Figure 1. Geometry model of the axial fan.

Table 1. Main structural parameters of the fan.

Structural Parameters	Value
Rotation speed (r/min)	1490
Hub ratio	0.668
Number of rotor blades	2 × 24
Number of guide blades	2 × 23
Inlet diameter (m)	2.312
Outlet diameter (m)	2.305

The abnormal stagger angle of rotating blades in the second stage is shown in the Figure 2 and the value, recorded as $\Delta\beta_y < 0°$, is negative when the blade is deflected in the clockwise direction. On the contrary, the value is positive when the direction is anticlockwise.

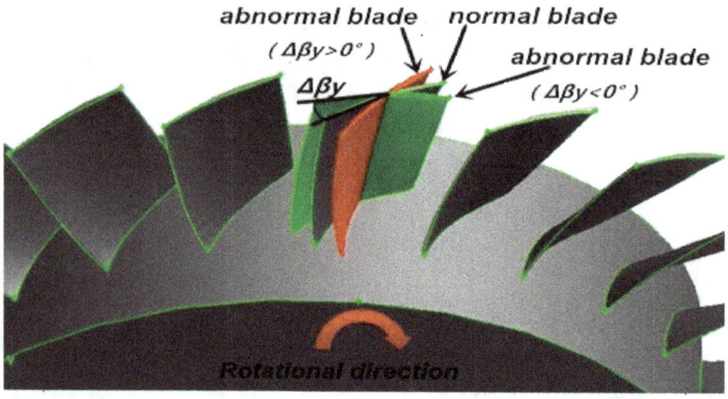

Figure 2. Diagram of the abnormal blade angle.

2.2. Mesh generation

The meshes were generated in the GAMBIT software (GAMBIT 2.4.6, ANSYS, Canonsburg, PA, USA). To satisfy the flow requirements in the impeller, the grids were generated using a T-Grid type and Tet/Hybrid elements. The leading and trailing edges and the tip clearance of blades are covered with grids intensively using a size stretching function. The specific parameters of the size function are as follows, the type is meshed, the source and attachment entities are the surfaces of all blades and the impeller volume, the growth rate is 1.2, and the maximum size is 20 mm. The types of elements employed in the mesh and their characteristic size in the different regions of the geometry are shown in Table 2. T-Grid type indicates that the fluid domain is divided into tetrahedron mesh, containing hexahedral mesh, pyramidal, and wedge mesh in the appropriate position. Cooper type means dividing the fluid domain according to the established source surface. A data exchange at the interfaces of different domains was accomplished using the Interface scheme. Figure 3 shows computational grid around the main blade edge and the local grid in rotor blade surface. The steady calculations use multiple reference frames (MRF) technique. The unsteady simulations use the moving mesh model. Blade surfaces were added on the boundary layer meshes. The boundary layer meshes are set to 15 rows with a growth factor of 1.2, and the first row is set to 0.02 mm. In this paper, the Y+ values are equal to 1.1 on the abnormal blades as well as on other blades.

In order to eliminate the influence of the number of elements of the mesh on results, this paper selected the grid number of elements of about 3.75 million, 5.87 million, 6.71 million, and 7.28 million to verify the grid independence with MRF technique and moving mesh technique, respectively. As shown in Figure 4a,b, when the grid numbers reach 6.71 million, the numerical results agree well with the experimental results obtained by the fan manufacturer. In the grid independence calculation, we obtained the number of elements by varying the number of internal points on the edge of the blades and the maximum size from 10 to 30 mm.

Table 2. The details of the mesh in the different regions.

Regions	Element	Type	Spacing
First stage guide vane	Tet/Hybrid	T-Grid	25
Second stage guide vane	Tet/Hybrid	T-Grid	25
Diffuser	Hex/Wedge	Cooper	40
Bell mouth	Tet/Hybrid	T-Grid	30
First stage rotor	Tet/Hybrid	T-Grid	20
Second stage rotor	Tet/Hybrid	T-Grid	20

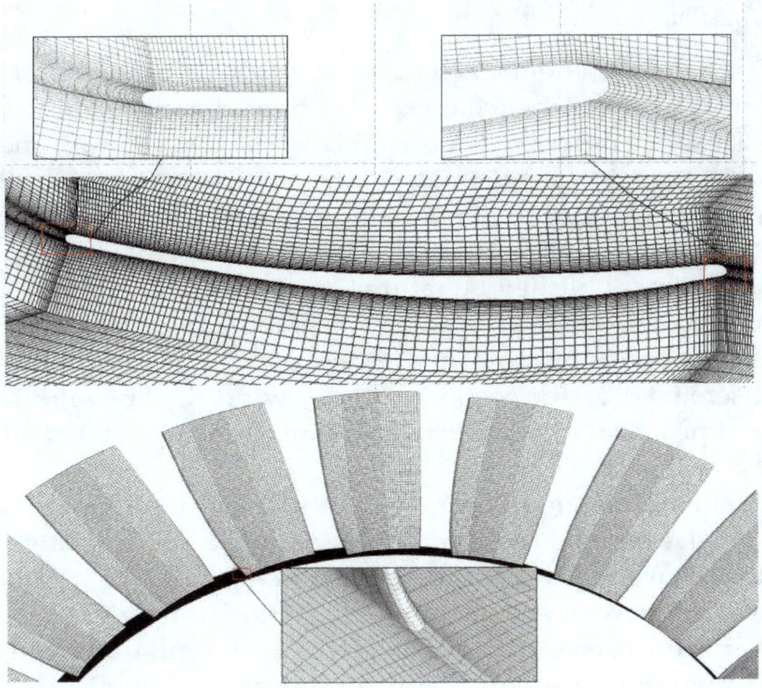

Figure 3. Computational grid around main blade edge and local grid in rotor blade surface.

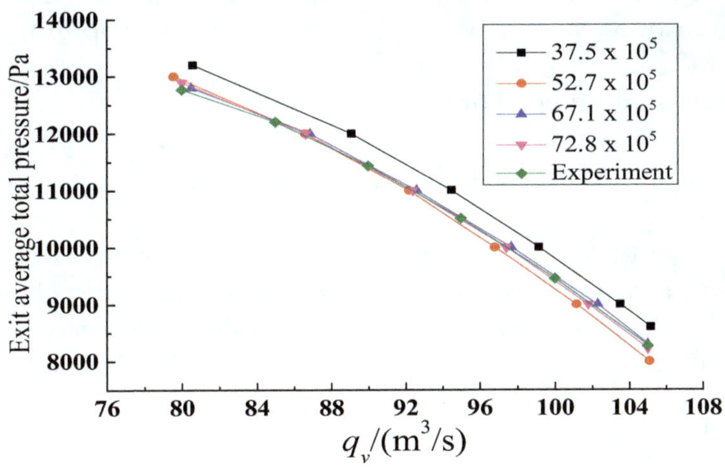

(**a**)

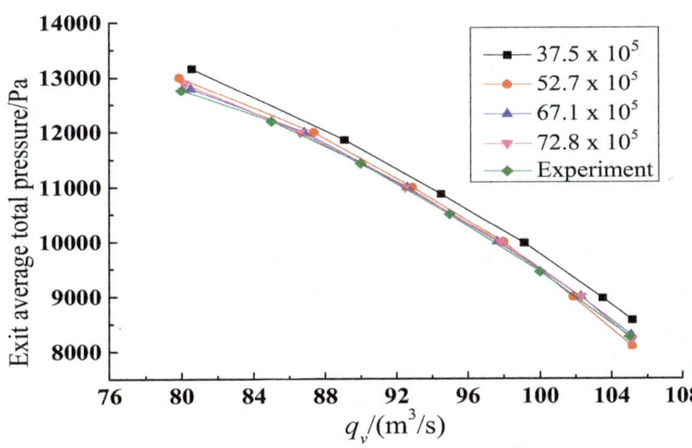

(**b**)

Figure 4. Grid independence calculation. (**a**) Grid independence calculation with multiple reference frames (MRF) technique; (**b**) Grid independence calculation with moving mesh technique.

2.3. Governing Equation and Boundary Conditions

A commercial CFD scheme of Fluent 15.0 is used to simulate the internal flow field of the axial fan. Rotating stall is an unsteady flow that changes with time, the unsteady Reynolds time-averaged Navier–Stokes equation is used to describe its flowing law. The temperature and density are constant, and the values are 288.16 K and 1.205 kg/m^3. Due to the fact that the internal flow of axial flow fans is extremely complex and flow separation phenomenon are found, the SST k-ω turbulence model is selected to predict the flow separation accurately, such a choice is useful in the application to instability rotating stall and were verified in published literature [28]. The wall function is used near the wall, the finite volume method is used to discretize the governing equations, the SIMPLE algorithm is used to address the pressure-velocity coupling. In order to improve the accuracy, the convection term and the diffusion term are discretized by the second-order upwind scheme and the center scheme, and the unsteady calculation time adopts an implicit dual time step approach, and the time precision is second order.

In the unsteady calculations, the time step is 0.000839 s in searching for the flow rate at the onset of rotating stall. The time step has been adjusted to 0.0000839 s after the flow rate at the onset of rotating stall is found. Meanwhile we have run simulation with time step 0.000112 s, that is the delta angle = 1°, and the results of the static pressure of exit are only 0.1% different than with the delta angle = 7.5°.

The collector inlet and diffuser outlet are selected as the inlet and outlet of the calculation regions, separately, and the boundary conditions are set as inlet total pressure and outlet static pressure, separately. The static pressure is given in steady calculations, the self-defined throttle valve model [29] is loaded at the outlet in unsteady calculations to carry out iterative calculations for each step. The accuracy of the throttle valve model has been verified by applying it to a centrifugal fan in literature [6].

The throttle value function can be expressed as follows

$$Ps_{out}(t) = Pi_{in} + \frac{1}{2}\frac{k_0}{k_1}\rho U^2, \tag{1}$$

where Ps_{out} and Pi_{in} represent the outlet back pressure and environmental pressure, k_0 and k_1 are constant and throttle value, respectively, ρ represents the air density, and U is the axial velocity at the outlet. The throttle valve opening is set to 1 at the initial, the flow rate decreases and approaches to the occurrence of rotating stall gradually by decreasing the value of k_0 until to zero.

3. Results and Discussions

3.1. Analysis of Static Pressure Characteristics of Fan Outlet

Curves of the outlet static pressure with time can help shed light on the time when stall inception occurs, which can induce steep drop of the pressure. Figure 5 shows the variation of the static pressure at the outlet of the fan with time under conditions of design stagger angle and one of blades in the second-stage impeller with abnormal angles −6° and +6° respectively. k_s and k_s + 0.001 represent the throttle value (k_1) when the fan operate under the conditions of near and induced stall respectively.

The static pressure at the outlet of fan is basically stable at 13230 Pa at the throttle value of k_s + 0.001 in Figure 5a. When the throttle value is reduced to k_s, the outlet static pressure begins to decrease approximately at T = 12 (in this paper, the rotation period is represented by T), and is followed by periodic and stable fluctuations with small ranges again at approximately T = 18 after a steep drop. That is to say, stall inceptions occur at T = 12 when the flow disturbance increases, with a throttle value at k_s in the present of the design installation angle in the axial fan. After approximately six rotor cycles, stall inceptions evolve into complete stall cells, and the static pressure at the outlet of the fan shows periodic fluctuations.

Figure 5b shows the outlet static pressure under the condition of an abnormal deviation equal to 6°. The static pressure at outlet is stable at about 13049 Pa at the throttle value of k_s + 0.001. The fluid disturbance in the fan occurs at about $T = 8$ at the throttle value of k_s, and then the generation and deterioration of stall inceptions cause a trend of steep drop in outlet static pressure. After approximately nine rotor cycles, it exhibits stable periodic fluctuations at approximately $T = 17$, resulting in stable stall cells. The outlet static pressure under the condition of an abnormal deviation angle equal to −6° is shown in Figure 5c. The static pressure is stable after a short period of time and is maintained at approximately 13047Pa afterwards at the throttle value of $k_s + 0.001$. When the throttle value is set to k_s, the outlet static pressure begins to decrease at appoximately $T = 2.5$. The deterioration of the flow field causes the occurrence of stall in the impeller. The steep drop trend ends at approximately $T = 16.5$, and the value of pressure is stable around 10930 Pa. What is more, the static pressure exhibits stable periodic fluctuations afterwards.

When the throttle value at $k_s + 0.001$, the steady value of outlet static pressure is lower than that when there is no abnormal deviation, by comparing diagrams of Figure 5 with the abnormal deviation angle being −6° or +6°. Moreover, it reaches a steady state at +6° in a short time. When the throttle value is set to k_s and stall is about to occur, the occurrence of stall inceptions is earlier in time than that with a normal stagger angle. Comparing Figure 5a,b, the static pressure drops rapidly and then presents a stable periodic fluctuation with large amplitude when the abnormal stagger angle is −6°, and the outlet static pressure exhibits a stable periodic fluctuation after a short-time steep drop, with amplitude greater than the one under the normal stagger angle when the abnormal deviation angle is −6°.

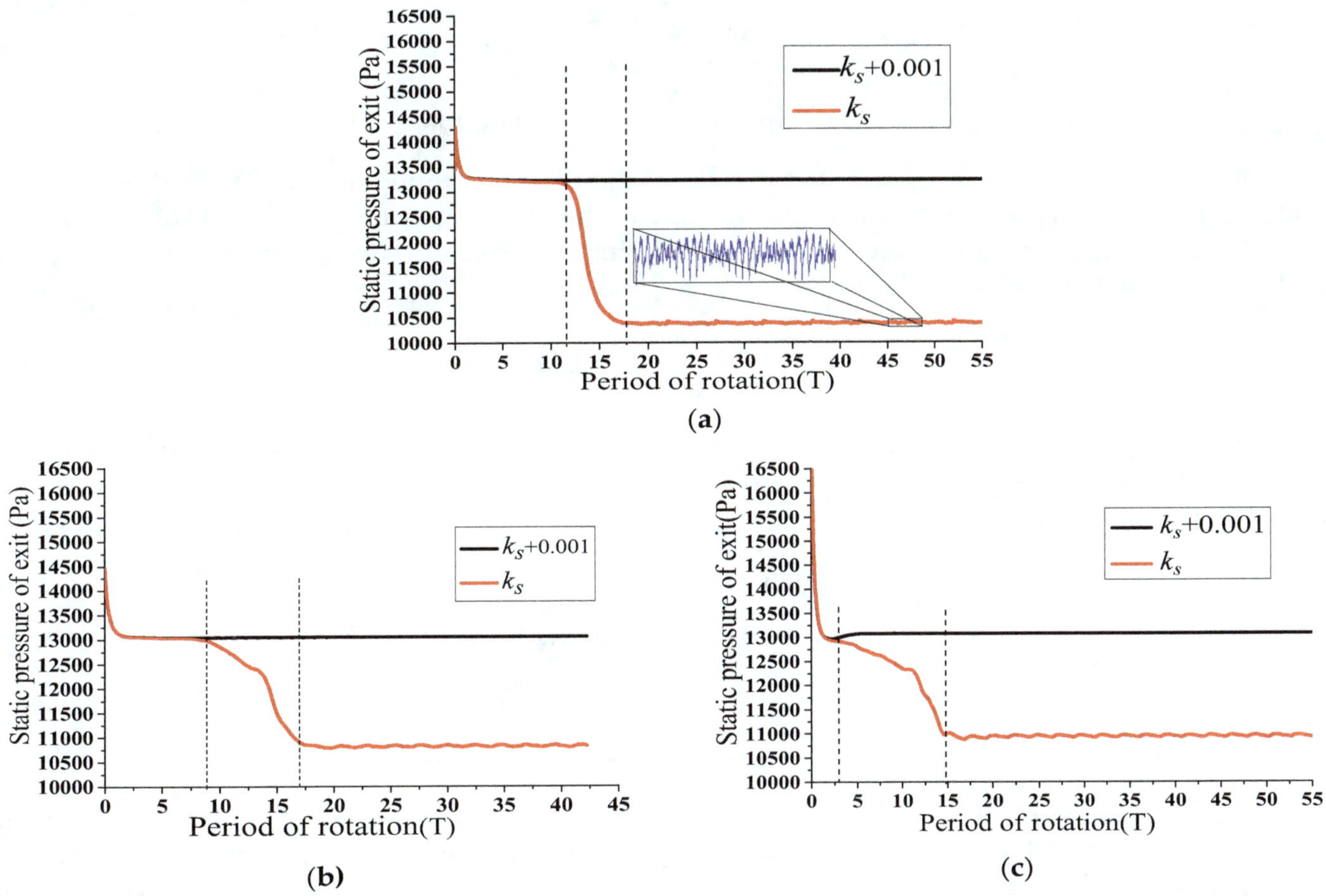

Figure 5. Variation of the outlet static pressure with time. (**a**) Normal blades; (**b**) an abnormal stagger angle of a blade equal to −6°; and (**c**) an abnormal stagger angle of a blade equal to +6°.

3.2. Analysis of Flow Rate at the Beginning of Stall and Stall Margin

The flow rate corresponding to the working conditions when stall inception occurs during simulations is called the flow rate at the beginning of stall, and is an important parameter to characterize the stable operating range of fan and calculate the stall margin. Figure 6 shows the flow rates at the beginning of stall under different stagger angles in the fan. The flow rates at the beginning of stall are 65.44m^3/s and 67.07m^3/s, in response to the single rotor blade with abnormal stagger angle of $-6°$ and +6°, respectively, and the difference is little. The flow rate is only 59.5 m^3/s under normal design angle. Compared with those at the design angle, the flow rates at the beginning of stall increase when the abnormal blade deviation is $-6°$ or +6°. The data above shows that the stable operating range of fan is narrowed when the single blade with abnormal stagger angle in the second stage rotor is $-6°$ or +6°. The stable operating range is narrower under an abnormal stagger angle of +6°, with a great impact on stall of the fan, compared with that under the abnormal stagger angle of $-6°$.

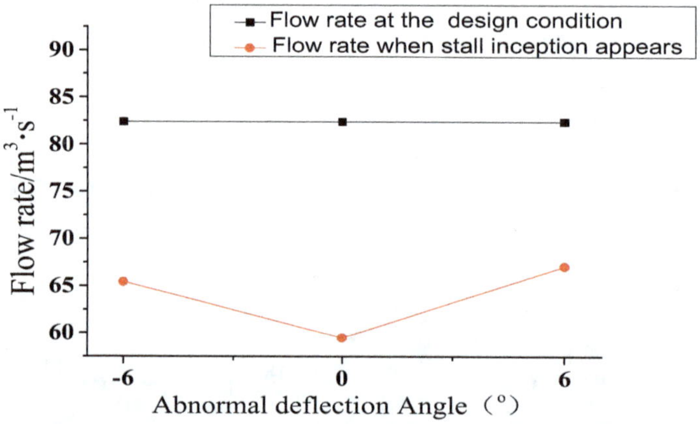

Figure 6. The flow rate when stall inception appears in.

The stall margin mainly represents the ratio of the flow rate range under the steady operation conditions to the design total flow, which can reflect the stable operating range of the fan clearly. The stall margin (Δm) under the abnormal deviation which can be calculated according to Equation (2), is calculated and analyzed in the following.

$$\Delta m = \frac{q_v{}^* - q_{vs}}{q_v{}^*} \times 100\%, \tag{2}$$

where $q_v{}^*$ and q_{vs} represent the flow rate under the design conditions, which is 82.4 m^3/s, and the flow rate of the starting stall, respectively. The calculation results are shown in Table 3 below, where 0° is the design angle of the blade.

Table 3. Stall margins.

Abnormal Angles	Stall Margins (%)
−6°	20.58
0°	27.79
+6°	18.60

It can be seen from Table 3 that the presence of the abnormal deviation of blades in the second stage, namely $-6°$ and +6°, leads to the drop of the stall margin in the fan. Whereas the angles above, namely $-6°$ and +6, have impacts on the stall margin in the fan, 7.21% and 9.19%, respectively, compared with that of the normal blades. And the difference of the value is only ~2%.

The curves above about the starting flow rate of stall and the outlet static pressure with time can be used to judge that stall occurs at a biggish flow rate, owing to the existence of the abnormal deviation of the blade in the second stage impeller, with values of -6° and +6, separately. They can also be used to analyze the moment when stall inception is generated, developed and evolve into complete stall cells, but cannot account for the reason why the changes occur completely. Therefore, the induced position and pattern of stall inception will be discussed below based on the relative velocities monitored at the monitoring points in the first and second rotors.

3.3. Analysis of Induced Position and Pattern of Stall Inception

The induced position of stall can be judged by the change of the relative speed with time in the first and second rotors, using the monitoring points. The numerical probes are placed at 50% and 90% of the radial height of the two-stage rotors and are circumferentially spaced with three impeller flow paths being between them. The numerical probes in the first rotor and the second rotor are labeled as m1-1, m1-2, m1-3, h1-1, h1-2, h1-3, and m2-1, m2-2, m2-3, h2-1, h2-2, h2-3, separately. The arrangement of the numerical probes in the first stage rotor is shown in Figure 7, and the probes in the second rotor are distributed in the corresponding passages according to placement in the first rotor.

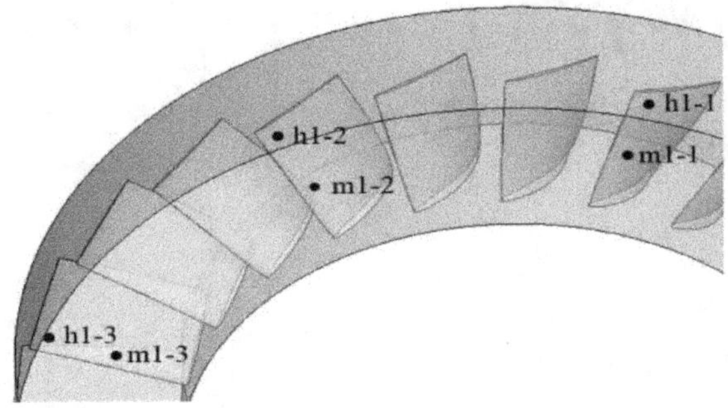

Figure 7. Distribution of the numerical probes in the first stage rotor.

Figure 8 shows curves of the relative velocities obtained by numerical probes in the first and second stage rotors when the throttle value is at k_s, wherein the curves of Figure 8a,b show the phenomenon under the design angle. The relative velocity values of h1-2 and h2-2 are shifted upwards by 30 units and the relative velocity values of h1-3 and h2-3 are shifted upwards by 60 units for comparison. In Figure 8a,b, it can be seen that, after ~13 rotor cycles, the relative velocity of each monitoring point begins to show fluctuations with large amplitude, stall inception induced in the impeller, and stall inception develop into stall cell after 3–4T with the relative velocity exhibiting periodic oscillation and the cycle is regarded as the rotating cycle of the stall cell. The relative velocity in the second stage rotor shows a significant jump at around 12T when stall inception is formed and the appearance time is about one rotation period earlier than that of the first stage rotor. Afterwards, the overall rule of the relative velocities in the process from the occurrence of the stall inception to the complete stall cell is approximately the same as that of the first stage rotor. It can be seen from the analysis above that stall inception first occurs in the second stage rotor, and then gradually affects the flow field in the first stage rotor, causing the flow field to deteriorate.

It can also be drawn from Figure 8a,b that the difference of the moments when stall inception occurs at the monitoring points h1-1 and h1-3 is ~0.334T at the range of 12 to 14T, and the value is 0.458T when stable stall cells are formed. The interval angle between the monitoring points h1-1 and h1-3 is 90° and the spread speed can be calculated as

$$\omega_s = \frac{P_m}{P_s}\omega_r = \frac{90}{0.334 \times 360}\omega_r = 0.750\omega_r, \tag{3}$$

$$\omega_T = \frac{P_m}{P_T}\omega_r = \frac{90}{0.458 \times 360}\omega_r = 0.549\omega_r, \tag{4}$$

The cycle of stall cell is $1.833T$ and the number of stall cell can be calculated as

$$N_c = \frac{P_T}{T_T \cdot P_m} = \frac{0.458}{1.833 \times \frac{90}{360}} \approx 1, \tag{5}$$

where ω_r stands for the speed of the rotor, P_m, P_s, P_T stand for the phase difference between two monitoring points, stall inceptions and stall cells, respectively. T_T is the period of the stall cell.

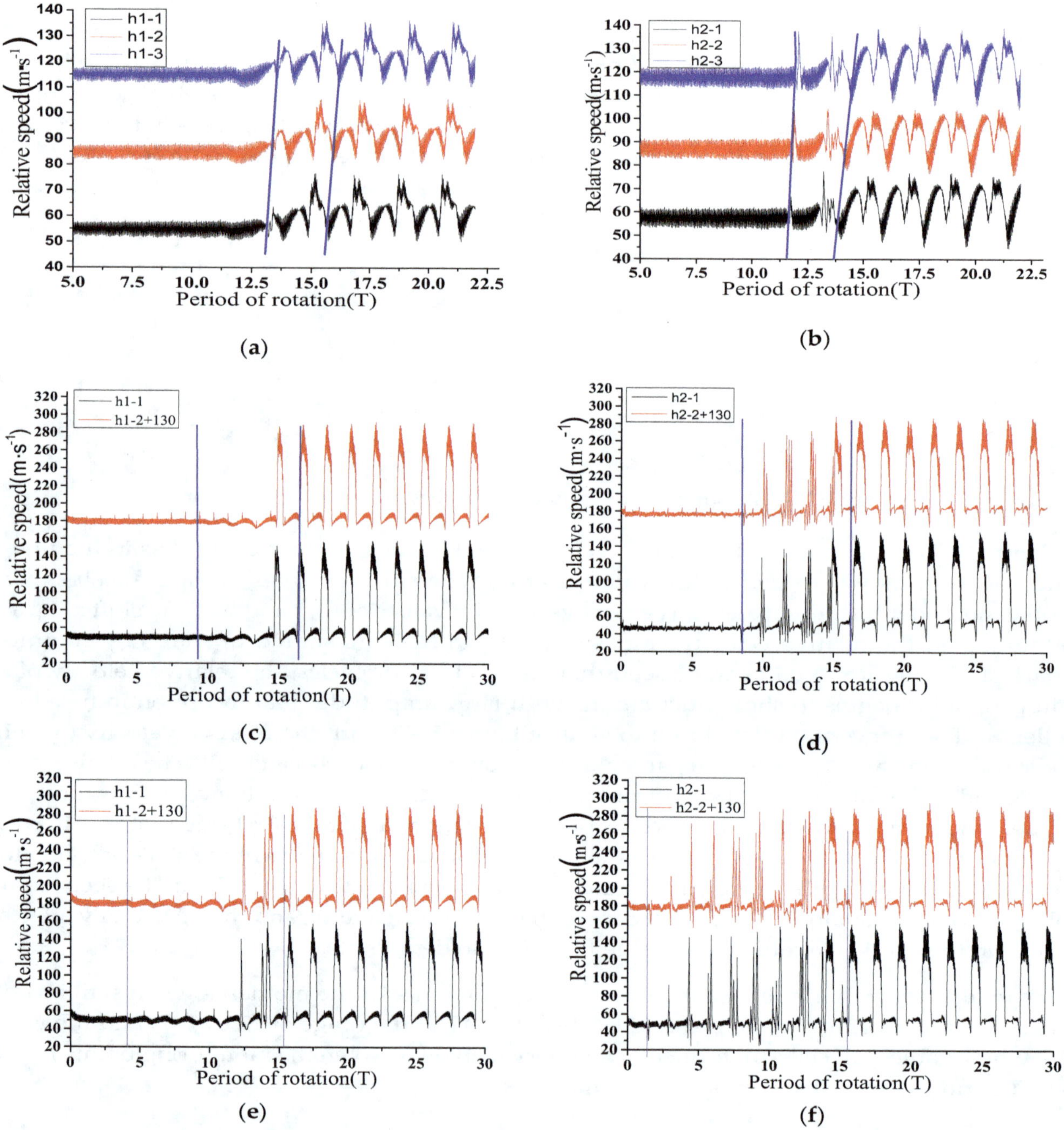

Figure 8. Curves of the relative speed at the monitoring points. (**a**) The first stage rotor under design angle. (**b**) The second stage rotor under design angle. (**c**) The first stage rotor when the abnormal deviation angle equal to $-6°$. (**d**) The second stage rotor when the abnormal deviation angle equal to $-6°$. (**e**) The first stage rotor when the abnormal deviation angle equal to $+6°$. (**f**) The second stage rotor when the abnormal deviation angle equal to $+6°$.

The values of spread speeds of stall inceptions and stable stall cells in the second stage rotor are calculated to be $0.75\omega_r$ and $0.549\omega_r$, respectively, and the number of stall cell is one, using the same method. It can be seen that the stall inception in the second-stage rotor occurs initially, the spread speed of stall inceptions in both rotors are the same, and the spread speeds of stall cells are also consistent after the development of stall cell is completed. Moreover, the stall inceptions appearing in the two stage impellers show identical characteristics: the unsteady fluctuations with small scale in impellers, the high speed of stall inceptions, the evolution into the stable stall cells within three rotor cycles, and the significant decrease of the spread speed of stall cells. Comparing the above characteristics with the researches on stall inceptions at home and abroad [30], the stall inception patterns in the two stage rotors belong all to the spike model under design blade angle condition.

The fluctuation ranges of the speed of monitoring points are large owing to the existence of blades with the abnormal stagger angle; therefore, only the relative speed of two adjacent monitoring points is analyzed. The relative velocity values of h1-2 and h2-2 are shifted by 130 units for comparison. The curves of the relative velocity of monitoring points are showed in the Figure 8c,d, with an abnormal deviation angle of $-6°$. It can be seen that stall inceptions occur in the second stage rotor at ~$8T$, and develop into stall cells after nine rotor cycles. Stall inceptions occur in the first stage rotor owing to the influence of the stall inception in the second sage rotor and turn into stable stall cells at $17T$ eventually. The curves of relative velocities of monitoring points under the abnormal stagger deviation angle of +6° are showed in the Figure 8e,f. It can be seen that stall inception first occur in the second rotor, at ~$2T$, and develop into complete stall cell after about 13 rotor cycles. The relative velocity of monitoring points in the first stage rotor fluctuates at $4.5T$ and appears periodic at ~$15T$. The analysis above shows that the occurrence of stall inception in the second stage rotor affects the flow field in the first stage rotor, causing flow instability at $4.5T$, which continues to deteriorate. Finally, stall inceptions in the first and second stage impellers evolve into complete stall cells at approximately the same time ($15T$).

The Equations (4)–(6) are used to calculate the circumferential propagation velocity of stall inception and stall cell under the abnormal stagger angle and the results are listed in the Table 4. R1 and R2 represent the first stage rotor and the second stage rotor separately.

Table 4. Circumferential propagation velocity of stall inceptions and stall cells.

Abnormal Stagger Angles	−6°		0°		+6°	
	R1	**R2**	**R1**	**R2**	**R1**	**R2**
Spread speed of stall inception (ω_r)	0.638	0.718	0.750	0.750	0.672	0.753
Spread speed of stall cell (ω_r)	0.61	0.584	0.549	0.549	0.584	0.658

The circumferential propagation velocity of the first and second stage impellers is ~70% of the rotor speed with a blade of an abnormal stagger angle of +6° or −6°, compared with that of the design blade angle. The propagation velocity decreases after developing into stable stall cell, and the propagation velocities in two rotors are different. There are two reasons for the asynchronization of speed: considering the cases of abnormal blades, the rotation of stall cell is not strictly periodic, and there may be some errors in data acquisition and calculation; otherwise, the existence of abnormal blades may also affect the propagation speed of stall inceptions and stall cells to a certain extent, leading to the difference of the propagation speed in different impellers.

3.4. Analysis of Dynamic in Rotating Impellers before and after Stall Induced

The evolution process of stall inception in the two rotors is basically the same under the design blade angle condition, so this paper analyzes the relative velocity and the turbulent kinetic energy contours of the middle span-wise cross section in the second stage rotor at typical moments, as shown in Figure 9.

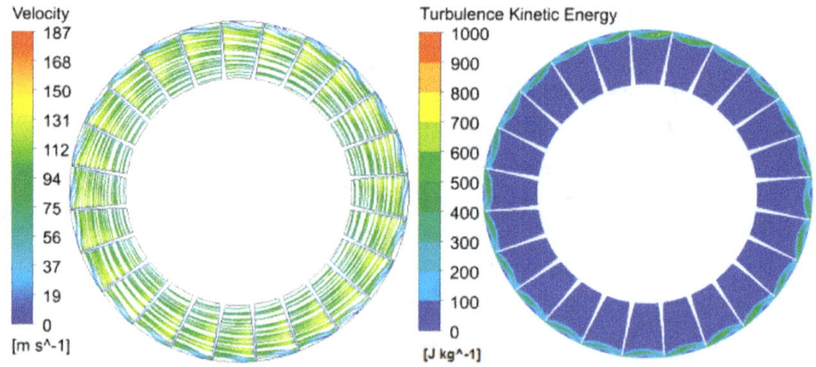

(**a**)

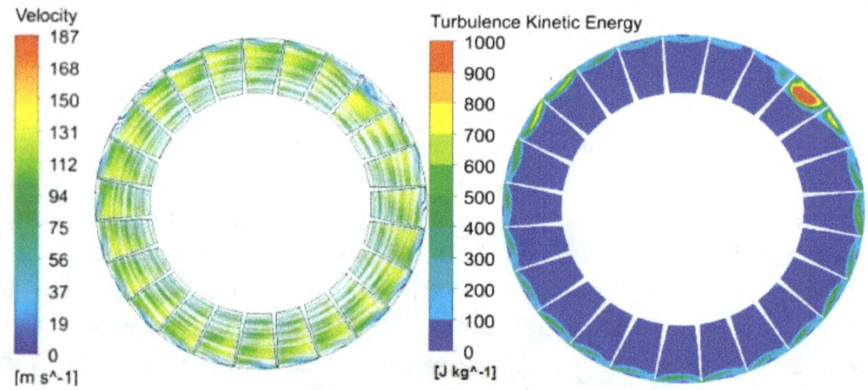

(**b**)

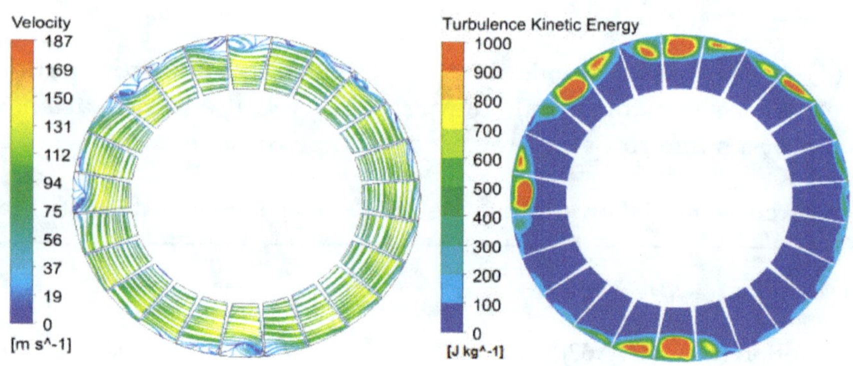

(**c**)

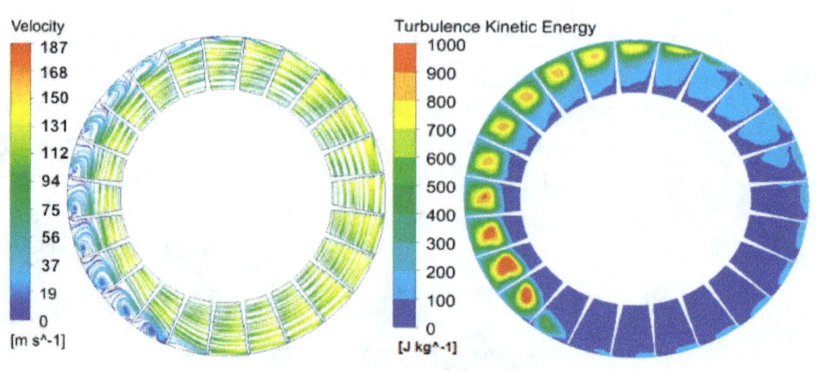

(**d**)

Figure 9. Streamline and the turbulent kinetic energy contour of the second rotor midsection. (**a**) Near stall inception at 11T, (**b**) stall inception stat at 12T, (**c**) development of stall inception at 15T, and (**d**) complete stall cell at 18T.

Figure 9a shows that the flow field in each passage is relatively uniform near the induced stall inception. However, there is a small radial velocity at the tip of the blade; the small fluctuation of the flow rate is caused by the enhancement of the tip leakage vortex. The kinetic energy is evenly distributed along the circumference, the turbulent kinetic energy is obviously increased near the tip of the blade, covering almost the entire tip clearance area. This is due to the positive angle that gradually increases with the decrease of flow rate, and the vortex area that increases, causing the improvement of turbulent pulsation. The low kinetic energy zone in the middle of the flow passages which is elliptical and has an increasing trend, develops toward the top of the blade.

As shown in Figure 9b, stall inception appears. The flow field disturbance is further enhanced, and large reflow and flow separation in the tip region are observed in three flow passages as the streamline diagram illustrates. The largest separation vortex occupies about two flow passages, the recirculation zone in the middle flow passages is large, and the recirculation zone in several flow passages adjacent along the counterclockwise direction is small, with an increasing trend; on the contrary, the area of the recirculation zone in the flow passages along the clockwise direction is also small with a tendency to decrease. The recirculation process in flow passages involves air flowing from the suction side of the vane to the pressure surface, and then flowing back to the suction side after the fluid is mixed.

The turbulent energy distribution corresponding to the flow field has a similar law. The region of turbulent kinetic energy occupies a large area near the tip of the blade to the 70% blade height, and turbulent energy increases. The area of turbulent kinetic energy in adjacent flow passages along the counterclockwise direction has an increasing tendency, with a decreasing tendency in the adjacent flow paths along the clockwise direction, owing to that the propagation speed of stall inception is smaller than the rotor speed. From the perspective of the relative coordinate system, stall inceptions rotate along the counterclockwise direction, resulting in the deterioration of the flow in the adjacent passages along the counterclockwise direction and the flow inside the adjacent passages along the clockwise direction. With the evolution of stall inception, the flow field disturbance is further enhanced. The separation vortex appears in the four regions of the flow diagram, the largest of which occupies approximately five flow passages, and the recirculation zone is larger in the middle flow passages. Corresponding to the turbulent kinetic energy contour, there is a similar distribution law, that is, the area of high kinetic energy in the middle flow passages is large, occupying the area from the tip to 60% of the blade height. There is an increasing trend in the adjacent flow passages along the clockwise direction, and a decreasing trend along the clockwise direction, as shown in Figure 9c.

The velocity and distribution of turbulent energy at a certain moment after stall inception evolve into stall cell are shown in Figure 9d, when stall cell rotate in the circumferential direction at a fixed speed. It can be seen that stall cell occupy about 11 passages along the circumferential direction, and the blocked area in five passages in the middle of stall cell is larger, occupying the area from the tip to 50% of the blade height. Compared with Figure 9a, the stall cell have a larger influence on the flow filed, and evolve from stall inceptions to stall cells in a short period of time, which is consistent with the spike-type stall inception. It can be seen from the Figure 8d that a zone of high kinetic energy appears in the region corresponding to the stall cell. Compared with Figure 9c, the maximum value of turbulent kinetic energy in passages is almost constant, whereas the area of the turbulent kinetic energy is significantly increased. As stall cell spreads along the circumferential direction, the region of turbulent kinetic energy in passages undergoes a dynamic process that first increases and then decreases in the radial direction.

The turbulent kinetic energy indicates the kinetic energy of the turbulent pulsation, it can reflect the magnitude of the normal value of the Reynolds stress and the intensity of the turbulent pulsation. Analyzing the contours of turbulent energy distribution under the design blade angle condition, the variation of regions where the kinetic energy is high in impellers can be used to clearly determine the generation and development process of stall inceptions. Therefore, taking on the change of kinetic energy in the first and second rotors as an example when the abnormal blade deviation angle is $-6°$, the evolution process of stall inception is analyzed.

Figure 10 shows the distribution of the turbulent energy in the middle section of the second stage rotor at four typical moments when the abnormal deviation angle is $-6°$. It can be seen that the kinetic energy in three passages corresponding to regions of stall inceptions is higher. The high kinetic energy fully occupies the tip area of the flow paths in the counter-clockwise direction of the abnormal deflection blade; the area of high kinetic energy in clockwise passages is particularly small. As the flow field evolves, as shown in Figure 10b, the region of high kinetic energy occupies about five impeller flow passages, with regions of the highest turbulent energy occupying at approximately from 50% to 90% of blade height. The development Figure 10b,c also shows that flow fields are affected by stall cell and zones of the high kinetic energy develop in radial direction from the tip of the blade to 50% blade height and in the direction along the circumference, more passages being influenced. The stall inception is further developed, eventually forming a stable stall cell occupying by about nine passages, as shown in Figure 10d.

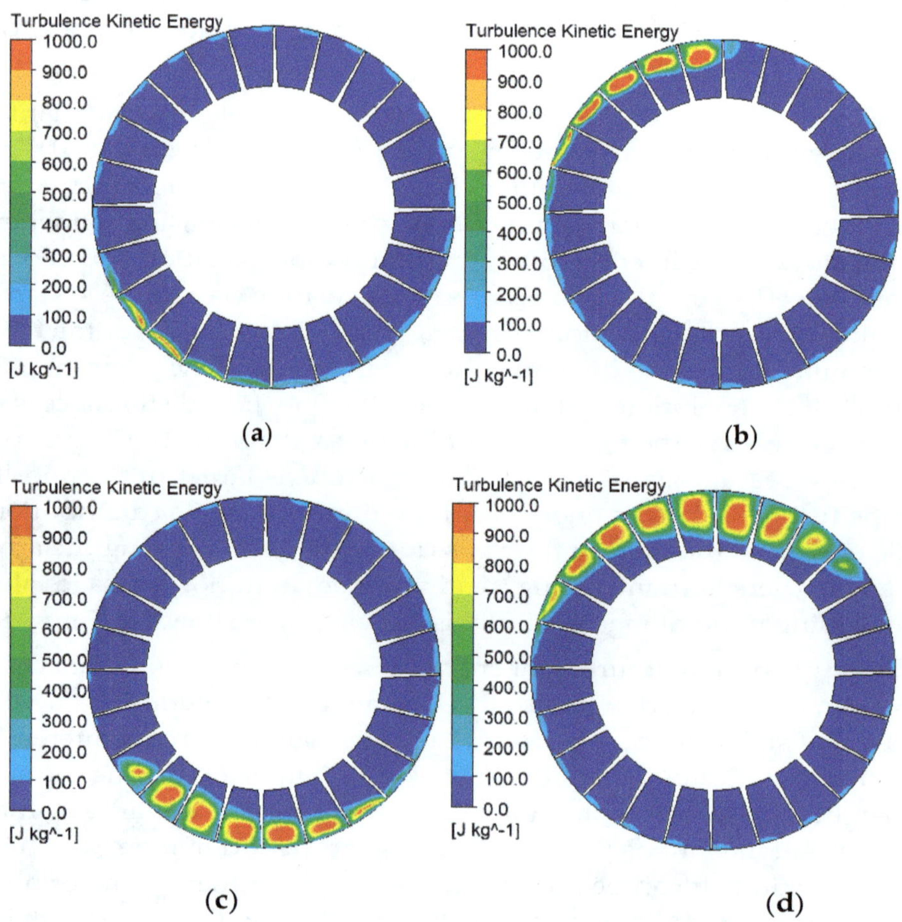

Figure 10. Turbulent kinetic energy contour of the first rotor midsection when the abnormal angle equal to $-6°$. (**a**) Near stall inception at $11T$, (**b**) stall inception start at $13T$, (**c**) development of stall inception at $15T$, and (**d**) complete stall cell at $17T$.

The distribution of the turbulent energy in the middle section of the second stage rotor at four typical moments is shown in Figure 10. Figure 11a shows the stage of stall inception generation, where the zone of high kinetic energy occupies about two impeller passages. Subsequently, four immature small stall cells are generated almost simultaneously in passages, and then develop stably and occupy two passages respectively, as shown in Figure 11b. Zones of sorghum kinetic energy develop from the tip to the 70% blade height, and three small stall cells are combined into a large one owing to the impact of the generation of stall in the first-stage rotating impeller, as shown in Figure 11c. It can be clearly seen from Figure 11d that the flow filed is further deteriorated, four small stall cells developing into a stable one and occupying about 11 passages. Regions of higher turbulent energy occupy the area from the 40% to the 90% of blade height, making passages easy to be blocked.

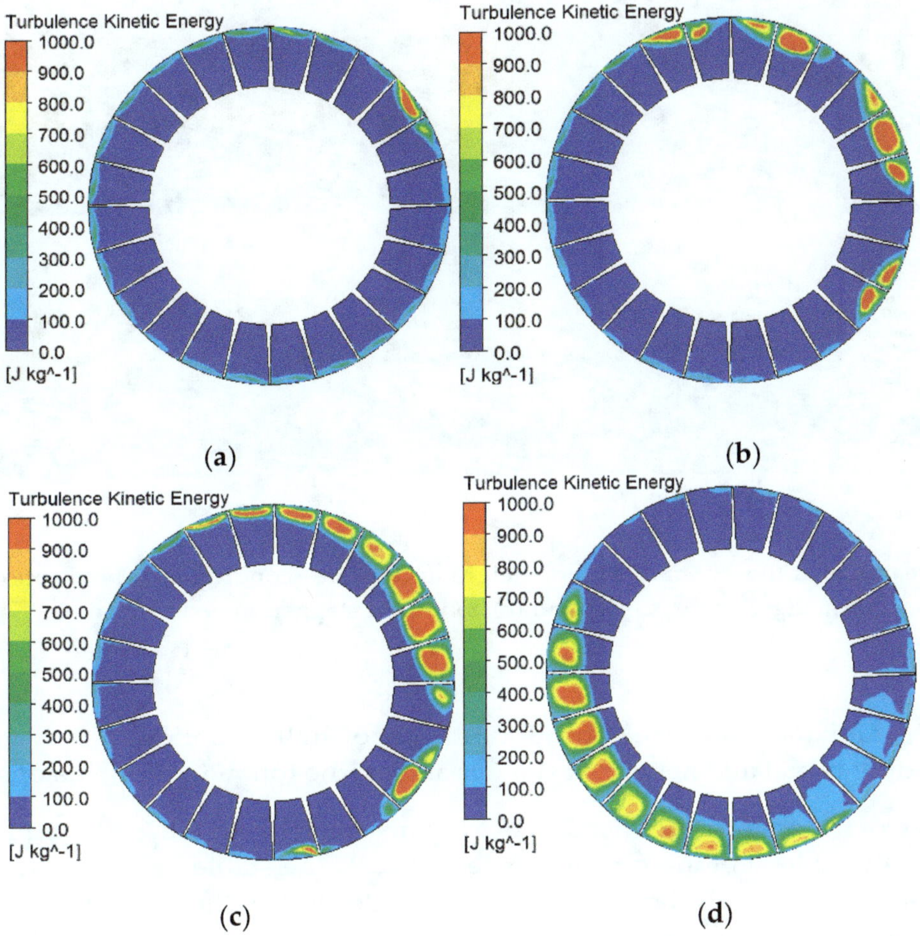

Figure 11. Turbulent kinetic energy contours of the second rotor midsection when the abnormal angle is $-6°$. (**a**) Near stall inception at $8T$. (**b**) Stall inception start at $11T$. (**c**) Development of stall inception at $15T$. (**d**) Complete stall cell at $17T$.

3.5. Analysis of the Streamline at 95% Radial Blade Height

The streamline diagram can clearly show the flowing law of the tip leakage flow to reveal the induced mechanism of rotating stall of the impeller. Figure 12 is a chart showing streamline diagrams at the 95% radial blade height in the second-stage rotor under three conditions. It can be seen that the streamline diagrams near the tip are not completely the same in the three cases, but the flowing laws are basically the same: a leakage vortex is generated near the leading edge of blade, and appears near suction surface, opposite to the main flow direction. A low velocity region is generated in the passage and changes the fluid direction to flow from the suction side of the blade to the pressure surface after mixing with the mainstream fluid. The reflow is therefore generated in the low velocity zone, causing deterioration of the flow field and the occurrence of rotating stall. In the case of abnormal and normal deviation angle of the blade, the same flowing law near the tip of the blade when the stall inception appears indicates that the inducement mechanism of stall is the same under the three conditions.

(**a**)

(**b**)

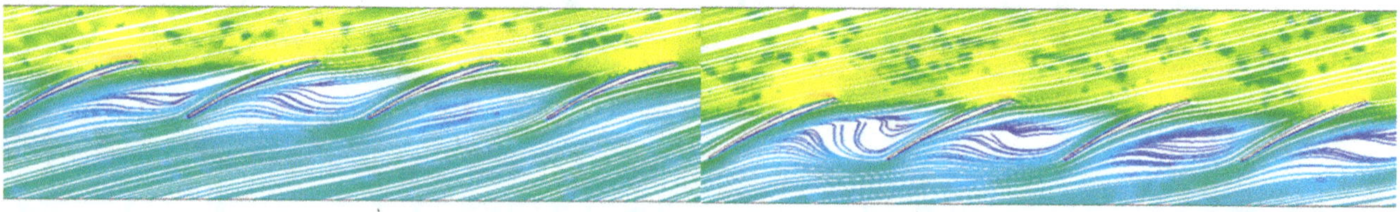

(**c**)

Figure 12. Streamline at the 95% radial blade height under three conditions: (**a**) No abnormal stagger angle; (**b**) abnormal stagger angle is −6°; and (**c**) abnormal stagger angle is +6°.

4. Conclusions

By simulating and analyzing the evolution process of stall inception in the two-stage variable pitch axial fan under normal and abnormal stagger angles, the following conclusions can be drawn.

(1) The existence of the abnormal deflection blade causes the fan to fall into rotating stall at a large flow rate, and the fan stall margin is reduced, that is, the stable operating range of the fan is reduced. A single blade with abnormal deviation angle induces the fan to get into stall in advance.

(2) Compared with the designed stagger angle condition, stall inception still first appears around the leading edge of the second-stage rotor blades, and the stall inception spreads at a high speed, about 70% of the rotor speed. After developing into a complete stall cell, the spread speed is significantly reduced, and shows characteristics of the spike-type stall inception. An abnormal deviation blade in the second stage rotor has little effect on induced position and type of stall inception.

(3) By observing and analyzing the flow diagram and the turbulent kinetic energy contours inside impellers, it was found that the number of stall cell and the evolutionary time change during the evolution process of stall inception. The existence of an abnormal deviation blade has great influence on the evolution process from stall inception to stall cell.

(4) By analyzing the streamlines at 95% radial blade height, it can be seen that the flowing laws in passages are basically the same, and inducement mechanisms of rotating stall are consistent under three conditions. An abnormal deviation blade in the second stage rotor does not change the inducement mechanism of rotating stall.

Author Contributions: L.Z. (Lei Zhang) and L.Z. (Liang Zhang) proposed the ideas of the research and analyzed and explained the data obtained by numerical simulation. Q.Z. carried out part of numerical simulation work, determined the writing method, and revised the paper. K.J. wrote the manuscript and made charts based on the numerical simulation data. Y.T. collected the relevant references and the data of the numerical simulation, and revised the paper. S.W. participated in part of the numerical simulation work in this study.

References

1. Zhang, L.; Yan, C.; He, R.Y.; Zhang, Q. Numerical Study on the Acoustic Characteristics of a Axial Fan under Rotating stall Condition. *Energies* **2017**, *10*, 1945. [CrossRef]
2. Niazi, S.; Stein, A.; Sankar, L. Numerical studies of stall and surge alleviation in a high-speed transonic fan rotor. *Mech. Eng.* **2000**, *225*, 38–39.
3. Emmons, H.W.; Pearson, C.E.; Grant, H.P. Compressor surge and stall propagation. *Trans. ASME* **1955**, *77*, 455–469.
4. Tang, Y.; Liu, Y.; Lu, L. Solidity Effect on Corner Separation and Its Control in a High-Speed Low Aspect Ratio Compressor Cascade. *Int. J. Mech. Sci.* **2018**, *142–143*, 304–321. [CrossRef]
5. Hou, J.H. *Research on Condition Monitoring and Fault Diagnosis of Fan Based on Multi-Parameter*; North China Electric Power University: Baoding, China, 2003.
6. Zhang, L.; Wang, S.L.; Zhang, Q.; Wu, Z.R. Dynamic Characteristics of Rotating Stall for Centrifugal Fans. *Proce. CSEE* **2012**, *32*, 95–102.
7. Kazutoyo, Y.; Hiroaki, K.; Ken-ichiro, I.; Furukawa, M.; Gunjishima, S. An Explanation for Flow Features of Spike-Type Stall Inception in an Axial Compressor Rotor. *J. Turbomach.* **2013**, *135*, 1–8.
8. Salunkhe, P.B.; Pradeep, A.M. Stall Inception Mechanism in an Axial Flow Fan under Clean and Distorted Inflows. *J. Fluids Eng.* **2010**, *132*, 1–11. [CrossRef]
9. Sheard, A.G.; Corsini, A.; Bianchi, S. Stall Warning in a Low-Speed Axial Fan by Visualization of Sound Signals. *J. Eng. Gas Turbines Power* **2011**, *133*, 1–9. [CrossRef]
10. Moore, F.K.; Greitzer, E.M. A theory of post-stall transients in axial compression systems. I: Development of equations. *J. Eng. Gas Turbines Power* **1986**, *108*, 68–76. [CrossRef]
11. Day, I.J. Stall inception in axial flow compressors. *J. Turbomach.* **1993**, *115*, 1–9. [CrossRef]
12. McDougall, N.M.; Cumpsty, N.A.; Hynes, T.P. Stall inception in axial compressors. *J. Turbomach.* **1990**, *112*, 116–125. [CrossRef]
13. Anish, S.; Sitaram, N.; Kim, H.D. A Numerical Study of the Unsteady Interaction Effects on Diffuser Performance in a Centrifugal Compressor. *J. Turbomach.* **2014**, *136*, 1–9. [CrossRef]
14. Biela, C.; Brandstetter, C.; Schiffer, H.P.; Heinichen, F. Unsteady wall pressre measurement in a one-and-a-half stage axial transonic compressor during stall inception. *Proc. Inst. Mech. Eng. Part A J. Power Energy* **2013**, *227*, 643–653. [CrossRef]
15. Bianchi, S.; Corsini, A.; Mazzucca, L.; et al. Stall Inception, Evolution and Control in a Low Speed Axial Fan with Variable Pitch in Motion. In Proceedings of the ASME 2011 Turbo Expo: Turbine Technical Conference and Exposition, Vancouver, BC, Canada, 6–10 June 2011; pp. 445–456.
16. Gaetani, P.; Persico, G.; Osnaghi, C. Effects of Axial Gap on the Vane-Rotor Interaction in a Low Aspect Ratio Turbine Stage. *J. Propuls. Power* **2015**, *26*, 325–344. [CrossRef]
17. Lakshminarayana, B.; Horlock, J.H. Review: Secondary flows and losses in cascades and axial-flow turbomachines. *Int. J. Mech. Sci.* **1963**, *5*, 287–307. [CrossRef]
18. Tong, Z.T. *Experimental Study on the Unsteady Correlation of Tip Leakage Vortex, Stall Inception and Micro Tip Injection*; Graduate School of the Chinese Academy of Sciences: Shenzhen, China, 2006.
19. Vo, D.H.; Tan, C.S.; et al. Criteria for Spike Initiated Rotating Stall. *J. Turbomach.* **2008**, *130*, 11–23. [CrossRef]
20. Torresi, M.; Camporeale, S.M.; Pascazio, G. Detailed CFD Analysis of the Steady Flow in a wells Turbine Under Incipient and Deep Stall Conditions. *J. Fluids Eng.* **2009**, *131*, 1–9. [CrossRef]
21. Tomita, I.; Ibaraki, S.; Furukawa, M. The Effect of Tip Leakage Vortex for Operating Range Enhancement of Centrifugal Compressor. *J. Turbomach.* **2013**, *135*, 1–8. [CrossRef]
22. Cameron, J.D.; Bennington, M.A.; Ross, M.H.; Morris, S.C. The Influence of Tip Clearance Momentum Flux on Stall Inception in a High-Speed Axial Compressor. *J. Turbomach.* **2013**, *135*, 1–11. [CrossRef]
23. Choi, M.; Vahdati, M.; Imregun, M. Effects of Fan Speed on Rotating Stall Inception and Recovery. *J. Turbomach.* **2011**, *133*, 1–8. [CrossRef]
24. Pavesi, G.; Cavazzini, G.; Ardizzon, G. Numerical Analysis of the Transient Behaviour of a Variable Speed Pump-Turbine during a Pumping Power Reduction Scenario. *Energies* **2016**, *9*, 534. [CrossRef]
25. Ye, X.M.; Li, C.X.; Yin, P. Effect of Abnormal Blade Reverse Deviation on Performance of the Axial Fan. *J. Chin. Soc. Power Eng.* **2013**, *9*, 702–710.
26. Li, C.X.; Lin, Q.; Ding, X.L.; Ye, X.M. Performance, aeroacoustics and feature extraction of an axial flow fan with abnormal blade angle. *Energy* **2016**, *103*, 322–339. [CrossRef]

27. Ye, X.M.; Ding, X.L.; Zhang, J.K.; Li, C. Numerical simulation of pressure pulsation and transient flow field in an axial flow fan. *Energy* **2017**, *129*, 185–200. [CrossRef]
28. Zhang, L.; Jiang, K.; Wang, S.L.; Zhang, Q. Effect of the First-stage of Rotor With Single Abnormal Blade Angle on Rotating Stall of a Two-stage Variable Pitch Axial fan. *Proc. CSEE* **2017**, *37*, 1721–1730.
29. Gourdain, N.; Burguburu, S.; Leboeuf, F.; Michon, G.J. Simulation of rotating stall in a whole stage of an axial compressor. *Comput. Fluids* **2010**, *39*, 1644–1655. [CrossRef]
30. Tan, C.S.; Day, I.; Morris, S.; Wadia, A. Spike-Type Compressor Stall Inception, Detection, and Control. *Annu. Rev. Fluid Mech.* **2010**, *42*, 275–300. [CrossRef]

11

Comparative Study of Shell and Helically-Coiled Tube Heat Exchangers with Various Dimple Arrangements in Condensers for Odor Control in a Pyrolysis System

Sun-Min Kim, Jun-Ho Jo, Ye-Eun Lee and Yeong-Seok Yoo *

Division of Environmental and Plant Engineering, Korea Institute of Civil Engineering and Building Technology, 283, Goyang-daero, Ilsanseo-gu, Goyang-si, Gyeonggi-do 10223, Korea; sunminkim@kict.re.kr (S.-M.K.); junkr@kict.re.kr (J.-H.J.); yeeunlee@kict.re.kr (Y.-E.L.)

* Correspondence: ysyoo@kict.re.kr.

Academic Editor: Bjørn H. Hjertager

Abstract: This study performed evaluations of the shell and helically-coiled tube heat exchangers with various dimple arrangements, that is, flat, inline, staggered, and bulged, at different Dean numbers (*De*) and inlet temperatures of a hot channel. Conjugated heat transfer was analyzed to evaluate the heat transfer performance of the exchangers through temperature difference between the inlet and outlet, Nusselt number inside the coiled tube, and pressure drop of the coiled tube by using 3-D Reynolds-averaged Navier–Stokes (RANS) equations with shear stress transport turbulence closure. A grid dependency test was performed to determine the optimal number of the grid system. The numerical results were validated using the experimental data, and showed good agreement. The inline and staggered arrangements show the highest temperature differences through all *De*. The staggered arrangement shows the best heat transfer performance, whereas the inline arrangement shows the second highest performance with all ranges of *De* and the hot channel's inlet temperature. The inline and staggered arrangements show the highest pressure drop among all inlet temperatures of the hot channel.

Keywords: shell and tube; heat exchanger; odor control; heat transfer; dimple; condenser; pyrolysis; RANS

1. Introduction

Pyrolysis is defined as one of the most effective and efficient processes of thermal decomposition to obtain energy in the form of char from resources in an anoxic condition. This process has the ability to convert sewage sludge, biomass, and other carbonaceous materials to reusable fuels [1]. Biomass pyrolysis results in three products: biochar, bio-oil, and gas [2]. Biochar is a solid carbon-rich byproduct obtained through the thermal stabilization of biomass or any other organic matter [3]. Bio-oil has a good feedstock for power generation as it contains high amounts of energy, which is, in some cases, comparable to that of the fossil fuels after upgrading [4]. Moreover, gas produced through pyrolysis has high energy, which is, in some cases, comparable to that of the coal used in industries as feedstocks for fuel [5,6]. Food wastes are known as good resources of pyrolysis among various raw materials.

Food wastes have a high energy content. Consequently, they offer a good potential as feedstocks for pyrolysis in power plants [7]. Therefore, many researchers and industry officials have focused on food-waste pyrolysis as a future source of energy supply. However, in the pyrolysis of food wastes, numerous types of odors are generated and the dust formed in the process, including the dry process, can be a factor of ignition [8,9]. One of the best and commonly used methods to solve this problem is condensation [10]. By condensing, the byproducts (such as the odors or dust) of the pyrolysis process are reduced and removed. The shell and tube heat exchanger (STHX) systems have played an important role to realize condensation [11], and it is important to increase the heat transfer performance for the best condensation conditions.

Therefore, many researchers focus on the STHX systems, especially on the heat transfer performance. Selbaş et al. [12] estimated the minimum heat transfer area required for a given heat duty, as it governs the overall cost of the heat exchanger, by using the genetic algorithms (GA) with varying design variables, such as the outer tube diameter, tube layout, number of tube passes, outer shell diameter, baffle spacing, and baffle cut. In addition, they concluded that the combinatorial algorithms, such as GAs, provide significant improvement in the optimal designs than the conventional designs. Babu and Munawar [13] applied differential evolution (DE) and its various strategies for the optimal design of STHXs. They determined that the DE, an exceptionally simple evolution strategy, is significantly faster than the GA and yields the global optimum for several key parameters. Xie et al. [14] experimentally investigated three heat exchangers by applying an artificial neural network (ANN) to the heat transfer analysis of STHXs with segmental or continuous helical baffles, and predicted outlet temperature differences on each side and overall heat transfer rates. As a result, the maximum deviation between the predicted results and experimental data was less than 2%, and the ANN showed superiority for prediction than correlation. Fesanghary et al. [15] explored the use of global sensitivity analysis (GSA) and a harmony search algorithm (HSA) for design optimization of STHXs from the economic viewpoint; total cost of STHXs are identified using GSA. They compared the HSA results with those obtained using GA, and revealed that the HSA can converge to an optimum solution with higher accuracy. Guo et al. [16] applied the field synergy principle to the optimization design of an STHX with segmental baffles by using the GA to solve the heat exchanger optimization problems by using multiple design variables. The comparison with the conventional heat exchanger optimization design approach, with the total cost as the objective function, shows that the field synergy number maximization approach is more advantageous. Zhang et al. [17] experimentally compared several STHXs: one with segmental baffles and four with helical baffles at helix angles of 20°, 30°, 40°, and 50°. The results show that, based on the same shell-side flow rate, the heat transfer coefficient of the heat exchanger with helical baffles is lower than that of the heat exchanger with segmental baffles, while the shell-side pressure drop of the exchanger with helical baffles is even lower.

As mentioned above, many researchers have focused on the STHX systems to enhance the heat transfer performance during condensation of odorized materials. However, there has been no attempt to develop a new combined system with the dimpled helical tube that enhances the heat transfer performance of the STHX systems. The tube was devised to intentionally enhance the heat transfer performance. Thus, the superiority of the dimpled helical tube was evaluated. However, the various arrangements of the dimples, such as staggered, in-lined, and alternating arrangements, must be assessed. Therefore, the present study suggests several arrangements of the dimples on the helical tube; the heat transfer performance was evaluated in terms of the Nusselt number. In addition, the effect of the Dean number (De) on the heat transfer performance was estimated. Furthermore, the performance of the proposed STHX systems was comparatively assessed using 3-D Reynolds-averaged Navier–Stokes (RANS) equations.

2. Configurations of the Shell and Helically-Coiled Tube Heat Exchanger Systems

Figure 1 presents dimensions of the shell and helically-coiled dimpled-tube heat exchanger system. The overall system has a length of 230 mm and diameter of 120 mm. The coolant enters through the inlet of 12 mm diameter, and exits through the same diameter outlet. The outer and inner diameters of the helical tube are 12 and 9 mm, respectively. The helical tube has 10 turns with the coil pitch of 17 mm and a curvature radius of $2R_c$, as shown in Figure 1a. The height (H_d) and diameter (D_d) of the dimple are 1.336 and 6.903 mm, respectively; the pitch of the dimple (p) is 5.628 mm, as shown in Figure 1b.

Figure 2 shows the schematic of the dimple arrangement. The dimples on the helical tube are arranged in the following three arrangements: staggered, inline, and bulged. Further, this tube's heat transfer performance was compared with that of the flat helical tube. The heat transfer enhancement structure consists of two dimples: concave and convex dimples. However, the bulged arrangement consists of convex dimples.

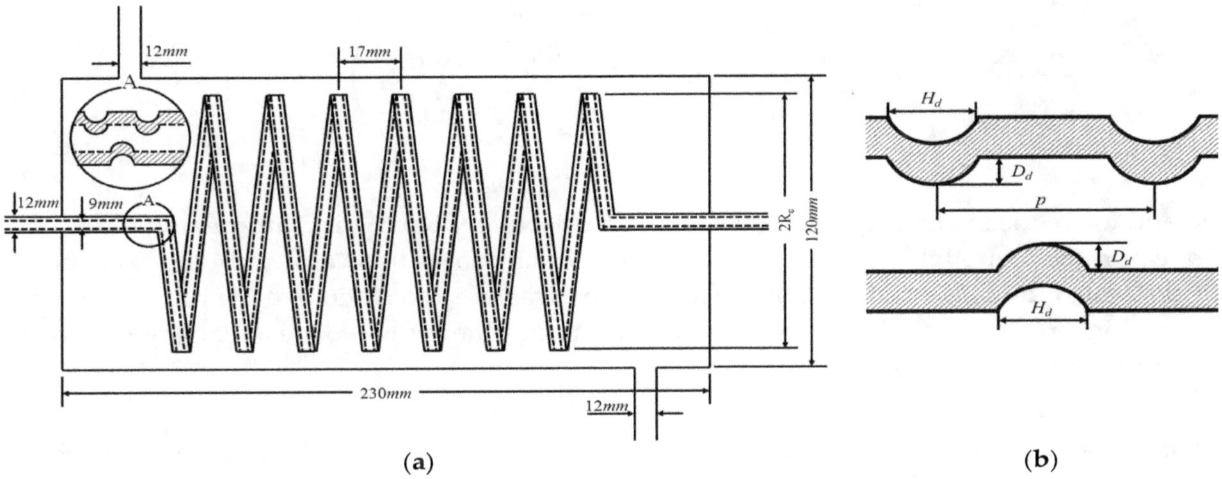

Figure 1. Dimensions of the shell and helically-coiled dimpled tube heat exchanger system: (**a**) overall system; and (**b**) dimple.

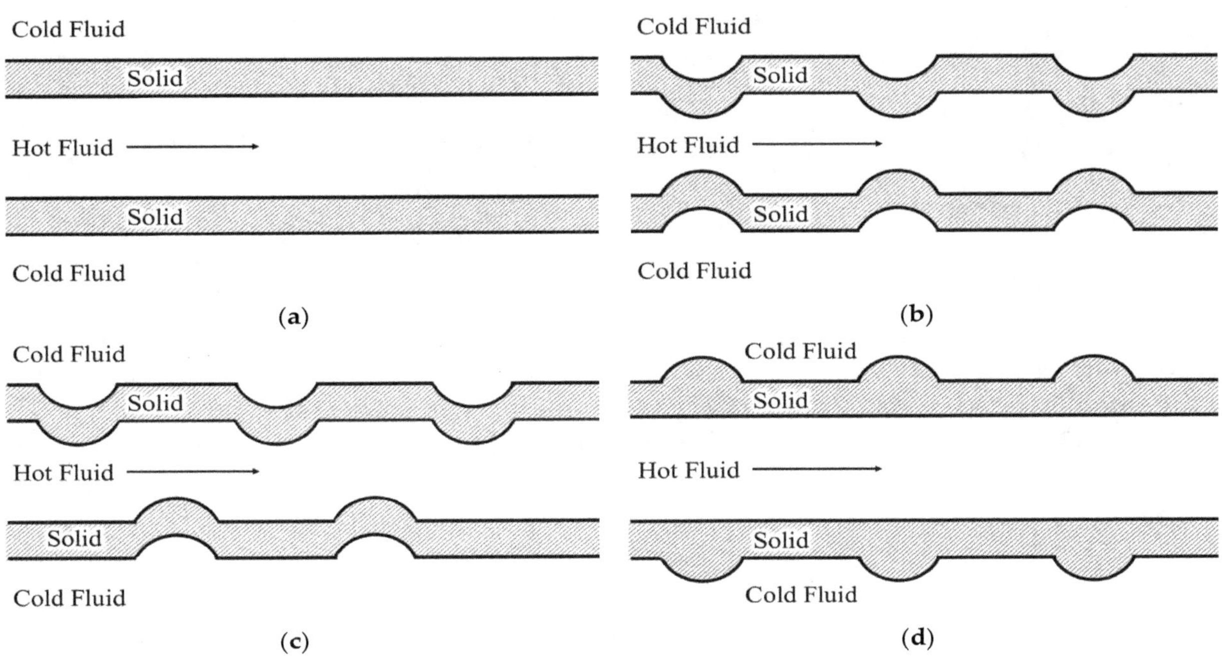

Figure 2. Schematic of the dimple arrangement: (**a**) flat; (**b**) inline; (**c**) staggered; and (**d**) bulged.

3. Numerical Analysis

The present numerical analysis was performed using commercial computational fluid dynamics code ANSYS CFX 17.0 (© 2016 ANSYS, Inc., Canonsburg, PA, USA) [18] to solve the governing equations for 3-D steady turbulent flow and conjugate heat transfer in the heat exchanger system.

For a steady, incompressible, turbulent flow, the continuity and Reynolds-averaged Navier-Stokes equations are given as:

$$\frac{\partial}{\partial x_i}(\rho u_i) = 0 \tag{1}$$

$$\frac{\partial}{\partial x_j}(\rho u_i u_j) = -\frac{\partial p}{\partial x_i} + \frac{\partial}{\partial x_j}\left[\mu\left(\frac{\partial u_i}{\partial x_j} + \frac{\partial u_j}{\partial x_i} - \frac{2}{3}\frac{\partial u_k}{\partial x_k}\delta_{ij}\right) - \rho\overline{u_i' u_j'}\right] + \overline{S}_i^u \tag{2}$$

where u_i and u_i' are mean and fluctuating velocities, respectively, and $\overline{S}_i^u$ is a production term. The shear stress transport (SST) model [19] with automatic wall treatment is used as a turbulence closure model. In addition, a second-order-accurate numerical scheme was used to discretize the governing equations. Moreover, a high-resolution scheme that uses a special nonlinear procedure at each node was used to discretize the advection terms.

Figure 3 shows the computational domain and boundary conditions. Water and Behran Hararat oil were respectively used as working fluids for the shell-side fluid and tube-side coolant. The thermophysical properties of the oil [20] were varied to state the dependency of the working fluid on temperature with an accuracy of 1.0%. The importance of temperature-dependent properties was already emphasized by many researchers [21,22]. The properties were defined as follows:

$$\mu = -1.31 \times 10^{-7}T^3 + 3.2813 \times 10^{-5}T^2 - 2.916875 \times 10^{-3}T + 9.9418750 \times 10^{-2}$$

$$\begin{aligned} C_p &= 3.7093T + 1814.3575 \\ k &= -8.7 \times 10^{-5}T + 0.164570 \\ \rho &= -0.725T + 877.350 \end{aligned} \tag{3}$$

where T (°C) is the caloric mean temperature of oil through the coiled heat. At the shell-side inlet, a mass flow rate of 0.06 kg/s was applied with a temperature of 287.15 K; at the tube-side inlet, a mass flow rate of 0.008–0.054 kg/s was assigned to maintain the De in the range of 17–195 with a temperature of 45 °C, 55 °C, and 75 °C. The De is defined as follows:

$$De = Re_i\left(\frac{d}{2R_c}\right)^{1/2} \tag{4}$$

where Re_i is the Reynolds number at the inside of the coiled tube, d is the diameter of the coiled tube, and R_c is the curvature radius. A relative pressure of 0 Pa was applied at both the shell- and tube-side outlets. At the contact surface between the solid and fluid domains, an interface condition was applied to estimate the conjugate heat transfer by using a general grid interface connection [18], which has been applied to many numerical analyses of different physical domains, such as rotating and stationary domains [23], or fluid and solid domains [24]. All of the wall surfaces on the outside of the heat exchanger and calculation domain were assumed to be nonslip and adiabatic.

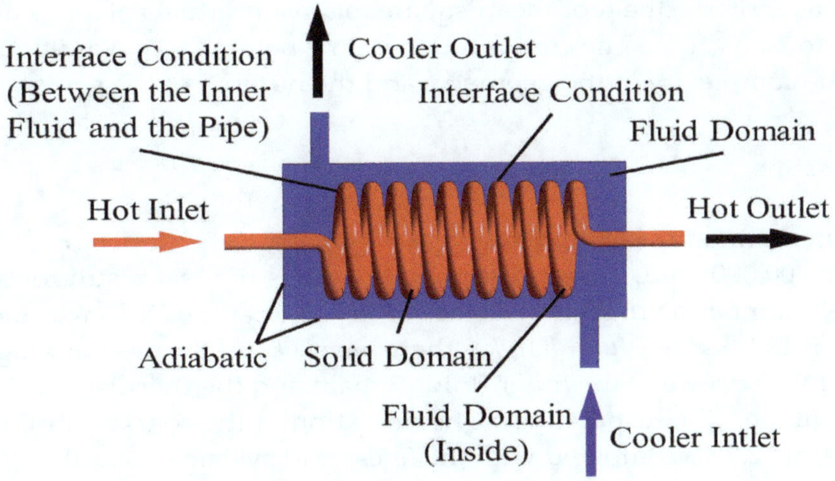

Figure 3. Computational domain and boundary conditions.

Figure 4 shows an example of the computational grid system, which comprises a tetrahedral mesh in most parts, and a hexahedral mesh near the inlet and outlet, to obtain more accurate inlet and outlet flow velocity profiles and reduce the overall number of grids. The mesh was finer in the helically-coiled tube and its surroundings for a complex flow phenomenon. A growth factor of 1.2 was adopted from the concentrated region to have a proper expansion factor for calculation stability.

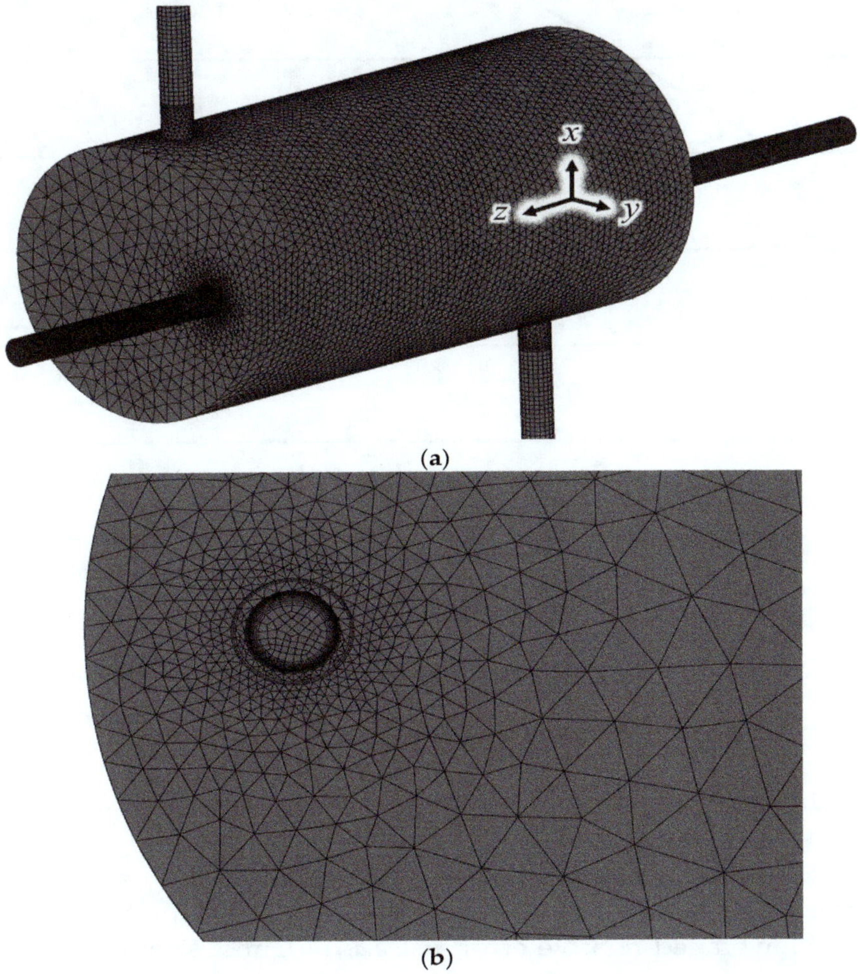

Figure 4. Example of the computational grid system: (**a**) overall and (**b**) wall treatment.

As the convergence criteria, the root-mean-square relative residual values of all the fluid and solid parameters were set to 1.0×10^{-5}. The computational time per analysis was about 12 h, and the time was dependent on the complexity of the geometry and the number of the grid system.

4. Result and Discussion

To determine the optimal number of the grid elements, three nodes of the grid system in the range of 900,000 to 3,300,000 were tested; Figure 5 illustrates the test result for a flat helical tube at $De = 111$. The Nusselt number on the inner surface of the coiled tube (Nu_i) was used as a criterion for the comparison and is defined as $Nu_i = h_i d_i / k$, where h_i and k are the area-averaged convective heat transfer coefficient at the surface of the inner fluid domain and the thermal conductivity of the inner fluid, as shown in Equation (3), respectively. The deviation of the coarsest grid system (0.9 million) shows a relative error of 12.22% compared with the finest grid system (3.3 million), while the deviation of the second finest grid system (1.9 million) shows a relative error of 2.19%. Resultantly, approximately 1.9 million nodes was decided to be the optimum number of nodes for calculation; however, the number of nodes varied with the shape of the coiled tube from 1.9 to 2.1 million.

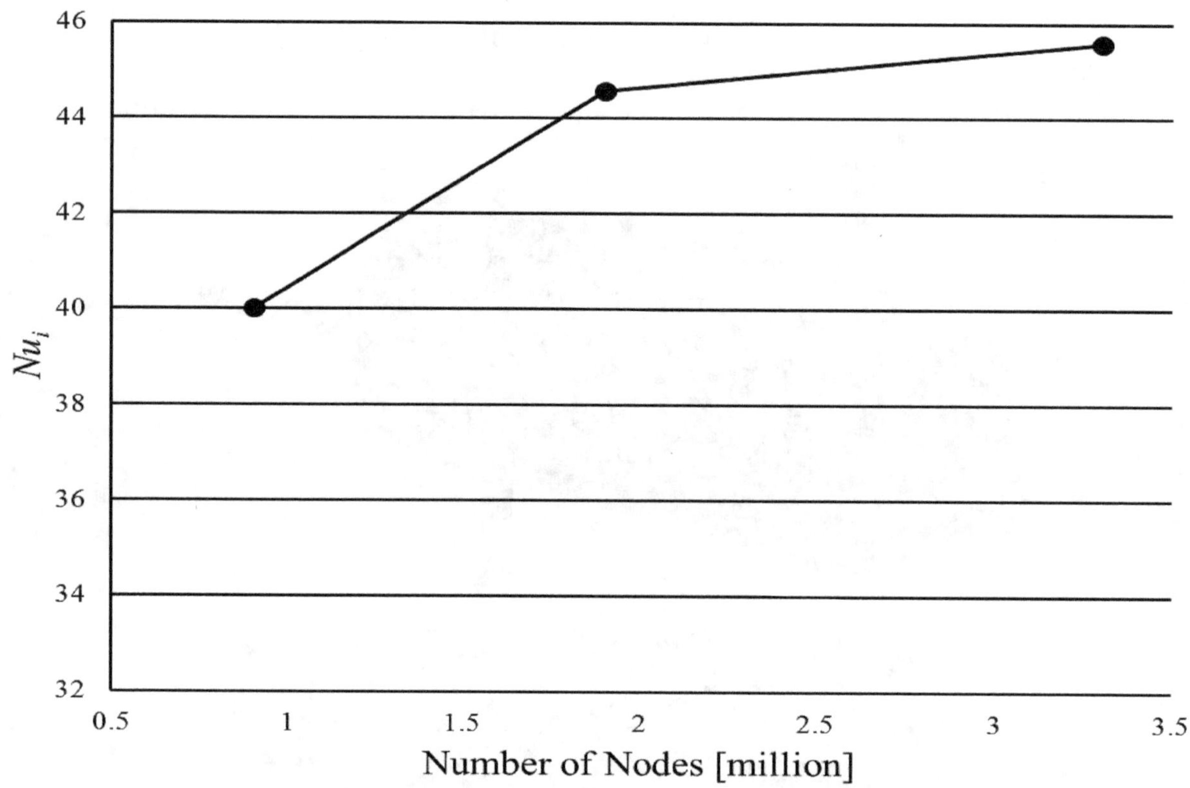

Figure 5. Grid dependency test.

To assess the validity of the numerical solutions, the results were compared with the experimental data of the flat coiled tube, as shown in Figure 6. The validation was performed for Nu_i with various De and two temperatures at the inlet of the tube-side. The results of $T_i = 55$ °C showed a deviation of 14.0% relative error at $De = 31.8$, and 1.7% averaged relative error at $De > 66.7$. The results of $T_i = 70$ °C showed 5.7% averaged relative error at $De > 140$, and 11.5% averaged relative error at $De < 98.3$. Overall, the present numerical results show good agreement with the experimental data.

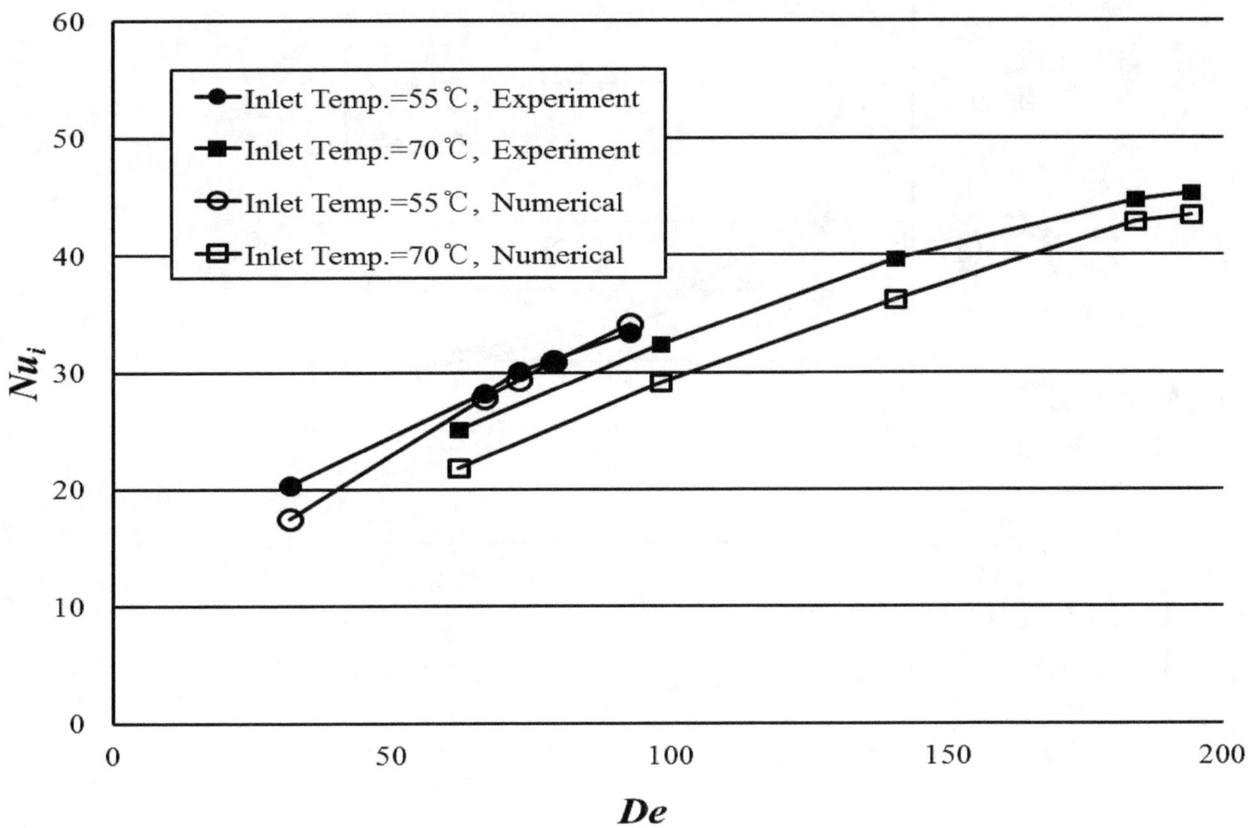

Figure 6. Validation of the numerical results with the experimental data [20] for the Nusselt number inside the helically-coiled tube.

Figure 7 presents the temperature difference charts of the hot channel with various *De*. The difference was calculated as $\Delta T = T_{hot,in} - T_{hot,out}$. The deviation in each case shows an almost similar trend. The inline and staggered arrangements show the highest temperature differences through all *De* values, whereas the staggered arrangement shows a slightly higher temperature difference. In addition, the bulged arrangement shows a higher temperature difference at lower *De* (<40) with $T_{hot,in}$ = 45 and 55 °C; however, as *De* increases, results similar with those of the flat case are observed. At $T_{hot,in}$ = 70 °C with a higher *De* (>100), the flat case shows a slightly higher temperature difference. It was obvious that increase in the residence time of the hot channel fluid increased the temperature difference of the hot channel.

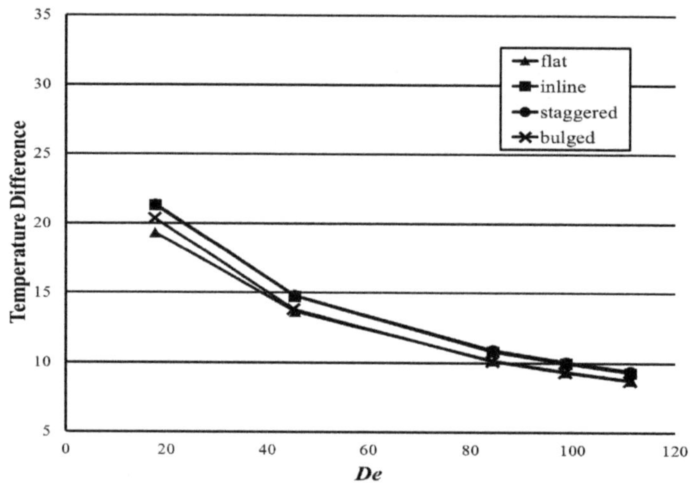

(a)

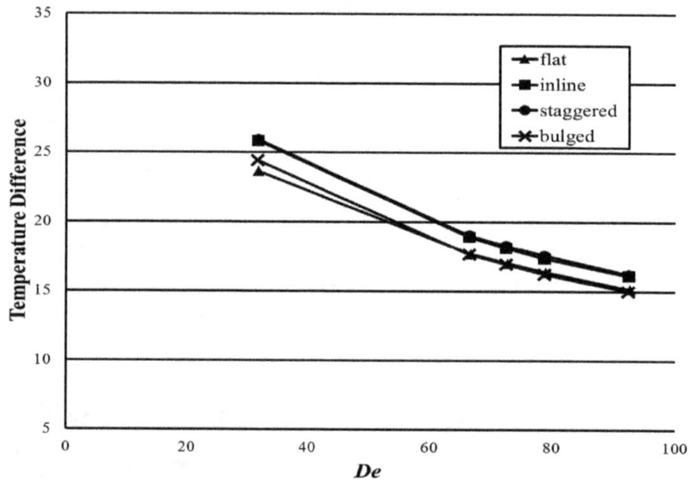

(b)

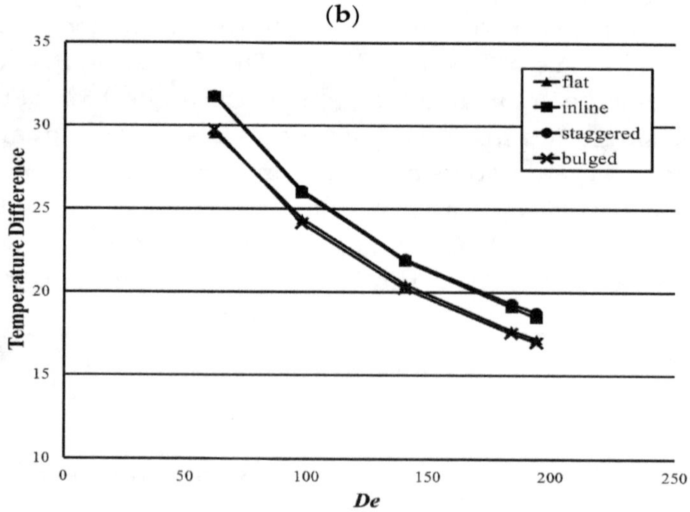

(c)

Figure 7. Temperature difference charts of the hot channel with various *De*: (**a**) $T_{hot,\,in}$ = 45 °C; (**b**) $T_{hot,in}$ = 55 °C; and (**c**) $T_{hot,in}$ = 70 °C.

Figure 8 shows temperature contours inside the helically-coiled tube with the velocity vector of the cold channel on the *y*–*z* plane at $x = 0$; $De = 17.87$ and inlet temperature of the hot channel is 45 °C. The results were focused on the temperature variation inside the coiled tube with a circumferential fluid flow on the shell-side. The hot fluid was affected by the centric fluid flow rather than outer fluid flow on the shell-side; therefore, the thermal layer inside the tube started at the inner surface for all cases. The high temperature fluid flows along the shape of the helically-coiled tube and the fluid that is closer to the center of the shell flows more slowly than the fluid that is farther from the center of the shell due to the centrifugal effect. Moreover, the fluid inside the shell-side flows faster at the inner area than at the outer (near wall) area. As the inline and staggered arrangements show rapid temperature variation in the coiled tube, the thermal layer was observed simultaneously from the inner and outer surfaces, and a lower temperature of the hot channel was observed near the exit, as shown in Figure 7.

Figure 9 presents the Nusselt number inside the helically-coiled tube with various *De* values and inlet temperatures of the hot channel. The staggered arrangement shows the best heat transfer performance, while the inline arrangement shows the second best performance for all ranges of *De* and inlet temperature. The bulged arrangement shows a higher than the flat arrangement at a lower *De* (<30); however, the flat arrangement shows a higher Nu_i value at a higher *De* (>40) at $T_{hot,in}$ = 45 °C. A similar result can be observed for $T_{hot,in}$ = 55 °C. The deviation between each case was increased with a higher *De* value at all inlet temperatures.

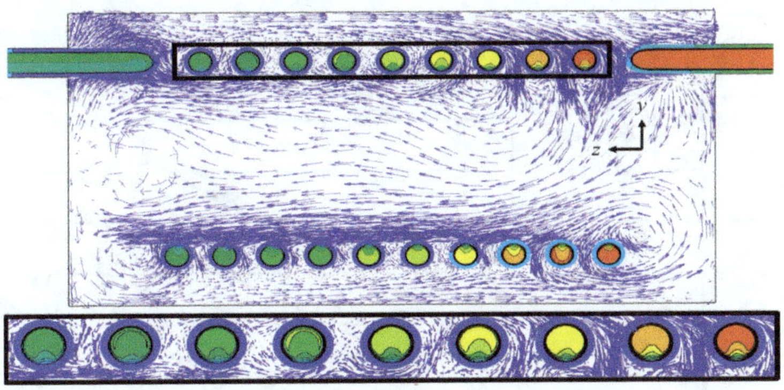

(a)

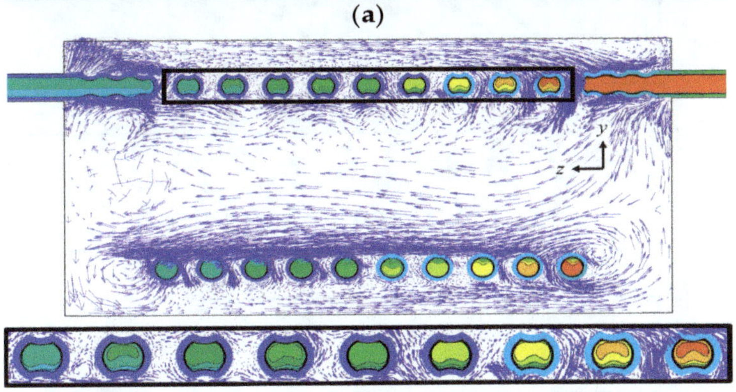

(b)

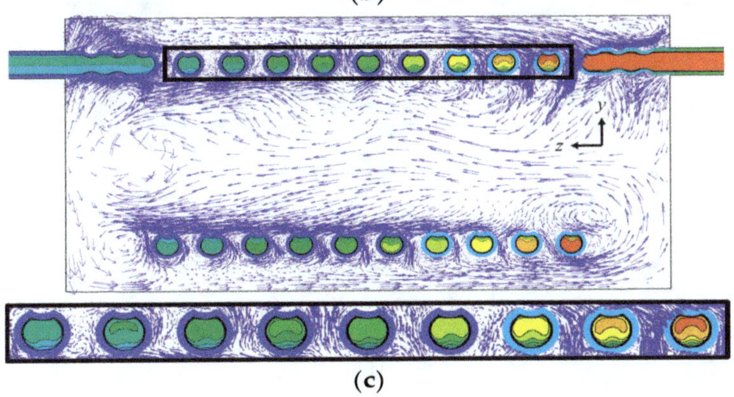

(c)

Figure 8. *Cont.*

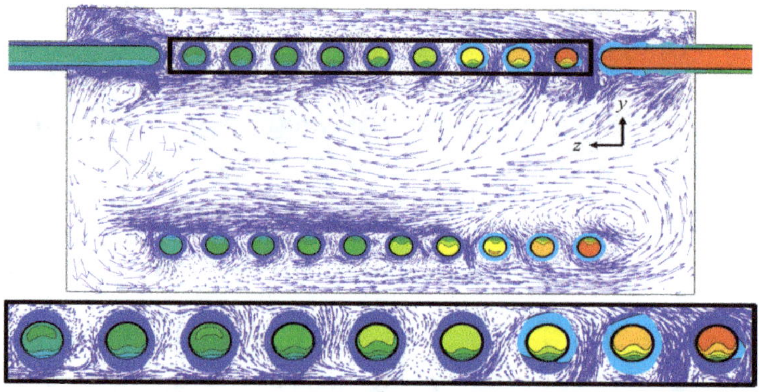

(d)

Figure 8. Temperature contours inside the helically-coiled tube with a velocity vector of the cold channel on the y–z plane at $x = 0$: (**a**) flat; (**b**) inline; (**c**) staggered; and (**d**) bulged.

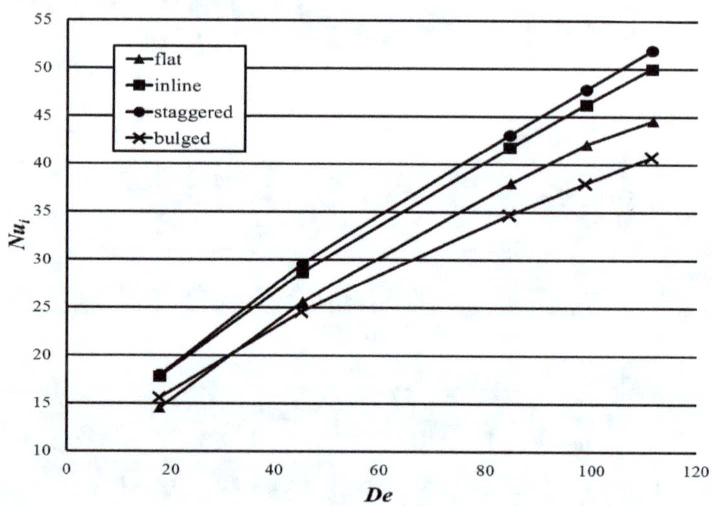

(**a**)

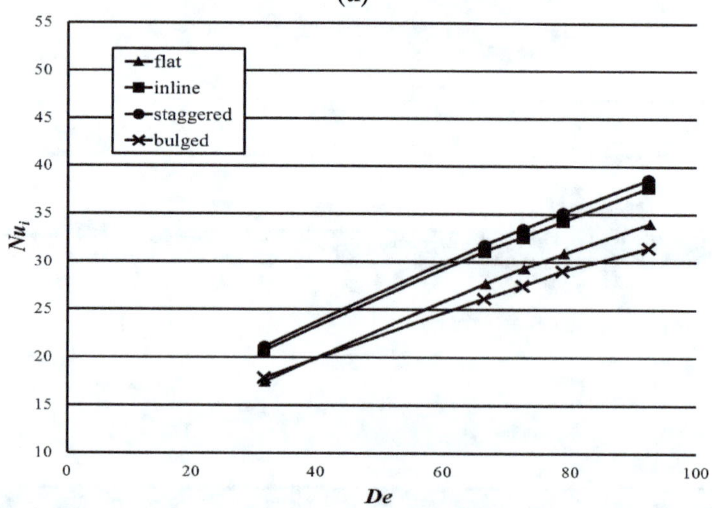

(**b**)

Figure 9. *Cont.*

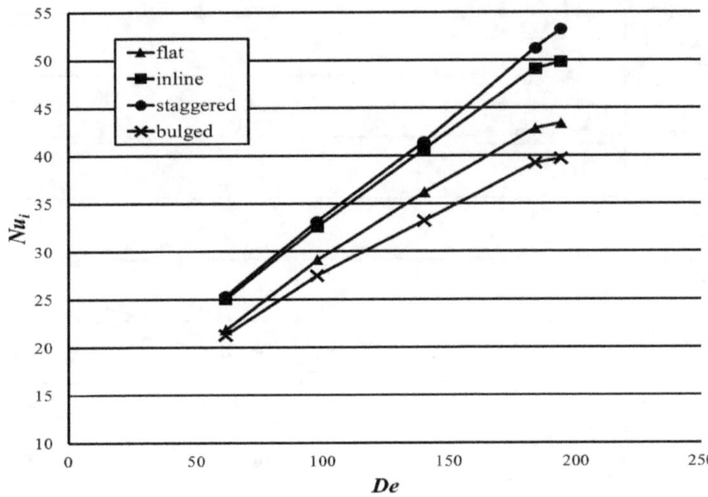

(c)

Figure 9. Nusselt number inside the helically-coiled tube with various *De*: (**a**) $T_{hot,\ in}$ = 45 °C; (**b**) $T_{hot,in}$ = 55 °C; and (**c**) $T_{hot,in}$ = 70 °C.

Figure 10 shows pressure drops at the helically-coiled tube with various *De*. The inline and staggered arrangements with the best heat transfer performance show the highest pressure drop for all inlet temperatures of the hot channel, and the flat arrangement shows the lowest pressure drop for all inlet temperatures. The deviations between each case increased with the *De* for all inlet temperatures.

Overall, introducing the staggered arrangement structure to the shell and helically-coiled tube system efficiently enhanced the heat transfer performance; however, the pressure drop increased through the system.

It is generally known that the performance of the dimple, as a device of heat transfer augment, can be influenced by the shape of the dimple, such as the dimple height, dimple width, and dimple periodic arrangement. In a further study, the effect of geometric variables of the dimple will be investigated to optimize the system and to generalize the results to other dimple systems.

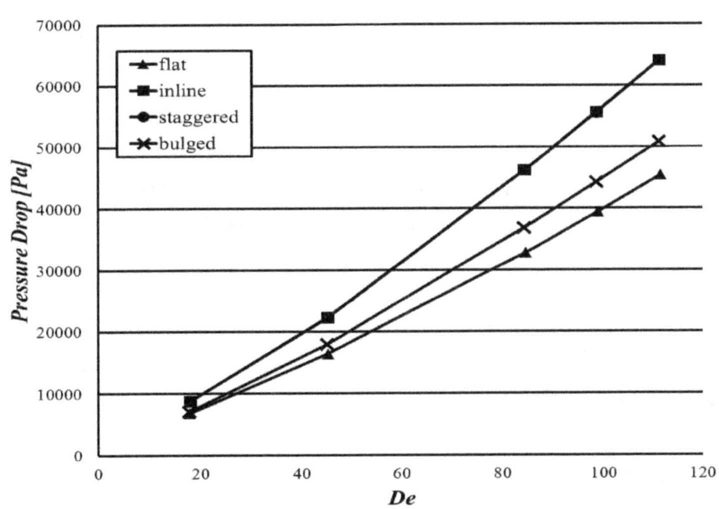

(a)

Figure 10. *Cont.*

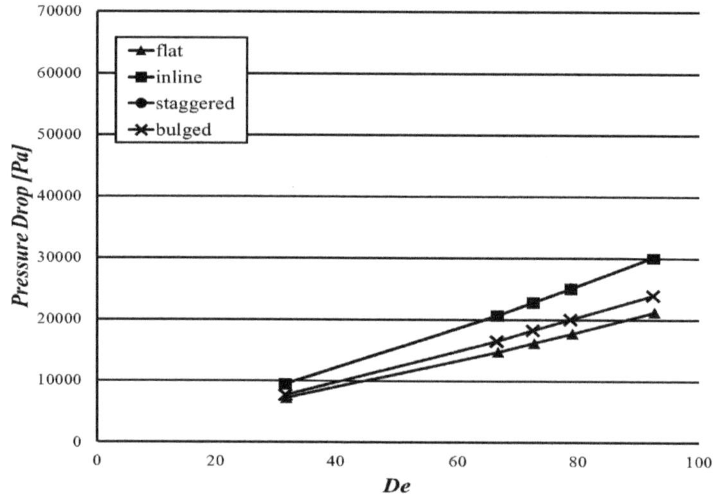

(b)

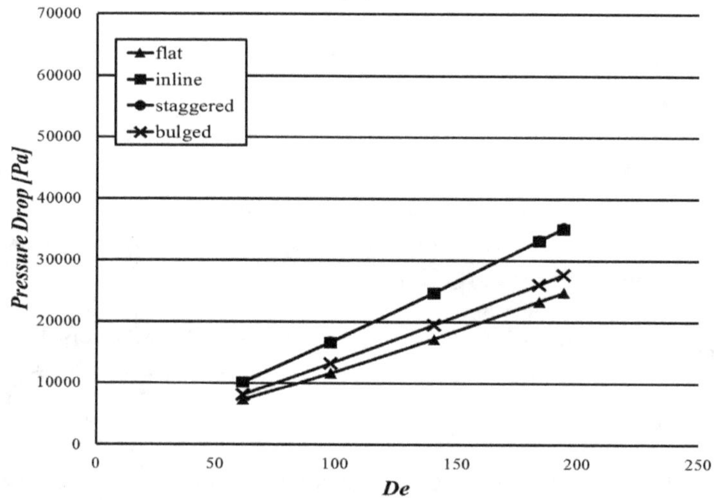

(c)

Figure 10. Pressure drop at the helically-coiled tube with various *De*: (**a**) $T_{hot,in}$ = 45 °C; (**b**) $T_{hot,in}$ = 55 °C; and (**c**) $T_{hot,in}$ = 70 °C.

5. Conclusions

This paper described the evaluations of the shell and helically-coiled tube heat exchangers with various dimple arrangements, flat, inline, staggered, and bulged, at different *De* and inlet temperatures of the hot channel. These evaluations were performed for determining the temperature difference between the inlet and outlet, Nusselt number inside the coiled tube, and pressure drop of the coiled tube by using 3-D RANS equations with SST turbulence closure. A grid dependency test was performed to determine the optimal number of the grid system; approximately 190 million nodes were found to be optimal for calculation. The numerical results were validated using the experimental data, and showed good agreement within the average relative error of 1.7%–5.7%. The inline and staggered arrangements show the highest temperature differences for all values of *De*, and the bulged and flat arrangements show lower temperature differences. The staggered arrangement shows the best heat transfer performance, followed by the inline arrangement for all ranges of *De* and inlet temperature of the hot channel. The inline and staggered arrangements show the highest pressure drop and the flat arrangement shows the lowest pressure drop through all inlet temperatures of the hot channel. The performance is expected to be enhanced further by optimizing the geometric configuration through a range of optimization techniques [25–28].

Acknowledgments: This research was supported by a grant (16AUDP-B083704-03) from Architecture & Urban Development Research Program funded by Ministry of Land, Infrastructure and Transport of Korean government.

Author Contributions: Jun-Ho Jo and Ye-Eun Lee supported the numerical investigation, and the revision of the manuscript. Yeong-Seok Yoo supervised the project, and provided professional technical guidelines and insights. Sun-Min Kim designed the project, analyzed the data, wrote the draft of the paper, and carried out the final revision of the paper.

Nomenclature

C_p	specific heat (J/kg·K)
d	diameter (m)
De	Dean number
h	convective heat transfer coefficient (W/m^2·K)
H	height (m)
Nu	Nusselt number
p	pitch (m)
R_c	curvature radius (m)
Re	Reynolds number
T	temperature (K)
x, y, z	Cartesian coordinates (m)

Greek symbols

μ	dynamic viscosity (Pa·s)
k	thermal conductivity
ρ	density (kg/m^3)
v	kinematic viscosity (m^2/s)
τ	shear stress (N/m^2)

Subscripts

d	dimple
hot	hot channel
$cold$	cold channel
in	inner
out	outer

References

1. Doshi, V.A.; Vuthaluru, H.B.; Bastow, T. Investigations into the control of odour and viscosity of biomass oil derived from pyrolysis of sewage sludge. *Fuel Process. Technol.* **2005**, *86*, 885–897. [CrossRef]
2. Tripathi, M.; Sahu, J.N.; Ganesan, P. Effect of process parameters on production of biochar from biomass waste through pyrolysis: A review. *Renew. Sustain. Energy Rev.* **2016**, *55*, 467–481. [CrossRef]
3. Mohanty, P.; Pant, K.K.; Naik, S.N.; Parikh, J.; Hornung, A.; Sahu, J.N. Synthesis of green fuels from biogenic waste through thermochemical route–The role of heterogeneous catalyst: A review. *Renew. Sustain. Energy Rev.* **2014**, *38*, 131–153. [CrossRef]
4. Mašek, O.; Budarin, V.; Gronnow, M.; Crombie, K.; Brownsort, P.; Fitzpatrick, E.; Hurst, P. Microwave and slow pyrolysis biochar: Comparison of physical and functional properties. *J. Anal. Appl. Pyrolysis* **2013**, *100*, 41–48. [CrossRef]
5. Park, H.J.; Dong, J.I.; Jeon, J.K.; Park, Y.K.; Yoo, K.S.; Kim, S.S.; Kim, J.; Kim, S. Effects of the operating parameters on the production of bio-oil in the fast pyrolysis of Japanese larch. *Chem. Eng. J.* **2008**, *143*, 124–132. [CrossRef]
6. Fu, P.; Hu, S.; Xiang, J.; Sun, L.; Su, S.; Wang, J. Evaluation of the porous structure development of chars from pyrolysis of rice straw: Effects of pyrolysis temperature and heating rate. *J. Anal. Appl. Pyrolysis* **2012**, *98*, 177–183. [CrossRef]

7. Ahmed, I.I.; Gupta, A.K. Pyrolysis and gasification of food waste: Syngas characteristics and char gasification kinetics. *Appl. Energy* **2010**, *87*, 101–108. [CrossRef]
8. Digman, B.; Kim, D.S. Review alternative energy from food processing wastes. *Environ. Prog.* **2008**, *27*, 524–537. [CrossRef]
9. Rappert, S.; Müller, R. Odor compounds in waste gas emissions from agricultural operations and food industries. *Waste Manag.* **2005**, *25*, 887–907. [CrossRef] [PubMed]
10. Burgess, J.E.; Parsons, S.A.; Stuetz, R.M. Developments in odour control and waste gas treatment biotechnology—A review. *Biotechnol. Adv.* **2001**, *19*, 35–63. [CrossRef]
11. Wang, Q.; Chen, G.; Chen, Q.; Zeng, M. Review of improvements on shell-and-tube heat exchangers with helical baffles. *Heat Transf. Eng.* **2010**, *31*, 836–853. [CrossRef]
12. Selbas, R.; Kızılkan, Ö.; Reppich, M. A new design approach for shell-and-tube heat exchangers using genetic algorithms from economic point of view. *Chem. Eng. Process.* **2006**, *45*, 268–275. [CrossRef]
13. Babu, B.V.; Munawar, S.A. Differential evolution strategies for optimal design of shell-and-tube heat exchangers. *Chem. Eng. Sci.* **2007**, *62*, 3720–3739. [CrossRef]
14. Xie, G.N.; Wang, Q.W.; Zeng, M.; Luo, L.Q. Heat transfer analysis for shell-and-tube heat exchangers with experimental data by artificial neural networks approach. *Appl. Therm. Eng.* **2007**, *27*, 1096–1104. [CrossRef]
15. Fesanghary, M.; Damangir, E.; Soleimani, I. Design optimization of shell and tube heat exchangers using global sensitivity analysis and harmony search algorithm. *Appl. Therm. Eng.* **2009**, *29*, 1026–1031. [CrossRef]
16. Guo, J.; Xu, M.; Cheng, L. The application of field synergy number in shell-and-tube heat exchanger optimization design. *Appl. Energy* **2009**, *86*, 2079–2087. [CrossRef]
17. Zhang, J.F. Experimental performance comparison of shell-side heat transfer for shell-and-tube heat exchangers with middle-overlapped helical baffles and segmental baffles. *Chem. Eng. Sci.* **2009**, *64*, 1643–1653. [CrossRef]
18. ANSYS Inc. *ANSYS CFX-Solver Theory Guide, ANSYS CFX Release 17.0*; ANSYS Inc.: Canonsburg, PA, USA, 2016.
19. Menter, F.R. Two-equation eddy-viscosity turbulence models for engineering applications. *AIAA J.* **1994**, *32*, 1598–1605. [CrossRef]
20. Salimpour, M.R. Heat transfer characteristics of a temperature-dependent-property fluid in shell and coiled tube heat exchangers. *Int. Commun. Heat Mass Transf.* **2008**, *35*, 1190–1195. [CrossRef]
21. Kim, S.M.; Kim, K.Y. Optimization of a Hybrid Double-Side Jet Impingement Cooling System for High-Power Light Emitting Diodes. *J. Electron. Packag.* **2014**, *136*. [CrossRef]
22. Husain, A.; Kim, S.M.; Kim, K.Y. Performance analysis and design optimization of micro-jet impingement heat sink. *Heat Mass Transf.* **2013**, *49*, 1613–1624. [CrossRef]
23. Heo, M.W.; Seo, T.W.; Shim, H.S.; Kim, K.Y. Optimization of a regenerative blower to enhance aerodynamic and aeroacoustic performance. *J. Mech. Sci. Technol.* **2016**, *30*, 1197–1208. [CrossRef]
24. Kim, S.M.; Kim, K.Y. Microcooling system with impinging jets and a stalactite structure. *Numer. Heat Transf. Part A Appl.* **2016**, *69*, 1376–1389. [CrossRef]
25. Kim, K.Y.; Lee, Y.M. Design optimization of internal cooling passage with V-shaped ribs. *Numer. Heat Transf. Part A Appl.* **2007**, *51*, 1103–1118. [CrossRef]
26. Lee, K.D.; Kim, K.Y. Shape optimization of a fan-shaped hole to enhance film-cooling effectiveness. *Int. J. Heat Mass Transf.* **2010**, *53*, 2996–3005. [CrossRef]
27. Li, P.; Kim, K.Y. Multiobjective optimization of staggered elliptical pin-fin arrays. *Numer. Heat Transf. Part A Appl.* **2008**, *53*, 418–431. [CrossRef]
28. Husain, A.; Kim, K.Y. Enhanced multi-objective optimization of a microchannel heat sink through evolutionary algorithm coupled with multiple surrogate models. *Appl. Therm. Eng.* **2010**, *30*, 1683–1691. [CrossRef]

An Experimental Study on the Radiation Noise Characteristics of a Centrifugal Pump with Various Working Conditions

Chang Guo [1], Ming Gao [1,*], Dongyue Lu [2] and Kun Wang [1]

1 School of Energy and Power Engineering, Shandong University, Jinan 250061, China; gg3263@163.com (C.G.); sduwang1993@163.com (K.W.)
2 State Nuclear Power Engineering Company, Shanghai 200233, China; ldy11energy@163.com
* Correspondence: gm@sdu.edu.cn.

Abstract: To investigate the radiation noise characteristics of a centrifugal pump under various working conditions, a noise measurement system is established; afterwards, the distribution of different points and intervals, as well as the overall level of noise, are studied. The total sound pressure level distribution for different points manifests the dipole and asymmetric directivity characteristics. Additionally, the acoustic energy is introduced to compare the noise of different intervals to reveal the asymmetric characteristics, and it is found that variation in working conditions has little impact on the acoustic energy distribution, and the ratio of the acoustic energy in the direction facing the tongue, as well as that in the direction against the tongue, to total acoustic energy fluctuate around 0.410 and 0.160, respectively, under various working conditions. Also, the A-weighted average sound pressure level (L_{pA}) is applied to describe the overall level of noise, and L_{pA} increases gradually with the growth of rotational speed, but the growth slope decreases. While in the operation of throttling regulation, L_{pA} shows the trend that first increases, then remains stable, and increases again with the growth of flow rate. This study could provide guidance for optimizing the operating conditions and noise control of centrifugal pumps.

Keywords: centrifugal pump; radiation noise; distribution characteristic; acoustic energy; experimental research

---

1. Introduction

Widespread concerns about environmental protection are accelerating noise abatement investment, accounting for 15–20% of the environmental protection investment [1]. Centrifugal pumps, widely applied in many fields of national economy [2,3], radiate a lot of noise during operation. The unexpected noise could affect the flow performance and deteriorate the working environment. To establish theoretical basis of radiation noise control technology for centrifugal pumps, this study reveals the changing rules of radiation noise under various working conditions.

Experimental studies could provide the most reliable results for scholars to detect the noise from centrifugal pumps. Choi et al. [4] conducted experiments and revealed that the main cause of radiation noise generated by an impeller without volute was the pressure fluctuation on the blade surface. Chu et al. [5,6] took the volute into consideration and attributed the primary noise to the interaction between the non-uniform outflux from the impeller and tongue [5,6], and further research showed that a slight increase of the gap between impeller and tongue would reduce the noise significantly [7]. Parrondo et al. [8] established an acoustic model to characterize the internal sound field at low frequency range, and concluded that the internal sound field could be characterized by a dipole-like source near the tongue. Cai et al. [9] measured the pressure fluctuation near the wall of the tongue under various rotational speed conditions, and found that the pressure fluctuation intensity increased more rapidly than that of rotational speed. The four-port model was introduced for research on internal sound fields in pipes, making it convenient to change the rules of noise in pipes under various working conditions [10–12]. Based on LabVIEW (LabVIEW8.6, National Instruments, Austin, Texas, USA), Yuan et al. [13] designed a measurement system for internal noise analysis of centrifugal pumps with synchronous measurement for noise signal, pressure and flow rate, and laid a foundation for the follow-up study about the influence of different structures [14–16] on sound pressure levels (SPL) at different frequencies. To characterize far field radiation noise, Ye et al. [17] measured the noise amplitude outside a centrifugal pump under various flow rate conditions, and found that noise amplitude reached maximum at the highest efficiency point.

Currently, as a useful research tool, numerical simulation compensated for weaknesses in the experiment. A hybrid method combining computational fluid dynamics (CFD) with Lighthill acoustic analogy was widely used to elucidate the acoustic generation [18,19]. Langthjem and Olhoff [20,21] applied the hybrid method for noise calculation in a two-dimensional centrifugal pump, and concluded that the main cause of noise was the dipole source. The dipole source was defined as unsteady fluid force acting on the wall surface, including the impeller and volute source, in centrifugal pumps. Huang et al. [22], Ma et al. [23] and Liu et al. [24] considered the two dipole sources to be the noise source for acoustic calculation, and compared the SPL at different orders of blade passing frequency (BPF) with different structures, and this research provided guidance for structural optimization of centrifugal pumps. In addition, Gao et al. [25] discovered that the radiation noise generated by an impeller demonstrated diploe directivity, while the volute-generated noise appeared to have asymmetric directivity, i.e., the noise level in the direction facing the tongue was higher than that of the direction against the tongue. Dong et al. [26] pointed out that the volute-generated noise increased monotonously and nonlinearly with the increase of rotational speed.

Obviously, the literature shows that previous experimental research mainly focused on the internal noise characteristics in pipes, while merely experimental research focused on far field radiation noise. During the numerical simulation, the previous research mainly considered the influence of the impeller dipole source and volute dipole source on the radiation noise separately, and most are the SPL at different orders of BPF. Actually, the radiation noise results from the interaction of the two dipole sources when pumps are running, so the interaction influence should be considered to obtain more accurate results. In addition, it is necessary to calculate the total sound pressure level (TSPL) of different monitoring points, calculated by the superposition of SPL at different characteristic frequencies, to acquire the directivity characteristic. Noise propagation is the spread of acoustic energy in the medium, so the acoustic energy changes with noise propagation process and the directivity characteristic can be detected. The overall level of radiation noise should also be conducted to evaluate the noise intensity under specific working conditions. In general, the pumps are always working at various conditions with various working demands, and the radiation noise changes accordingly, and it is also indispensable to carry out the various working condition study for radiation noise to figure out these changing rules.

In this paper, a centrifugal pump radiation noise measurement system is established. Then, the TSPL and acoustic energy distribution, as well as the overall level of radiation noise, are analyzed, which may provide guidance for optimizing the working conditions and noise control of centrifugal pumps.

2. Experimental Facility and Procedure

2.1. Parameters of the Test Pump

In this study, IS-80-50-250 (IS-80-50-250, Shanghai Pump Manufacture Co., Ltd, Shanghai, China) is chosen as the test pump, and the prototype figure is shown in Figure 1. Water at normal temperature is used as medium, and the geometric and performance parameters of the test pump are listed in Table 1.

Figure 1. Prototype figure of the test pump.

Table 1. Geometric and performance parameters of the test pump.

Parameter	Value
Inlet diameter, mm	80
Impeller diameter, mm	250
Outlet diameter, mm	50
Nominal flow rate, m^3/h	50
Best efficiency	63% (50 m^3/h)
Design head, m	80
Nominal rotational speed, rpm	2900
Blade number	6

2.2. Radiation Noise Measurement System

The experimental apparatuses used in the system are shown schematically in Figure 2, which include a soundproof room, water circulation system, circuit control system, and data acquisition and storage system. In the measurement system, to reduce the influence of the surrounding environment and motor operation on the measurement results, the centrifugal pump and motor are insulated by a soundproof room; the interior and exterior walls of the soundproof room, along with the motor, are surrounded by soundproof cotton. The pump is driven by the YVF2180L-2 type three-phase asynchronous motor (YVF2180L-2, Shanghai Nama Electric Co., Ltd, Shanghai, China), and the rotational speed is regulated by the Y0300G3 type frequency converter (Y0300G3, Shanghai Nama Electric Co., Ltd, Shanghai, China). The flow rate, inlet pressure, outlet pressure, as well as the radiation noise level are measured and recorded by corresponding instrument listed in Table 2.

During the operation, when the operation system reaches stable, the flow rate, pressure, radiation noise level are measured sequentially and stored in the computer terminal. After that, the rotational speed and flow rate are adjusted via frequency converter and artificial regulation, respectively, then these parameters under various working conditions are recorded for analysis. Figure 3 shows the structure diagram of the measurement system.

Table 2. Measurement characteristic of instruments.

Instruments	Type	Application	Measuring Range	Accuracy or Sensitivity
Flow meter	SLDG-800 (SLDG-800, Nanjing Shunlaida Measurement and Control Equipment Co., Ltd., Nanjing, Jiangsu, China)	Measuring flow rate	0–100 m^3/h	0.2% (accuracy)
Pressure transducer	MIK-300 (MIK-300, Hangzhou Meikong Automation Technology Co., Ltd., Hangzhou, China)	Measuring inlet pressure	−100–0 kPa (inlet pipe)	0.5% (accuracy)
		Measuring outlet pressure	0–1600 kPa (outlet pipe)	0.5% (accuracy)
Pressure recorder	RX-200D (RX-200D, Hangzhou Meikong Automation Technology Co., Ltd., Hangzhou, China)	Recording pressure	/	/
Microphone	AWA14423L (AWA14423L, Hangzhou Aihua Instruments Co., Ltd., Hangzhou, China)	Measuring radiation noise	10–20 kHz	50 mV/Pa (sensitivity)
Two channel signal analyzer	AWA6290M+ (AWA6290M+, Hangzhou Aihua Instruments Co., Ltd, Hangzhou China)	Recording radiation noise	/	/

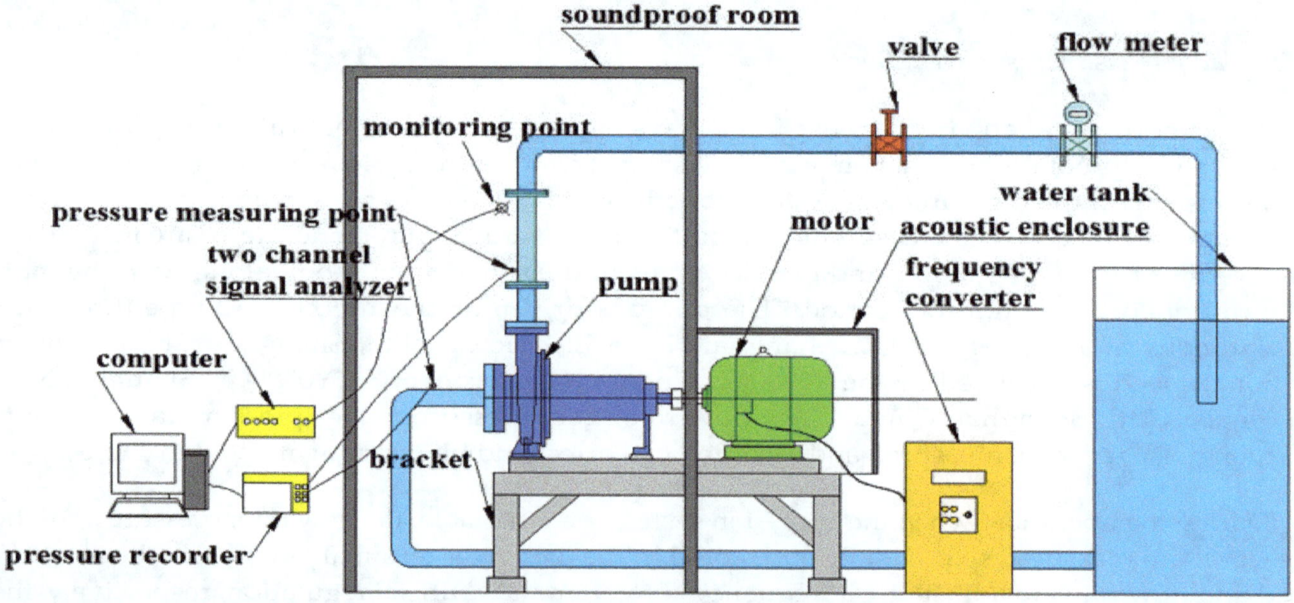

Figure 2. Layout and instrumentation of measurement system.

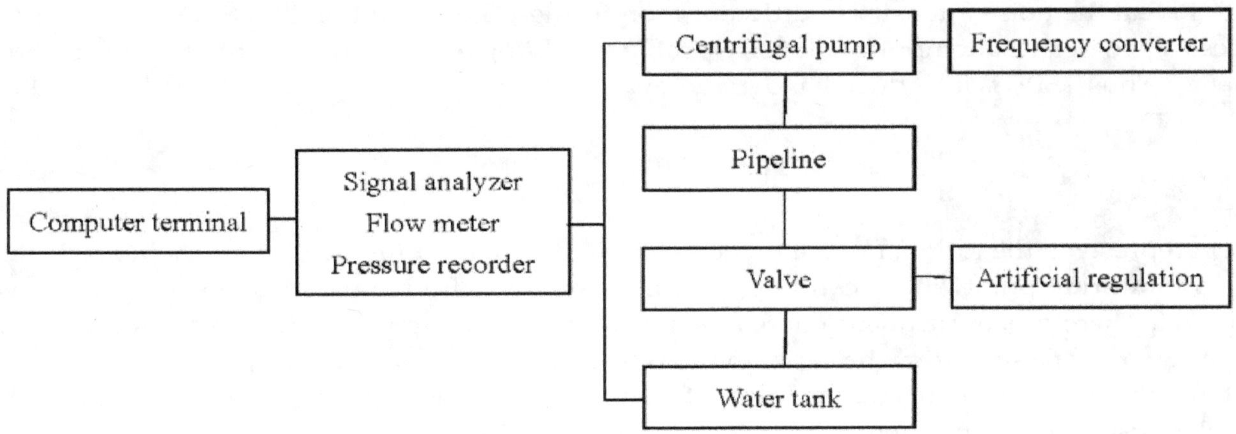

Figure 3. Structure diagram of measurement system.

2.3. Arrangement of the Monitoring Points

To acquire the distribution characteristic of radiation noise, 16 monitoring points are arranged on the measurement surface around the pump. As shown in Figure 4, the monitoring points are 1000 mm away from the center of impeller, and arranged evenly in circumferential directions [27]. During the measurement process, SPL characteristic of every point is measured sequentially by a microphone. Here, SPL is defined as:

$$\mathrm{SPL} = 20\log\frac{P_e}{P_{ref}} \tag{1}$$

$$P_e = \sqrt{\frac{1}{T}\int_0^T p'^2 dt} \tag{2}$$

$$T = \frac{t}{l} \tag{3}$$

where P_{ref} is the reference acoustic pressure (2×10^{-5} Pa in air), P_e is the effective value of acoustic pressure, p' is instantaneous acoustic pressure, t is the time of one revolution of the impeller, and l is the number of blades (l = 6 in this paper).

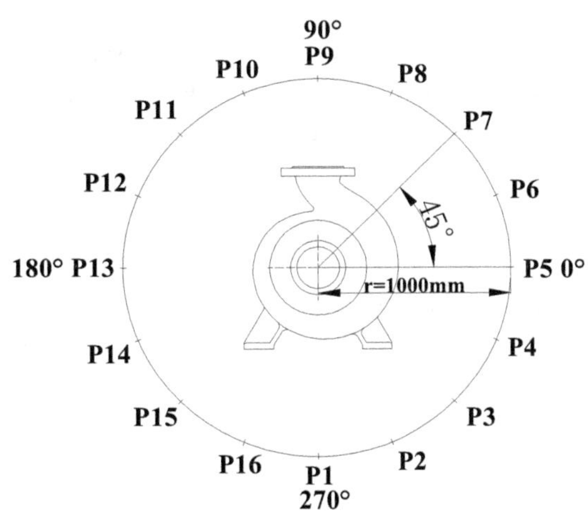

Figure 4. Arrangement of monitoring points in circumferential direction.

To reveal the noise intensity of different monitoring points and the directivity characteristic, it is necessary to derive a temporal intensity profile involving a superposition of SPL at each Fourier frequency. Thus TSPL is introduced and given by:

$$\text{TPSL} = 10\lg\sum_{i=1}^{n} 10^{\text{SPL}_i/10} \tag{4}$$

where n represents the number of frequencies. However, TSPL only represents the noise level of a certain monitoring point, and it can't be superimposed by arithmetic to describe the noise level of an interval. Therefore, the acoustic energy is more suitable for comparing the noise level of different monitoring intervals since it can be superimposed by arithmetic, moreover, the propagation of sound is essentially the propagation of energy. Briefly, the application of acoustic energy could reveal the noise level relationship between different intervals intuitively, so the average acoustic energy density [28] is analyzed, and it is defined as:

$$\varepsilon = \frac{P_e^2}{\rho c^2} \tag{5}$$

where ε, ρ and c represent the acoustic energy density, medium density (1.29 kg/m^3 in air) and the sound speed in medium (343 m/s in air), respectively.

To further evaluate the overall level of radiation noise on measurement surface, the average A-weighted sound pressure level (L_{pA}) of the measurement surface is introduced and expressed as:

$$L_{pA} = 10\lg\left(\frac{1}{m}\sum_{i=1}^{m} 10^{\text{TSPL}_i/10}\right) \tag{6}$$

where m represents the number of monitoring points, and the L_{pA} of 16 monitoring points on the measurement surface around the pump is calculated for analysis.

3. Radiation Noise Characteristic under Various Rotational Speed Conditions

The variable speed regulation has no throttling loss, and is an ideal adjustment method. In this section, on the basis of keeping the flow valve installed downstream on the outlet pipe opened fully and unchanged, then the radiation noise characteristic under various rotational speed conditions are studied. According to the similarity law, the relationship between flow rate and rotational speed is defined as:

$$\frac{Q_1}{Q_2} = \frac{n_1}{n_2} \tag{7}$$

where Q and n represent the flow rate and rotational speed, while the subscript 1 and 2 represent two different working conditions. It can be found that the flow rate is proportional to the rotational speed, so the flow rate of the centrifugal pump increases with the rise in rotational speed. To ensure the safety of the running system, seven different rotational speeds that are less than or equal to the nominal rotational speed are considered. The rotational speeds and corresponding flow rates are shown in Table 3.

Table 3. Rotational speeds and corresponding flow rates.

Rotational Speed, rpm	Flow Rate, m^3/h
1700	56.9
1900	64.8
2100	69.6
2300	74
2500	78.2
2700	82.1
2900	86

3.1. Directivity Characteristic of Radiation Noise under Various Rotational Speed Conditions

By measuring the SPL characteristic and calculating the TSPLs of 16 different points in the circumferential direction, the directivity of radiation noise is obtained under various rotational speed conditions.

As shown in Figure 5, it can be found that the TSPL increases from 80 to 100 dB with the increase of rotational speed. As a result of the symmetric characteristic of impeller, the directivity profile diagram of TSPL demonstrates the dipole symmetry, and one sees that the two TSPL valleys appear at 0° and 180°. In addition, apparent asymmetric characteristics also can be found due to the asymmetric structure of the volute, more concretely, the noise level in the direction facing the tongue (in the interval from 90° to 157.5°) is higher than that in the direction against the tongue (in the interval from 292.5° to 0°), which means that the radiation noise is the interaction result of impeller and volute dipole source.

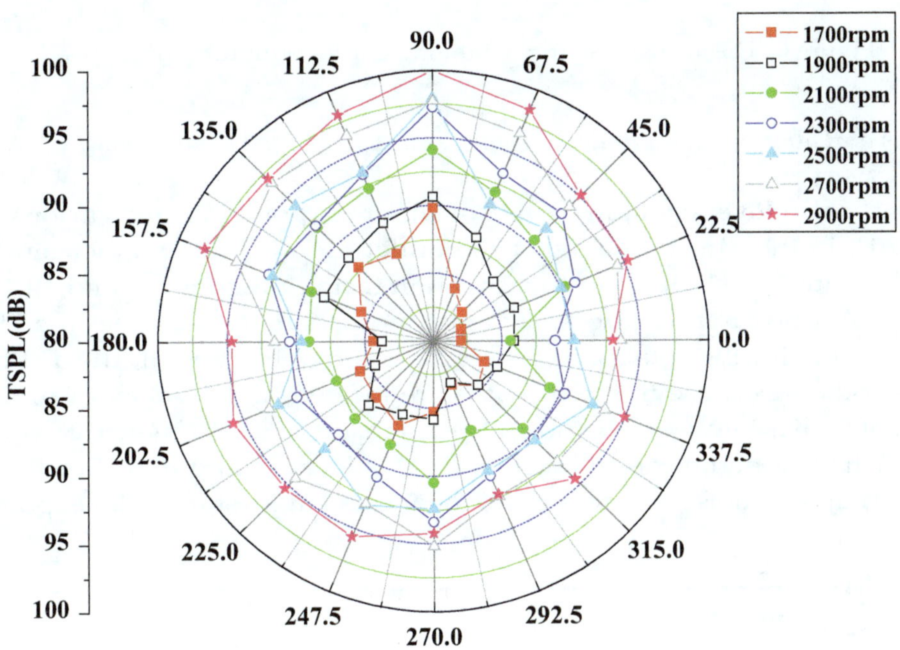

Figure 5. Directivity characteristic of radiation noise under various rotational speed conditions.

To further reveal the asymmetric characteristics under various rotational speed conditions quantitatively, the ratio of the acoustic energy in the interval from 90° to 157.5° (ε_1), as well as that in the interval from 292.5° to 0° (ε_2) to the total acoustic energy (ε_t) are calculated. As shown in Figure 6, the values of $\varepsilon_1/\varepsilon_t$ and $\varepsilon_2/\varepsilon_t$ change little with the change of rotational speed, and the value of $\varepsilon_1/\varepsilon_t$ fluctuates around 0.413, while the value of $\varepsilon_2/\varepsilon_t$ fluctuates near 0.157. This proves that the change of rotational speed would affect the acoustic energy, but it has little impact on the acoustic energy distribution. According to the past research [5–7], the volute tongue is the major noise source, and the measurement interval in the direction facing the tongue is closer to the tongue than other intervals. On the other hand, acoustic energy weakens gradually as the distance from the sound source increases, so the acoustic energy has less attenuation in the direction facing the tongue than that in the direction against the tongue, causing a much higher ratio of acoustic energy in the direction facing the tongue.

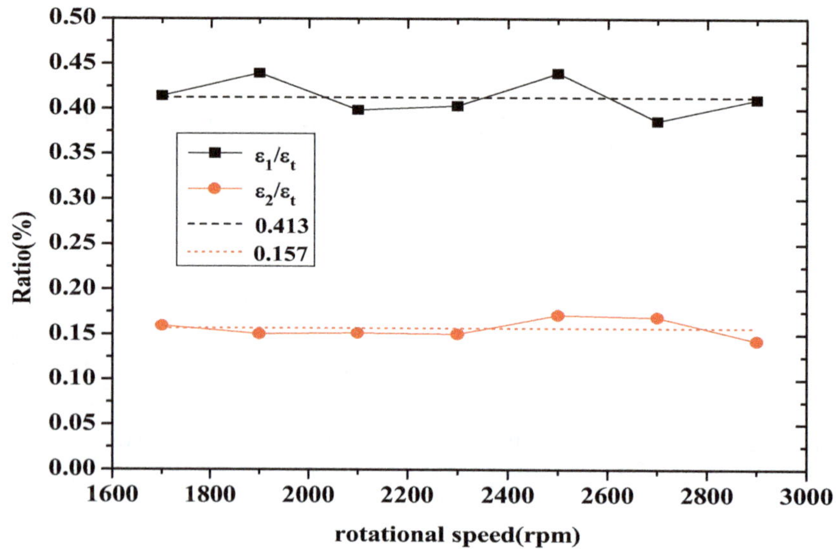

Figure 6. The acoustic energy changing curves with rotational speed.

3.2. Changing Rules of Radiation Noise under Various Rotational Speed Conditions

To analyze the noise level changing rules of different points under various rotational speed conditions, P1 (270°, in the direction against the outlet), P5 (0°, the valley point), P9 (90°, in the direction facing the outlet), P11 (135°, in the direction facing the tongue) and P13 (180°, the valley point) are selected. As shown in Figure 7, it can be more visually seen that the TSPLs of different monitoring points rise with the rotational speed increasing. In general, the TSPLs of P5 and P13 are lower than the others because the two points are located at the valley of dipole characteristic, meanwhile, TSPL of P1 lies between P5 and P11. As a result of the influence of volute tongue noise source, P9 and P11 have a higher noise level, and P9 is located in outlet direction, which is more affected by the internal flow noise in outlet pipe, so P9 is the highest noise level point.

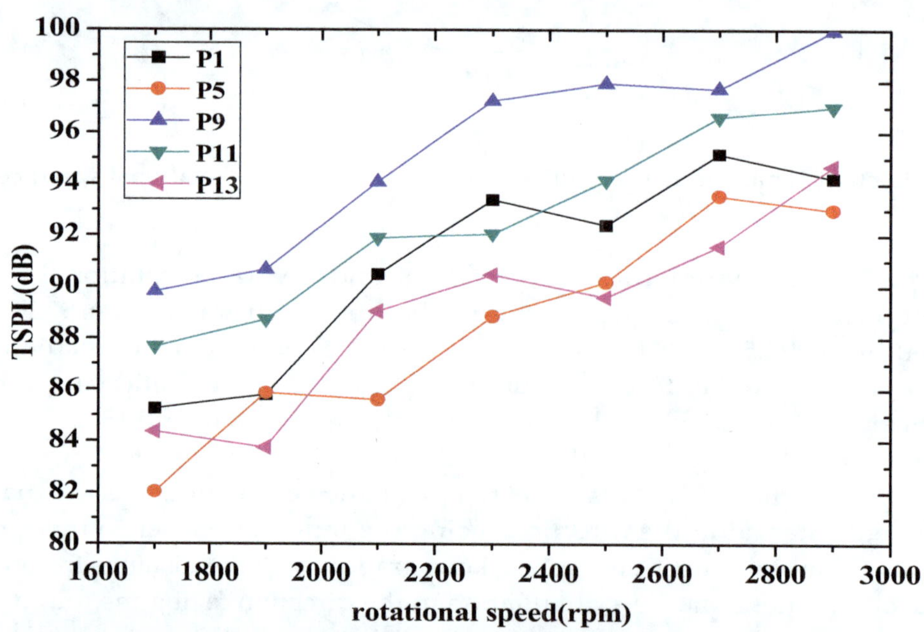

Figure 7. TSPL changing curves with rotational speed.

To evaluate the overall level of radiation noise, the L_{pA} of the 16 monitoring points is compared in Figure 8 under various rotational speed conditions. It can be observed that L_{pA} climbs gradually with the increase of rotational speed, and reaches to maximum at 2900 rpm. When rotational speed increases from 1700 to 2900 rpm, the L_{pA} increases by 12.40%. However, the growth slope decreases gradually with the increase of rotational speed generally, in particular, the L_{pA} grows rapidly in the interval between 1700 and 2300 rpm, with an average increase of 1.12 dB per 100 rpm, while in the range from 2300 to 2900 rpm, the growth rate of L_{pA} slows down, with an average increase of 0.65 dB per 100 rpm, which is less than that in the range from 1700 to 2300 rpm. It could be explained by that with the increase of rotational speed, the pressure fluctuation on wall surface also increases, but the increment rate of the variance of pressure fluctuation decreases [29], and causes a slow increase in acoustic pressure around the pump, and as a result, the growth slope of noise level decreases steadily.

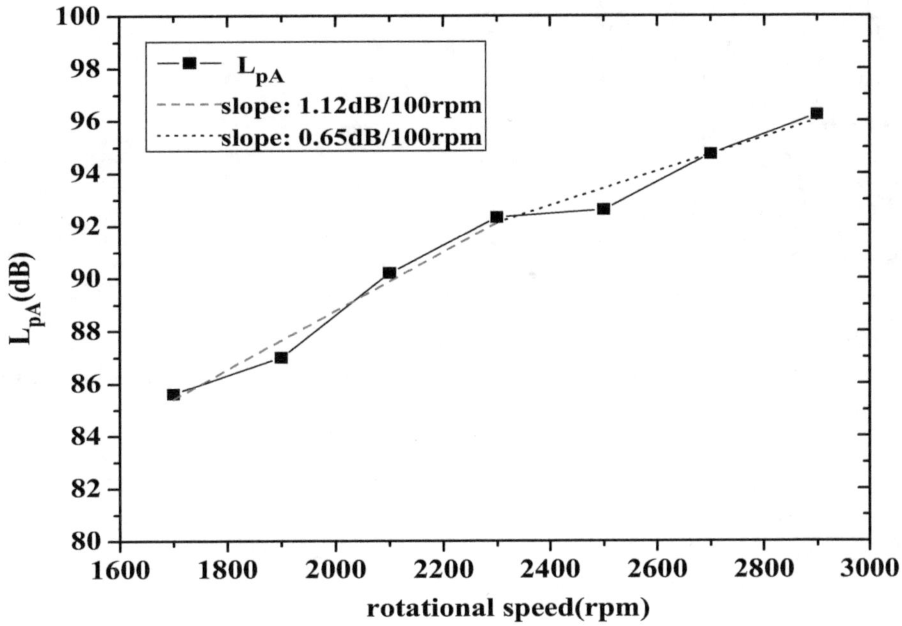

Figure 8. L_{pA} changing curve with rotational speed.

4. Radiation Noise Characteristics under Various Flow Rate Conditions

Compared with the variable speed regulation, the throttling regulation has the advantages of easy operation and low cost. Considering the two factors mentioned, throttling regulation is widely adopted at some occasions. In this section, the throttling regulation is realized by adjusting the valve opening on outlet pipe, then five different flow rates, i.e., 37.5, 50 (nominal flow rate), 62.5, 75 and 86 m^3/h (valve open fully), are studied to explore the radiation noise characteristic under various flow rate conditions when rotational speed is set as 2900 rpm.

4.1. Directivity Characteristic of Radiation Noise under Various Flow Rate Conditions

By calculating the TSPLs of 16 different points in the circumferential direction, then the directivity characteristic of radiation noise is analyzed under various flow rate conditions.

It is apparent in Figure 9 that TSPL changes between 86 dB and 100 dB, not only the dipole characteristic, but the asymmetric characteristic are presented in circumferential direction. And then the ratio of the acoustic energy in the interval from 90° to 157.5° (ε_1), as well as that in the interval from 292.5° to 0° (ε_2) to the total acoustic energy (ε_t) are compared under various flow rate conditions. As shown in Figure 10, the values of $\varepsilon_1/\varepsilon_t$ and $\varepsilon_2/\varepsilon_t$ show the fluctuation trend in the vicinity of 0.410 and 0.166, respectively. And these results are closer to those under various rotational speed conditions. It also can be found that the deviation between the value of $\varepsilon_1/\varepsilon_t$ and average value (0.410) is larger at 50 m^3/h. It could be explained by that in the measurement process, the measurement results are affected by the severe disturbance of the surrounding environment, but it does not affect the conclusion that the change of flow rate also has little impact on the acoustic energy distribution.

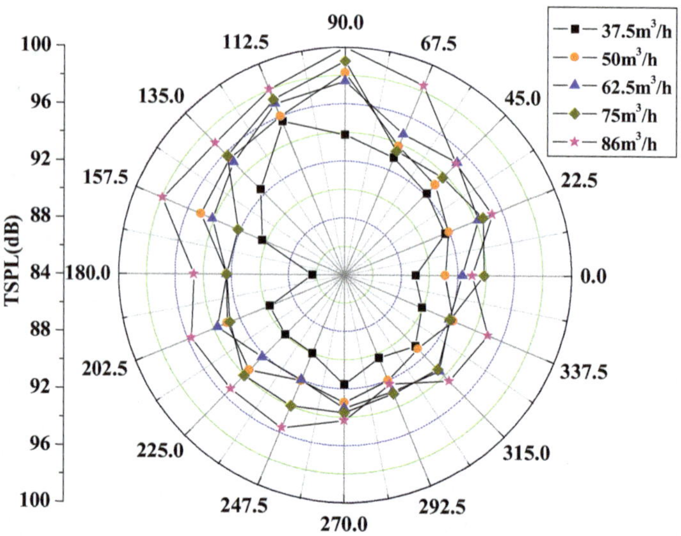

Figure 9. Directivity characteristics of radiation noise under various flow rate conditions (2900 rpm).

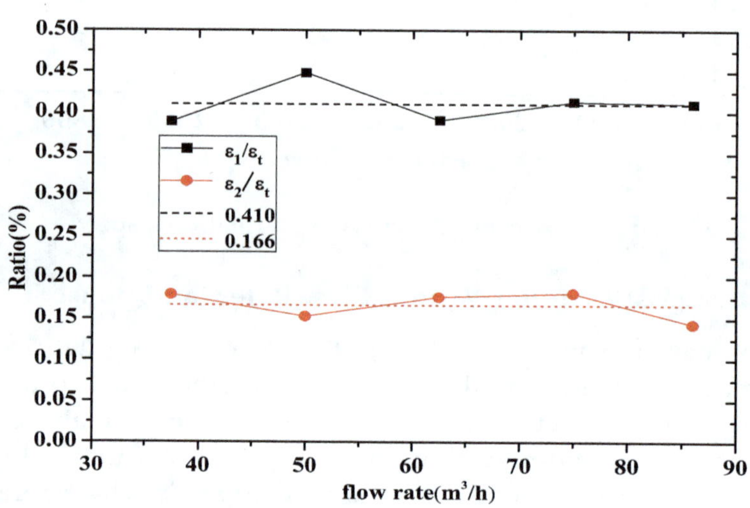

Figure 10. The acoustic energy changing curves with flow rate (2900 rpm).

4.2. *Changing Rules of Radiation Noise under Various Flow Rate Conditions*

In Figure 11, TSPLs of P1, P5, P9, P11 and P13 are also analyzed to reveal the noise level changing rules under various flow rate conditions. It is revealed that TSPLs of different points show similarly changing rules, specifically, the TSPL increases rapidly in small flow range from 37.5 to 50 m^3/h, then basically levels off in the interval from 50 to 75 m^3/h, and continues to increase when flow rate is higher than 75 m^3/h. What's more, the TSPL distribution characteristic of different points under various flow rate conditions also shows that the noise level at P5 and P13 are lower than others, while TSPLs of P9 and P11 are higher than others, and P9 is highest noise level point, which coincides with the distribution under various rotational speed conditions.

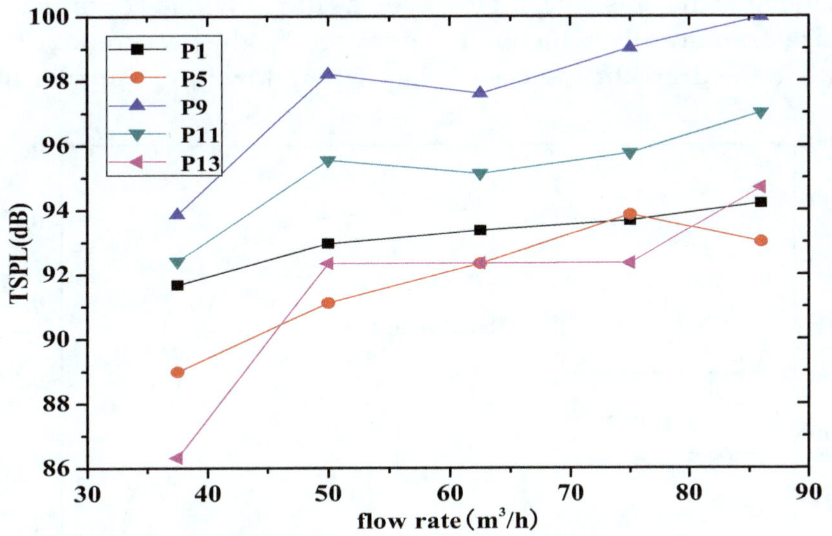

Figure 11. TSPL changing curves with flow rate (2900 rpm).

Furthermore, L_{pA} of 16 monitoring points is also studied under various flow rate conditions. It can be found in Figure 12 that the L_{pA} also increases sharply initially at small flow rate, then maintains stable, and reaches to maximum value at large flow rate interval. Additionally, L_{pA} changes a little with the increase of flow rate, which increases by 5.10% when flow rate grows from 37.5 to 86 m^3/h.

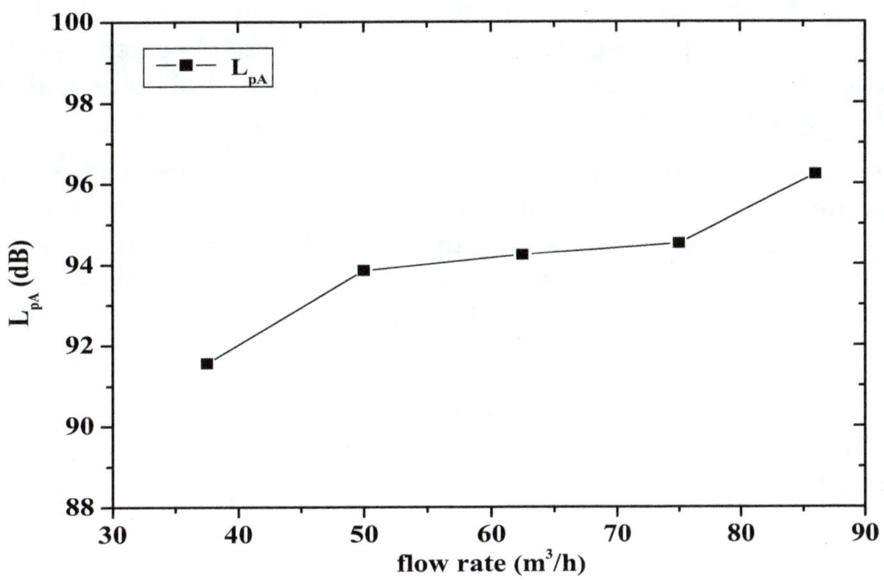

Figure 12. L_{pA} changing curve with flow rate (2900 rpm).

To explore the reasons for the changing rules of L_{pA} under various flow rate conditions, the head and efficiency are calculated, respectively. The changing rules of the two parameters are shown in Figure 13, it can be found that with the increase of flow rate, the head decreases gradually. In addition, the efficiency shows the tendency that increases from 37.5 to 50 m^3/h, and reaches to the maximum at 50 m^3/h, then remains essentially unchanged in the range from 50 to 75 m^3/h, which is the best efficiency range, and it keeps decreasing subsequently with the flow rate continues to grow. At low flow rate conditions, the pressure fluctuation inside the pump is low and causes low noise levels. With the increase of flow rate, the pressure fluctuation inside the pump increases accordingly, while in the best efficiency range, i.e., in the range from 50 to 75 m^3/h, the pressure becomes stable [17], and leads to little change for the noise level. However, as the flow rate continues to increase, the head and efficiency decrease dramatically, which may be caused by the occurrence of cavitation inside the pump, and contribute to the dramatic increase of L_{pA} when flow rate is greater than 75 m^3/h.

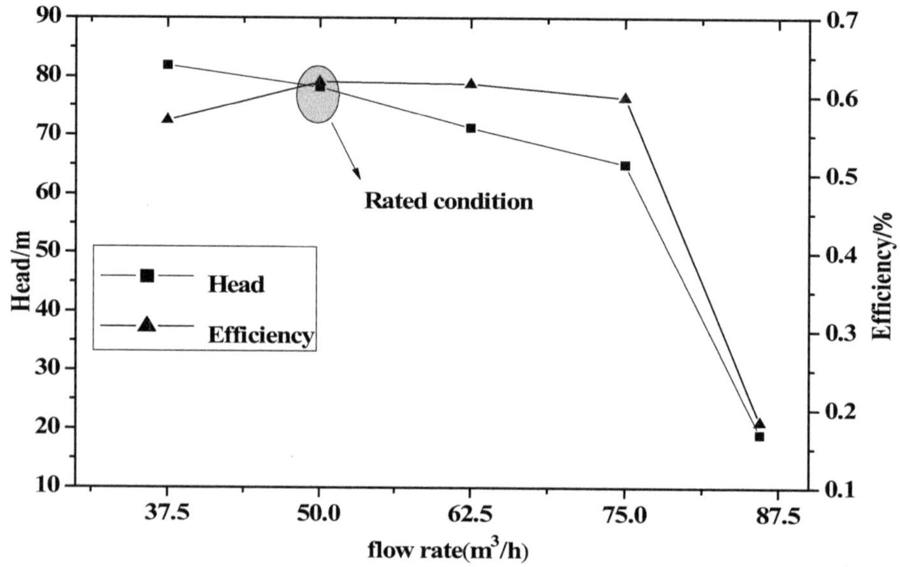

Figure 13. Head and efficiency changing curve with flow rate (2900 rpm).

5. Conclusions

In this paper, a centrifugal pump radiation noise measurement system is established, then the directivity distribution characteristics and overall level of radiation noise are analyzed under various working conditions. The main conclusions are drawn as follows:

(a) Under various working conditions, the total sound pressure level distribution for different monitoring points is dipolar; specifically, the two valley values appear at 0° (in the direction against the tongue) and 180°, and the minimum valley values are presented at the minimum rotational speed and minimum flow rate condition. Additionally, asymmetry is also validated, i.e., the noise level in the direction facing the volute tongue is higher than that in the direction against the tongue, and the monitoring point in outlet direction is the highest noise level point.

(b) The change in working conditions has little impact on the acoustic energy distribution of different intervals, and the ratios of the acoustic energy in the direction facing the tongue (ε_1), as well as that in the direction against the tongue (ε_2), to the total acoustic energy (ε_t) fluctuate around 0.410 and 0.160, respectively.

(c) In the operation of variable speed regulation, the average A-weighted sound pressure level (L_{pA}) increases gradually with the increasing of rotational speed, and it increases by 12.40% when rotational speed increases from 1700 to 2900 rpm, but the growth slope decreases gradually with the rise of pump rotational speed. While in the operation of throttling regulation, L_{pA} first increases, then remains stable, and continues to increase with the increase in flow rate, and it increases by 5.10% when flow rate grows from 37.5 to 86 m^3/h.

Acknowledgments: This paper is supported by National Natural Science Foundation of China (No. 51776111), Shandong Province Natural Science Foundation (No. ZR2016EEM35), and National Development and Reform Commission Foundation (No. 2013-1819).

Author Contributions: Chang Guo, Dongyue Lu designed the study, conducted the experiment and collected the experimental data; Kun Wang analyzed the experimental data; Chang Guo wrote the manuscript; Ming Gao reviewed and edited the manuscript.

References

1. Mao, D.X.; Hong, Z.H. *Environmental Noise Control Engineering*; Higher Education Press: Beijing, China, 2002.
2. Shah, S.R.; Jain, S.V.; Patel, R.N.; Lakhera, V.J. CFD for centrifugal pumps: A review of the state-of-the-art. *Procedia Eng.* **2013**, *51*, 715–720. [CrossRef]
3. Wu, D.F.; Liu, Y.S.; Li, D.L.; Zhao, X.F.; Li, C. Effect of materials on the noise of a water hydraulic pump used in submersible. *Ocean Eng.* **2017**, *131*, 107–113. [CrossRef]
4. Choi, J.S.; Mclaughlin, D.K.; Thompson, D.E. Experiments on the unsteady flow field and noise generation in a centrifugal pump impeller. *J. Sound Vib.* **2003**, *263*, 493–514. [CrossRef]
5. Chu, S.; Dong, R.; Katz, J. Relationship between unsteady flow, pressure fluctuation, and noise in a centrifugal pump—Part A: Use of PDV data to compute the pressure field. *J. Fluid Eng.* **1995**, *117*, 24–29. [CrossRef]
6. Chu, S.; Dong, R.; Katz, J. Relationship between unsteady flow, pressure fluctuation, and noise in a centrifugal pump—Part B: Effects of blade-tongue interactions. *J. Fluid Eng.* **1995**, *117*, 30–35. [CrossRef]
7. Dong, R.; Chu, S.; Katz, J. Effect of Modification to Tongue and Impeller Geometry on Unsteady Flow, Pressure Fluctuations, and Noise in a Centrifugal Pump. *J. Turbomach.* **1997**, *119*, 506–515. [CrossRef]
8. Parrondo, J.; Pérez, J.; Barrio, R.; González, J. A simple acoustic model to characterize the internal low frequency sound field in centrifugal pumps. *Appl. Acoust.* **2011**, *72*, 59–64. [CrossRef]
9. Cai, J.C.; Pan, J.; Andrew, G. Experimental study of the pressure fluctuation around the volute tongue of a centrifugal pump at variable rotating speed. *Fluid Mach.* **2015**, *43*, 13–16.
10. Rzentkowski, G.; Zbroja, S. Experimental characterization of centrifugal pumps as an acoustic source at the blade-passing frequency. *J. Fluid Struct.* **2000**, *14*, 529–558. [CrossRef]
11. Wang, Q.Y. Numerical and Experimental Research on the Flow-Induced Noise of Centrifugal Pump in the Seawater Pipe. Master's Thesis, Harbin Engineering University, Harbin, China, 2010.
12. Si, Q.R.; Yuan, S.Q.; Yuan, J.P. Experimental study on the influence of impeller-tongue gap on the performance and flow-induced noise characteristics of centrifugal pumps. *J. Vib. Shock* **2016**, *35*, 164–168.
13. Yuan, S.Q.; Yang, Y.; Yuan, J.P.; Luo, Y. Measurement system design of flow-induced noise in centrifugal pumps. *Drain. Irrig. Mach.* **2009**, *27*, 10–14.
14. Liu, H.L.; Wang, Y.; Yuan, S.Q.; Tan, M.G. Effects of impeller outlet width on the vibration and noise from centrifugal pumps induced by flow. *J. Huazhong Univ. Sci. Technol.* **2012**, *40*, 123–127.
15. Tan, M.G.; Wang, Y.; Liu, H.L.; Wu, X.F.; Wang, W. Effects of number of blades on flow induced noise vibration and noise of centrifugal pumps. *J. Drain. Irrig. Mach. Eng.* **2012**, *30*, 131–135.
16. Wang, Y.; Liu, H.L.; Liu, D.X.; Wang, J.; Wu, X.F. Effects of vane wrap angle on flow induced vibration and noise of centrifugal pumps. *Trans. Chin. Soc. Agric. Eng.* **2013**, *29*, 72–77.
17. Ye, X.M.; Pei, J.J.; Li, C.X.; Liu, Z. Experimental Study on Nosie Characteristics of Centrifugal Pump Based on Near-Field Acoustic Pressure Method. *Chin. J. Power Eng.* **2013**, *33*, 375–380.
18. Wang, M.; Freund, J.B.; Lele, S.K. Computational prediction of flow-generated sound. *Annu. Rev. Fluid Mech.* **2015**, *38*, 483–512. [CrossRef]
19. Liu, H.L.; Dai, H.W.; Ding, J.; Tan, M.G.; Wang, Y.; Huang, H.Q. Numerical and experimental studies of hydraulic noise induced by surface dipole sources in a centrifugal pump. *J. Hydrodyn.* **2016**, *28*, 43–51. [CrossRef]
20. Langthjem, M.A.; Olhoff, N. A numerical study of flow-induced noise in a two-dimensional centrifugal pump. Part I. Hydrodynamics. *J. Fluid Struct.* **2004**, *19*, 349–368. [CrossRef]
21. Langthjem, M.A.; Olhoff, N. A numerical study of flow-induced noise in a two-dimensional centrifugal pump. Part II. Hydroacoustics. *J. Fluid Struct.* **2004**, *19*, 36–386. [CrossRef]
22. Huang, J.X.; Geng, S.J.; Wu, R.; Liu, K.; Nie, C.Q.; Zhang, H.W. Comparison of noise characteristics in centrifugal pumps with different types of impellers. *Acta Acoust.* **2010**, *35*, 113–118.
23. Ma, Z.L.; Chen, E.Y.; Guo, Y.L.; Yang, A.L. Numerical Simulation of the Influence of the Diameter at the Outlet of an Impeller on the Noise Level Induced by the Flow Inside a Centrifugal Pump. *J. Eng. Therm. Energy Power* **2016**, *31*, 93–98.
24. Liu, H.L.; Ding, J.; Tan, M.G.; Cui, J.B.; Wang, Y. Analysis and experimental of centrifugal pump noise based on outlet width of impeller. *Trans. Chin. Soc. Agric. Eng.* **2013**, *29*, 66–73.
25. Gao, M.; Dong, P.X.; Lei, S.H.; Turan, A. Computational Study of the Noise Radiation in a Centrifugal Pump When Flow Rate Changes. *Energies* **2017**, *10*, 221. [CrossRef]

26. Dong, P.X.; Gao, M.; Guan, H.J.; Lu, D.Y.; Song, K.Q.; Sun, F.Z. Numerical simulation for variation law of volute radiated noise in centrifugal pumps under variable rotating speed. *J. Vib. Shock* **2017**, *36*, 128–133.
27. Tao, J.Y.; Lu, X.N.; Wang, L.L. *Methods of Measuring and Evaluating Noise of Pumps*; Chinese Quality Supervision Bureau: Beijing, China, 2013.
28. Du, G.H.; Zhu, Z.M.; Gong, X.F. *Fundamentals of Acoustics*; Nanjing University Press: Nanjing, China, 2001.
29. Dong, P.X. Computational Study on the Noise Radiation of a Centrifugal Pump Based on 3-D Flow field with Varying Working Condition. Master's Thesis, Shandong University, Jinan, China, 2016.

13

A Performance Prediction Method for Pumps as Turbines (PAT) using a Computational Fluid Dynamics (CFD) Modeling Approach

Emma Frosina *, Dario Buono and Adolfo Senatore

Department of Industrial Engineering, University of Naples Federico II, Via Claudio, 21-80125 Naples, Italy; darbuono@unina.it (D.B.); senatore@unina.it (A.S.)

* Correspondence: emma.frosina@unina.it.

Academic Editor: Bjørn H. Hjertager

Abstract: Small and micro hydropower systems represent an attractive solution for generating electricity at low cost and with low environmental impact. The pump-as-turbine (PAT) approach has promise in this application due to its low purchase and maintenance costs. In this paper, a new method to predict the inverse characteristic of industrial centrifugal pumps is presented. This method is based on results of simulations performed with commercial three-dimensional Computational Fluid Dynamics (CFD) software. Model results have been first validated in pumping mode using data supplied by pump manufacturers. Then, the results have been compared to experimental data for a pump running in reverse. Experimentation has been performed on a dedicated test bench installed in the Department of Civil Construction and Environmental Engineering of the University of Naples Federico II. Three different pumps, with different specific speeds, have been analyzed. Using the model results, the inverse characteristic and the best efficiency point have been evaluated. Finally, results have been compared to prediction methods available in the literature.

Keywords: energy saving; PAT; urban hydraulic network; numerical modeling

1. Introduction

Electricity generation presents many issues and is studied with different techniques in order to reduce its production cost and environmental impact. Conventional production with fossil fuels presents problems associated with the high cost, rapid depletion and detrimental environmental effects of these fuels. Renewable energy is probably the best solution for environmental issues, and many solutions have been developed since the last century, such as hydropower, hydrogen, fuel cells, biofuels, and solar power generation.

Among the renewable sources, small hydropower represents a very attractive source of energy generation. In many countries, small and micro hydropower systems are an important means of electricity generation. An efficient solution, from the point of view of energy efficiency, is the adoption of a turbine, but the purchase and maintenance costs of turbines make their implementation economically unattractive, especially for small hydropower [1–6].

Reverse-running centrifugal pumps (also called pumps as turbines or PAT) are a solution for generating and recovering power in small and micro hydropower situations. Pumps are relatively simple machines, inexpensive (compared to a hydraulic turbines), and readily available worldwide. It has been estimated that the capital payback period of a reverse-running pump in the range of 5–50 kW is less than two years [7,8]. Moreover, the use of PAT could be suitable because manufacturers of turbines worldwide are less numerous than pump producers, the market for turbines is smaller compared to pumps, and pumps are mechanically simple and require less maintenance. Moreover, an integral pump and electric motor can be purchased for use as a turbine and generator set; pumps are available in a wide range of heads and flows and in a large number of standard sizes. Generally, pumps have short delivery times, spare parts (such as seals and bearings) are easily available and the installation can be done using standard pipes and fittings.

The use of a pump running in reverse mode to generate electricity is not new; the first applications started almost 80 or 90 years ago and many theoretical and experimental studies have been done [2–6,9,10]. Much research is still being conducted, especially to predict the operating conditions and the efficiency of centrifugal pumps running in reverse [11].

The selection of a proper PAT for an existing site represents a critical issue because pump manufacturers do not supply the characteristic of the pump running in reverse. Many methods have been used to predict the inverse characteristic of a pump, based on numerical models, experiments, or theoretical procedures [4–10].

This research has demonstrated that these methods can be used only for a limited set of pumps. None of them, in fact, allows prediction for the reverse running conditions for all geometries and over a wide range of pump specific speeds. Several studies, based on a modeling approach with CFD code, are available and generally show good correspondence with the available experimental data [4,10].

A study [4] carried out with a computational model of a PAT is based on the concept called "flow zone". The flow regime within a PAT is divided into four major flow regions (volute casing, impeller, casing outlet and draft tube). A comparison has been made between the experimental and numerical results of a single stage end suction centrifugal pump that was operated in turbine mode at a speed of 800 rpm. CFD predictions of the hydraulic parameters were in good correspondence with the experimental results, but deviations (within 5% to 10%) have been found at certain load regions.

Nautiyal et al. [5,6] carried out a study on the application of CFD and its limitations for PAT using cases reported by previous researchers [4,9,10]. The study reported that CFD analysis was an effective design tool for predicting the performance of centrifugal pumps in turbine mode and for identifying the losses in turbo-machinery components such as the draft tube, impeller and casing, but there was some deviation between the experimental results and the CFD modeling results. Barrio et al. [11] carried out a numerical investigation on the unsteady flow in commercial centrifugal pumps operating in direct and reverse mode with the help of CFD code. The results of their simulation were in good correspondence with the experimental results. The study revealed that in the reverse mode, the flow only matched the geometry of the impeller at nominal conditions; re-circulating fluid regions developed at low flow rates (near the discharge side of the blades) and high flow rates (near the suction side).

Many correlations based on theoretical approaches are available to predict the performance of a PAT. Several researchers (Stepanoff, Childs, Sharma, Wong, Williams, Alatorre-Frenk, and others) have presented correlations for predicting the performance of a pump-as-turbine [5]. These correlations were based upon either pump efficiency or specific speed. However, deviations of more than 20% have been found between the experimental and predicted reverse operation of standard pumps [12]. The objective of these correlations is to calculate the best efficiency point (BEP) of pumps for operation in turbine mode by using the pump operation data provided by the manufacturer.

In 1962, Childs [13] presented a PAT prediction method based on the efficiency of the pump. A similar approach was then presented by McClaskey and Lundquist [14] and Lueneburg and Nelson [15] in 1976 and 1985, respectively.

Hancock [16] stated that for most pumps the turbine BEP lies within 2% of the pump mode BEP. Grover and Hergt [17,18] proposed a PAT prediction method based on specific speed for the turbine mode (obtained similar to the specific speed for a pump). Grover's method is applicable for the turbine mode specific speed range between 10 and 50 [17]. A comparison between experimental results and the methods proposed by the above researchers show relatively large deviations; therefore, the use of these formulae must be confined to an approximate selection of PATs.

Finally, a large number of experimental studies can be used to evaluate the inverse characteristics. These are often limited to the specific pumps tested, so that they cannot serve as a valid tool for pump selection, but are very useful for tuning and validating theoretical and modeling analyses.

In this paper, authors present a methodology for obtaining the reverse characteristics of a pump, starting from the results of three-dimensional CFD models. After a description of all prediction methods available in the literature, in the third section the adopted modeling approach is described. Three pumps have been studied and modeled using the three-dimensional CFD commercial code PumpLinx®, developed by Simerics Inc.® (1750 112th Ave NE, Ste C250, Bellevue, WA, USA). In the fourth section, the test bench layout is shown with all the transducers' characteristics.

Numerical models have been validated with experimental data obtained on a dedicated test bench installed in the Department of Civil Construction and Environmental Engineering of the University of Naples Federico II. Simulations have been run in both direct (as pump) and reverse (as turbine) modes with good accuracy.

In the fifth section, the models' results have been used to predict the efficiency curves of the three analyzed pumps in both modes. At the end of the paper, results of the proposed methodology have been compared to the prediction methods described in the second section. Analysis has demonstrated that the existing prediction methods underestimate or overestimate the real operation.

In this paper, the authors have shown only the first step of a research done on PAT. Other pumps are under study using the same modeling approach in order to realize a macro database for the prediction of pump performance as turbines. The final aim is the identification of a new prediction method more accurate than the others already available in literature. Using the new prediction method, a reduction of the experimentation will then be realized allowing an easy and fast choice of the PAT for each application.

2. Literature Overview on Prediction Methods

A methodology to calculate the inverse characteristic of a commercial pump is presented in this paper. The proposed approach is based on results of CFD modeling using a commercial code developed to simulate centrifugal machines. Therefore, it is important to describe prediction methods already available in literature. In fact, these methods, will be used in the last section of the paper to analyze the proposed methodology and to discuss our results. The following equations summarize different methods to predict the pump inverse characteristics. They are based on theoretical or experimental analyses [13–24]. Stepanoff [19] calculates the head, flow rate at the BEP in reverse mode using the efficiency, head and flow rate value at the BEP in direct mode. All relations between head and flow rate are reported in Equation (1).

$$\frac{H_t}{H_p} = \frac{1}{\eta_p}; \quad \frac{Q_t}{Q_p} = \frac{1}{\sqrt{\eta_p}}; \quad \eta_T = \eta_P; \quad N_{st} = N_s \eta_p \tag{1}$$

Alatorre-Frenk [20] calculates the head, flow rate and efficiency at the BEP in reverse mode using the efficiency, head and flow rate value at the BEP in direct mode. Correlations are presented by following:

$$\frac{H_t}{H_p} = \frac{1}{0.85\eta_p{}^5 + 0.385}; \quad \frac{Q_t}{Q_p} = \frac{0.85\eta_p{}^5 + 0.385}{2\eta_p{}^{9.5} + 0.205}; \quad \eta_t = \eta_p - 0.03 \tag{2}$$

The prediction method developed by Sharma [21] calculates the head and flow rate at the BEP in reverse mode using the efficiency, head and flow rate value at the BEP in direct mode:

$$\frac{H_t}{H_p} = \frac{1}{\eta_p{}^{1.2}}; \quad \frac{Q_t}{Q_p} = \frac{1}{\eta_p{}^{0.8}}; \quad P_t = P_p; \quad \eta_t = \eta_p \tag{3}$$

Schmiedl [22] as Sharma [21] calculates the head and flow rate at the BEP in reverse mode using the efficiency, head and flow rate value at the BEP in direct mode. Correlations are reported by following:

$$\frac{H_t}{H_p} = -1.4 + \frac{2.5}{\eta_p}; \quad \frac{Q_t}{Q_p} = -1.5 + \frac{2.4}{\eta_p{}^2} \frac{\eta_t}{p} = 1.158 - 0.265N_{st} \tag{4}$$

Head and flow rate at the BEP in reverse mode can be evaluated with correlations of Grover [17] using the specific speed value of the PAT at the BEP ($N_{st} = N_s\eta_p$). All relations between head and flow rate are listed by following.

$$\begin{cases} \frac{H_t}{H_p} = 2.693 - 0.0229N_{st} \\ \frac{Q_t}{Q_p} = 2.379 - 0.0264N_{st} \\ \frac{\eta_T}{\eta_P} = 0.893 - 0.0466N_{st} \end{cases} \tag{5}$$

As Grover [17], knowing that $N_{st} = N_s\eta_p$, Hergt [18] calculates the head and flow rate at the BEP in reverse mode using the specific speed value of the PAT at the BEP:

$$\frac{H_t}{H_p} = 1.3 - \frac{6}{N_{st}-3}; \quad \frac{Q_t}{Q_p} = 1.3 - \frac{1.6}{N_{st}-5} \tag{6}$$

Relations of Childs [13] evaluate the head and the flow rate at the BEP in reverse mode using the efficiency, head and flow rate value at the BEP in direct mode. Equations are reported by following:

$$\frac{H_t}{H_p} = \frac{1}{\eta_p}; \quad \frac{Q_t}{Q_p} = \frac{1}{\eta_p}; \quad \eta_t = \eta_p \tag{7}$$

Moreover, Derakhshan and Nourbakhsh [24] introduced a method based on theoretical analysis to evaluate the BEP of an industrial centrifugal pump. This method is based on the geometrical and hydraulic characteristics of the pump in direct mode. The final formula to evaluate the turbine's maximum efficiency is:

$$\eta_t = \frac{P_{nt}}{\gamma \times Q_t \times H_t} = \frac{\gamma \times Q_t \times H_t - P_{vt} - P_{lt} - P_{et} - P_{it} - P_{mt}}{\gamma \times Q_t \times H_t} \tag{8}$$

All of presented methods are based on different hypotheses. In Section 6 of this paper, all methods will be compared with the results of the proposed modeling methodology. The comparison has confirmed that each prediction method can be used only for a limited set of pumps. None of them, in fact, allows the prediction in the reverse running conditions for all geometries and over a wide range of pump specific speeds. In some cases, performance is underestimated or overestimated by as much as 30%.

3. Simulation Model

Three different centrifugal pumps have been modeled in order to obtain the necessary data to predict the performance of the pumps by the described procedures. The analyzed pumps are commercial ones and have three different specific speeds. The main characteristics are summarized in Table 1. It was decided to use pumps with different heads (from 3.9 to 60 m) and flow rates (from 45.4 to 148 m^3/s) to have different geometries and operating conditions to better test the prediction method.

Table 1. Pumps characteristics.

	Impeller Diameter (mm)	Delivery Outlet Diameter (mm)	H_{bep} (m)	Q_{bep} (m^3/h)
(*Ns* 37.6)	190	80	39	148
(*Ns* 20.5)	200	70	60	45.4
(*Ns* 64.0)	120	80	3.9	54

On the top side of Figure 1, the disassembled pump with a $N_S = 37.6$ is shown. This pump is a shrouded one with one- channel impeller and six blades and is called Pump 1. Starting from the real geometry in a .step format, the fluid volume has been extracted. In Figure 1, the fluid-volume is colored in green while the solid impeller is in blue and the solid rotor in red. In the same way, fluid volumes of others two pumps have been extracted and then modelled using a 3D-CFD approach.

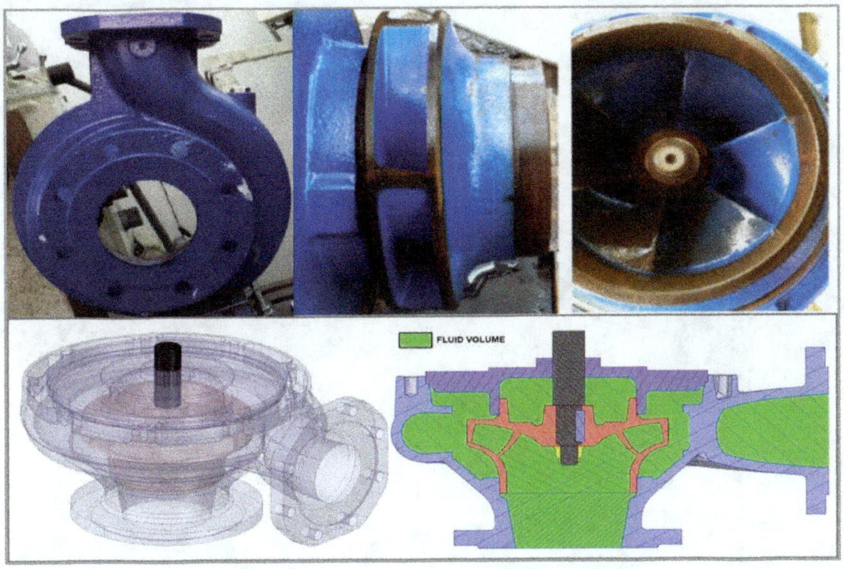

Figure 1. Geometry and fluid volume.

The study has been approached using the commercial code PumpLinx®. PumpLinx® is a three-dimensional CFD software developed by Simerics Inc. (1750 112th Ave NE, Ste C250, Bellevue, WA 98004, USA) [25–29]. It numerically solves the fundamental conservation equations of mass, momentum and energy and includes robust models of turbulence and cavitation.

The fluid volume of each pump has then been meshed with the PumpLinx® grid generator using a body-fitted binary tree approach. These grids have been demonstrated to be extremely accurate and efficient [25–29]. In fact, the parent-child tree architecture allows for an expandable data structure with reduced memory storage, the binary refinement is optimal for transitioning between different length scales and resolutions within the model, the majority of cells are cubes, and, since the grid is created from a volume, it can tolerate inaccurate CAD surfaces with small gaps and overlaps. It is important to underline that, cells are hexagonal not deformed therefore the skewness is zero [30].

Figure 2 shows the binary tree mesh of three pumps under study: Pump 1 (*Ns* = 37.6), Pumps 2 (*Ns* = 20.5) and Pump 3 (*Ns* = 64).

It is important to underline that using the binary tree approach in the boundary layer can easily increase the grid density on the surface without excessively increasing the total cell count. In this way, the grid has been subdivided and cut to conform it to the surface in regions of high curvature and small details [27]. Impellers and rotors fluid volumes of each model have been meshed separately. A maximum cell size of 0.025 has been chosen, where no cell in the volume can have a cell side larger than the maximum cell size. The minimum cell size has been fixed at 0.0001. The minimum cell size is a parameter used to limit how small cells can be in attempting to resolve the geometry using the general mesher. No cell in the volume can have a cell side smaller than the minimum cell size. The cell size on surfaces has also been fixed at 0.00625. This parameter is used to control the size of the cells for all surfaces of a mesh volume. Using a pure Eulerian approach the mesh of rotors is deformed by squeezing and expanding the cells. The rotor mesh is colored in green in Figure 2.

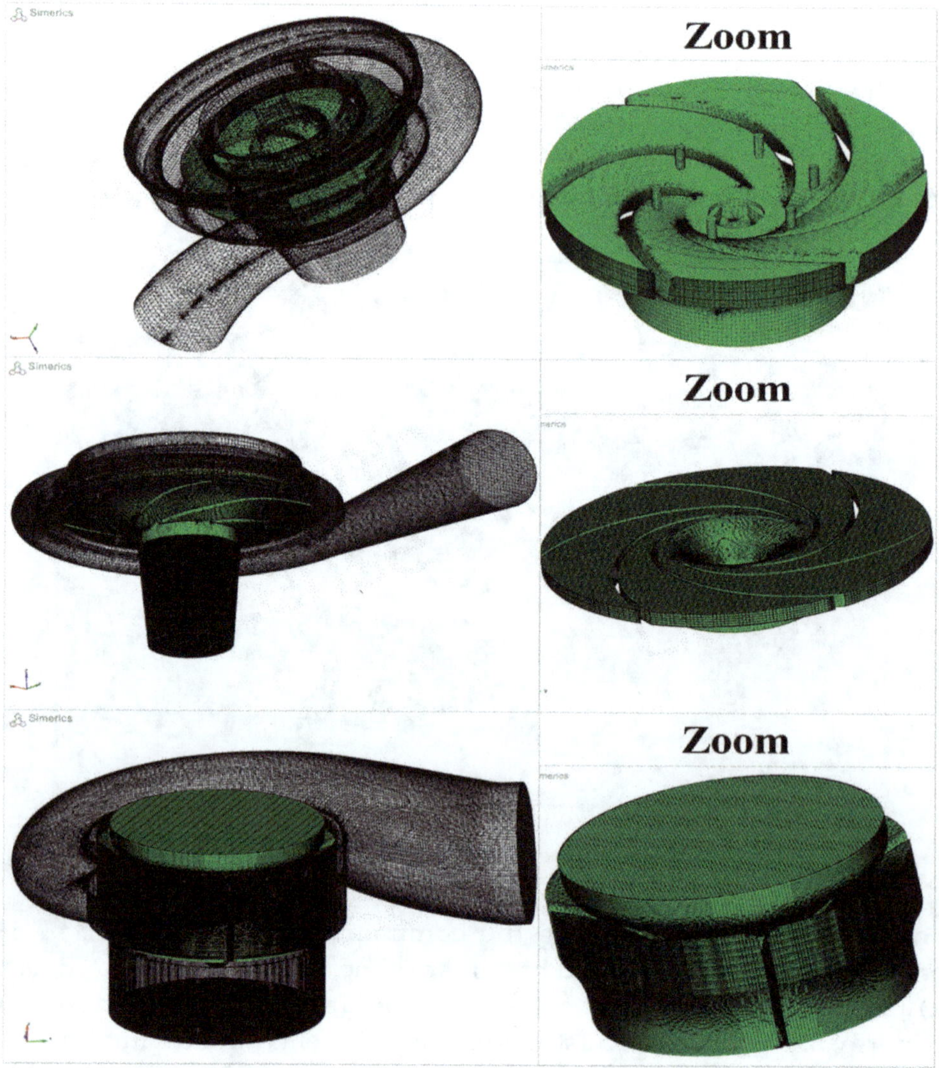

Figure 2. Binary tree mesh for three pumps.

A mesh sensitivity, as well known, is fundamental in studying any new problem with a CFD solver. The target is to allow excellent accuracy (in comparison with experimental tests) and computational efficiency. Pumps fluid volumes have been meshed increasing and decreasing the already described parameters: maximum cell size, minimum cell size and the cell size on surfaces. For each case the best compromise between results accuracy and computational time has been found and by following all mesh characteristic are listed:

PUMP 1 ($N_S = 37.6$): Total number of cells: 851.673

Total number of faces: 3.383.745

Total simulation time: 8.9 h as Pump, 9 h as PAT

PUMP 2 ($N_S = 20.5$): Total number of cells: 1.039.450

Total number of faces: 3.926.412

Total simulation time: 4.2 h as Pump, 4.8 h as PAT

PUMP 3 ($N_S = 64$): Total number of cells: 324.596

Total number of faces: 3.348.318

Total simulation time: 4.5 h as Pump, 4.8 h as PAT

Simulations have been run with an Intel(R) Xeon(R) CPU 2.66 GHz (two processors). It is important to underline that models are transient, and during simulations there is a simultaneous treatment of moving and stationary fluid volumes. In particular, each volume connects to the others via an implicit interface. Mismatched grid interfaces can identify overlap areas and match them without interpolation. These faces, during the simulation process, are treated no differently than an internal face between two neighboring cells in the same grid domain.

Using the models described it has been possible to study the internal fluid dynamics of each pump working in the direct (as pump) and reverse (as turbine) modes. A mature turbulence model has been implemented. It has been demonstrated that for these applications one of the more accurate model to study turbulence is the k–ε model and RNG k–ε model. Other turbulence models, such as LES and RNG k–ε model, might also provide good results but the adoption of higher order turbulence models would have increased the computational time with no relevant improvement of the results. Authors have also applied the presented strategy confirming the solution accuracy [25–29,31–33] in other analyses.

Therefore, in this research, the k-epsilon model has been used providing a good accuracy and computationally efficiency. This model is used by the adopted CFD code since it has been available for more than a decade and has been widely demonstrated to provide good engineering results for a wide range of applications [29,31–33]. The standard k–ε model, used for the simulations presented in this paper is based on the following two equations [25–29,31,32]:

$$\frac{\partial}{\partial t}\int_{\Omega(t)}^{0} \rho k d\Omega + \int_{\sigma}^{0} \rho((v - v_\sigma) \times n) k d\sigma = \int_{\sigma}^{0} \left(\mu + \frac{\mu_t}{\sigma_k}\right)(\nabla k \times n) d\sigma + \int_{\Omega}^{0} (G_t - \rho\epsilon) d\Omega \quad (9)$$

$$\frac{\partial}{\partial t}\int_{\Omega(t)}^{0} \rho\epsilon d\Omega + \int_{\sigma}^{0} \rho((v - v_\sigma) \times n) \varepsilon d\sigma = \int_{\sigma}^{0} \left(\mu + \frac{\mu_t}{\sigma_\varepsilon}\right)(\nabla \epsilon \times n) d\sigma + \int_{\Omega}^{0} \left(c_1 G_t \frac{\varepsilon}{k} - c_2 \rho \frac{\varepsilon^2}{k}\right) d\Omega \quad (10)$$

where $c_1 = 1.44$, $c_2 = 1.92$, $\sigma_k = 1$, $\sigma_\varepsilon = 1.3$, where σ_k and σ_ε are the turbulent kinetic energy and the turbulent kinetic energy dissipation rate Prandtl numbers.

The turbulent kinetic energy, k, is defined as [25–29,31–33]:

$$k = \frac{1}{2}(v' \cdot v') \quad (11)$$

where v' is the turbulent fluctuation velocity, and the dissipation rate, ε, of the turbulent kinetic energy is defined as [25–29]:

$$\varepsilon = 2\frac{\mu}{\rho}\overline{\left(S'_{ij} S'_{ij}\right)} \quad (12)$$

In which the strain tensor is [29,31]:

$$S'_{ij} = \frac{1}{2}\left(\frac{\partial u'_i}{\partial x_j} + \frac{\partial u'_j}{\partial x_i}\right) \quad (13)$$

with u_i' ($i = 1, 2, 3$) being components of v'.

The turbulent viscosity μ_t is calculated by [16,23,24,32,33]:

$$\mu_t = \rho C_\mu \frac{k^2}{\epsilon} \quad (14)$$

with $C_\mu = 0.09$.

The turbulent generation term G_t can be expressed as a function of velocity and the shear stress tensor as [16,23,24]:

$$G_t = -\rho\overline{u'_i u'_j}\frac{\partial u'_i}{\partial x_j} \tag{15}$$

where $\tau'_{ij} = \rho\overline{u'_i u'_j}$ is the turbulent Reynolds stress, which can be modelled by the Boussinesq hypothesis [15,16,23,24,32,33]:

$$\tau'_{ij} = \mu_t\left(\frac{\partial u_i}{\partial x_j} + \frac{\partial u_j}{\partial x_i}\right) - \frac{2}{3}\left(\rho k + \frac{\partial u_k}{\partial x_k}\right)\delta_{ij} \tag{16}$$

With the built models, simulations have first been run comparing the results in the direct and reverse working modes. Model results for Pump 1 at 2900 rpm are shown in Figure 3; in Figure 3a the pressure distribution at the BEP in the 0–9 bar pressure range is presented. The fluid properties of water have been used in all the simulations.

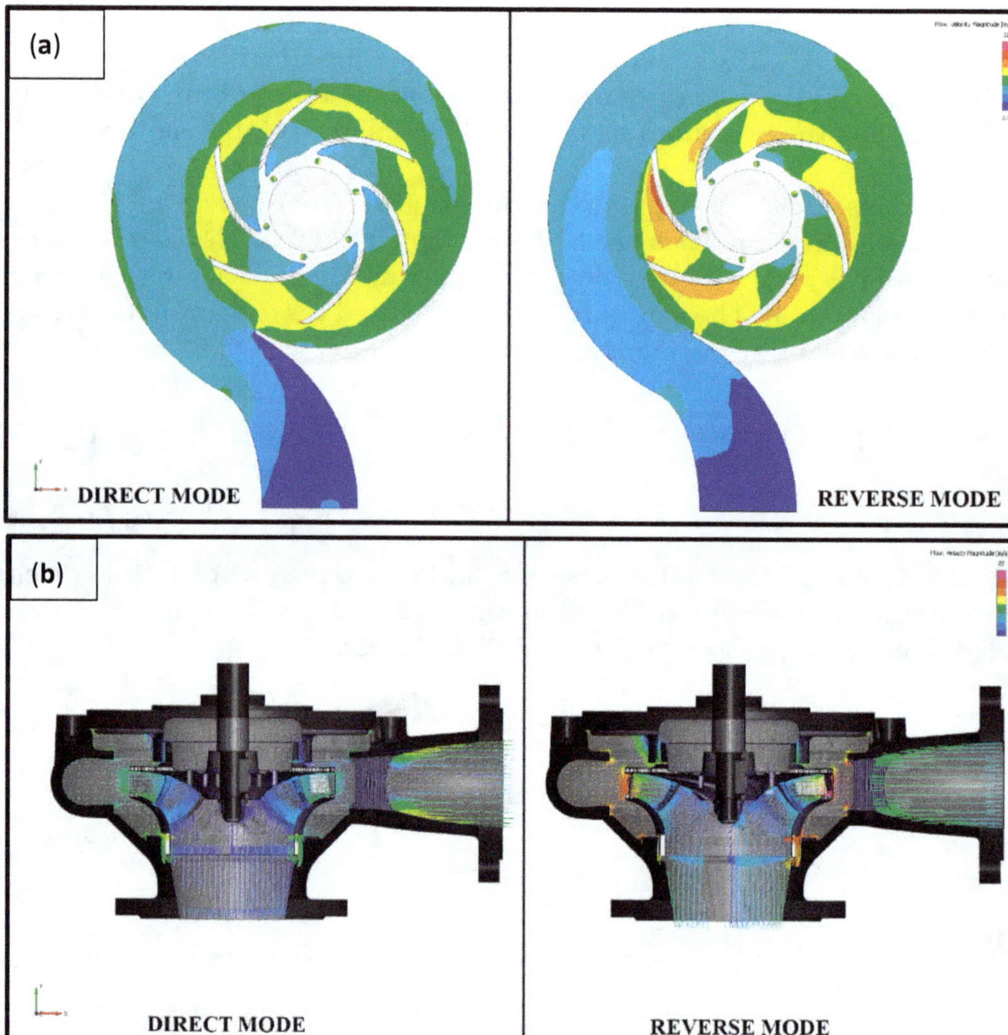

Figure 3. Model results for Pump 1 (Ns = 37.6) at 2900 rpm. (**a**) Pressure distribution; (**b**) velocity vectors.

Figure 3b shows the velocity vectors in the fluid volume; the velocity range (0–32 m/s) is the same for the direct and reverse modes. In this picture, it is possible to visualize the flow evolution inside the machine and the acceleration/deceleration of the fluid. Both figures confirm that the velocity is higher in the reverse mode.

Similarly, Figure 4 shows the pressure distributions for pump 2 (N_S = 20.5) and pump 3 (N_S = 64) at 2900 rpm.

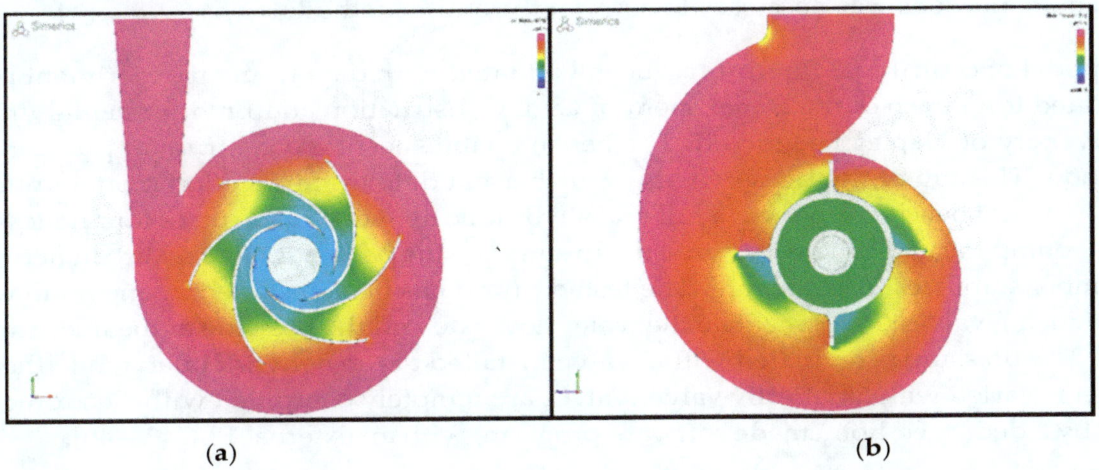

Figure 4. Pressure distribution in the fluid volume of pumps 2 and 3 at 2900 rpm. (**a**) Ns = 20.5; (**b**) Ns = 64.

For the direct mode, CFD models have been validated using the data supplied by the pump manufacturers. In Figure 5, the head vs. flow rate plots (as the blue curves) are shown. Across the range of flow rates (30–207) m^3/h, the head varies from 47 m to 3 m. In the plots in Figure 5, the model results are shown in red. The comparison in Figure 5 demonstrates the accuracy of the adopted methodology; in fact, the percentage error is always less than 4% while for many points the error is near zero.

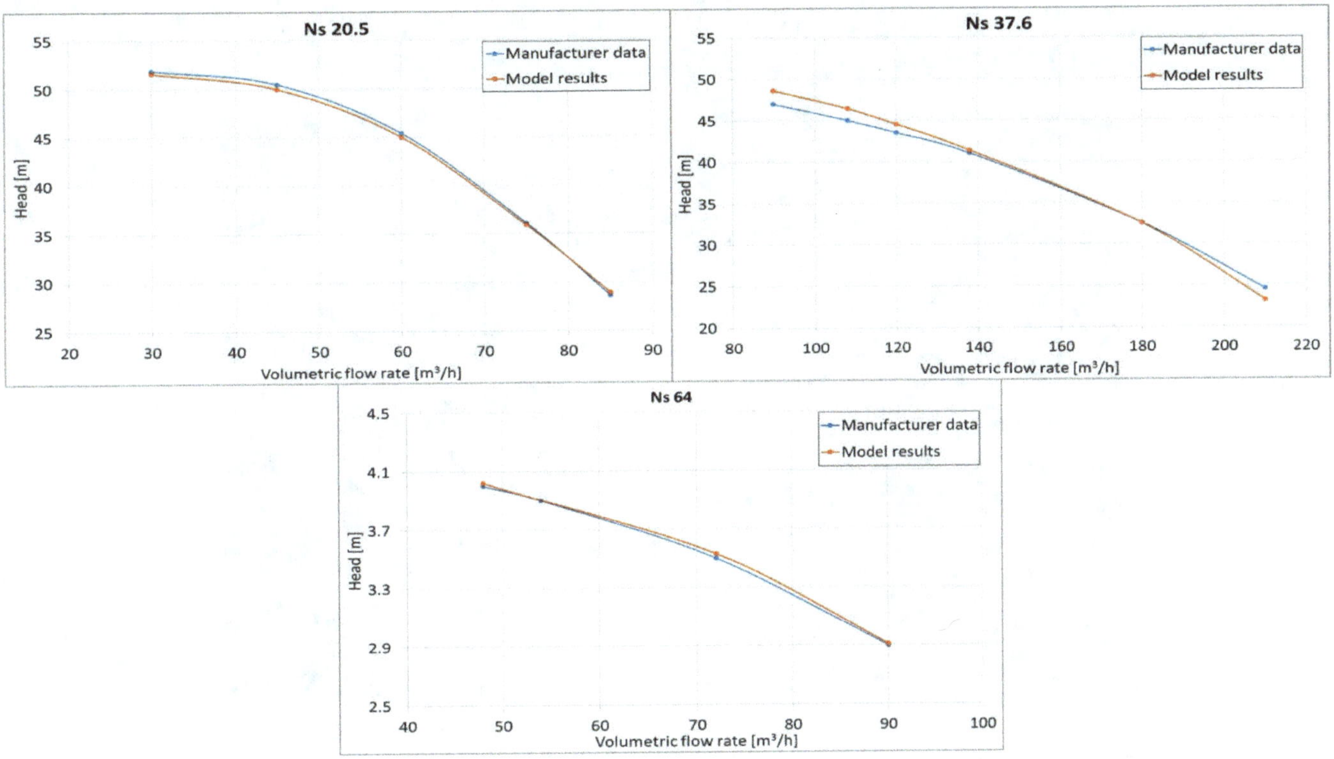

Figure 5. Model validation: Comparison between manufacturer's data and model results.

4. *Ns* 37.6 Pump Model Validation with Experimental Data

Once the model had been validated in the pumping mode, it was decided to also validate it in the reverse mode, to assess whether the model reproduces the turbine mode well. Because the proposed methodology is based only on the results of the CFD model under reverse conditions, the validation under reverse conditions was necessary to confirm the entire methodology.

The model of centrifugal Pump 1 has been validated with data from an experiment performed on a dedicated test bench of the Department of Civil Construction and Environmental Engineering of the University of Naples Federico II. The bench enables testing a centrifugal pump running in reverse mode. The aim of this activity was to further validate the simulation model under reverse conditions. The test bench reproduces a full-scale hydraulic network, made up of four nodes (Figure 6). An external pump increases the water pressure to simulate the behavior of a real urban network while an air chamber stabilizes the flow rate. The tested pump has been installed in one node where two pressure-reducing valves (PRV) regulate the water flow rate and the pressure at the inlet and outlet of the pump. The pressure is regulated with a valve installed at the network inlet, while the flow rate at the outlet is varied with a butterfly valve. Valves are remotely controlled with electronic actuators controlled by a dedicated homemade software program, with an external PLC [34–40].

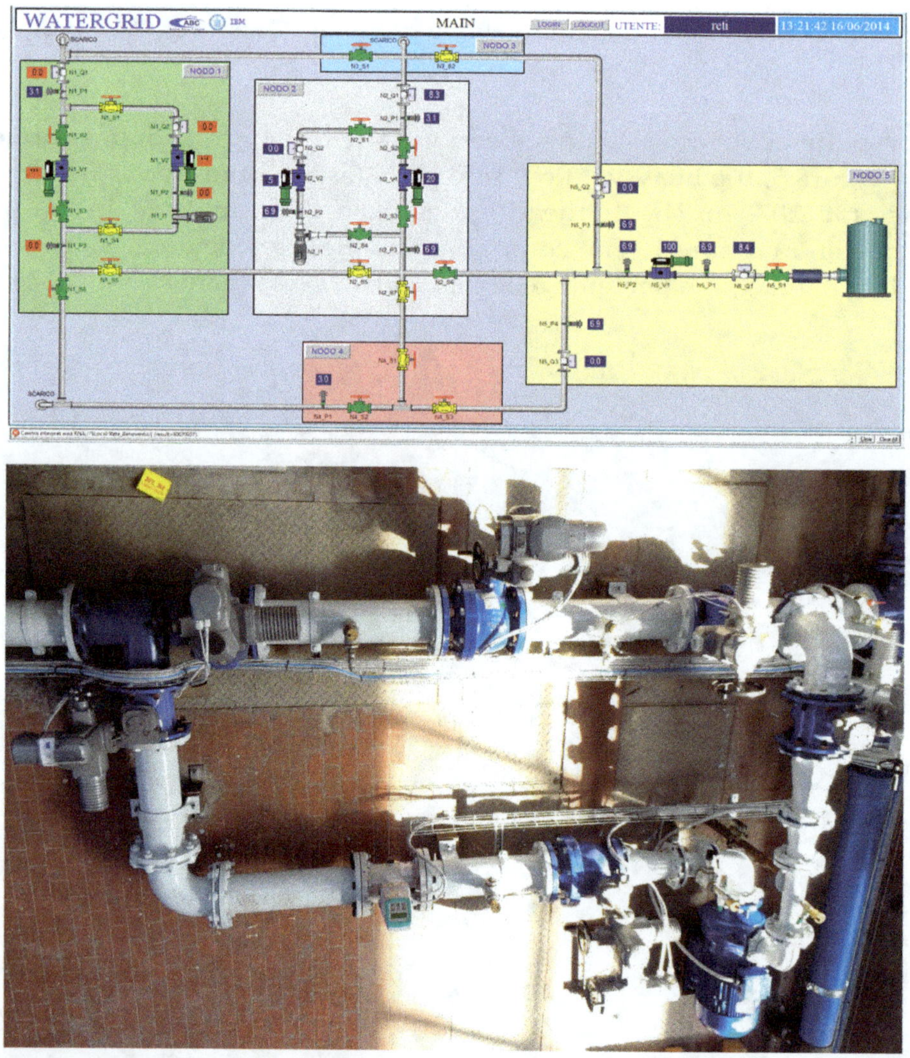

Figure 6. Test bench—water grid of the Department of Civil Construction and Environmental Engineering of the University of Naples Federico II.

The electric motor of the pump is linked to an inverter and the produced electrical power is connected to the urban power grid. In the node, two pressure transducers, P_1 and P_2 (Burkert® model 8314), and a flow meter Q (Siemens® mag 500) have been installed. All test bench data have been acquired by a homemade acquisition system. Furthermore, a 360-tooth encoder has been installed on the electrical motor to acquire the shaft speed.

The hardware system is based on a data acquisition board NI DAQ Card (12-bit ADC converter resolution, 16 input channel, two 24 bit counters), a 68-pin shielded desktop connector block (NI TBX-68). NI LabView performs a homemade software.

The pressure is acquired with Buckert transducers. Sensors are installed upstream and downstream of the PAT, as it is shown in Figure 6. The characteristics of the pressure transducers are as follows:

- Ceramic technology
- 0–10 bar pressure range
- ±0.25% accuracy
- 2 ms response time

The electric motor shaft speed is acquired with a BAUMER BHK 16.05A.360-I2-5 incremental encoder, with 360 teeth, while, as already said, the flow rate is measured with an electromagnetic Siemens mag 500 transducer, (accuracy 0.25%, response time 1 s). The sample frequency was 1 Hz.

The experiments have been performed only in steady-state conditions, varying the water flow rate and the pressure at the inlet of the pump, for different shaft speeds. In particular, the flow rate has been varied between 8 and 21 L/s, and the shaft speed between 300 and 2200 rpm. During the test, pressure at the inlet and outlet of the pump have been acquired at a sample frequency of 1 Hz.

As stated above, the tests have been done in steady-state conditions, running the pump in reverse mode. The flow rate, the pressures at the inlet and outlet of the pump and the shaft rpm were measured. In Figure 7, all results of the experimental campaign are shown.

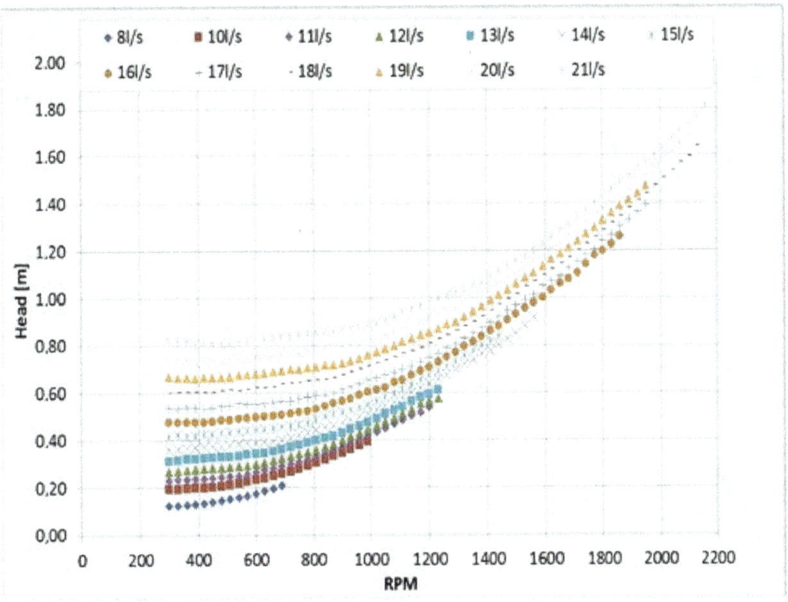

(a)

Figure 7. *Cont.*

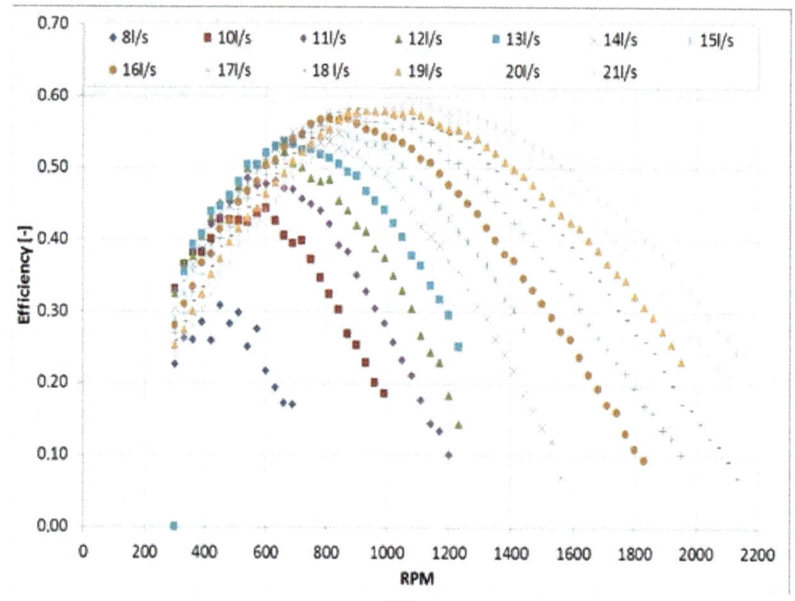

(**b**)

Figure 7. Experimental results. (**a**) Head; (**b**) Efficiency.

To examine the PAT performance, the total head (m) versus shaft rpm is reported, varying the water flow rate for all the examined conditions. Results confirm what is known from the literature: the PAT head increases with the rpm and with the flow—rate, and it can be easily noted that, for the tested conditions, the head varies between 0.1 and 1.8 m.

In Figure 8, the whole validation of the simulation model is presented: it shows that the model reproduces the experimental data well with very small differences between the experimental and the model results for all the running conditions that were analyzed.

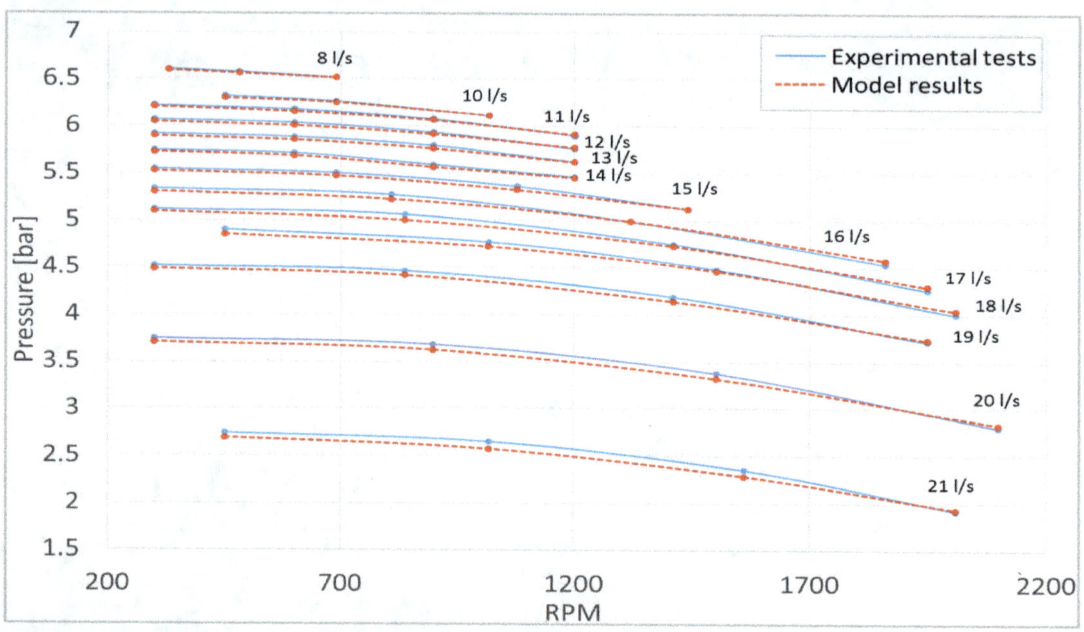

Figure 8. Model validation.

5. Model Results

After the validation phase in pump and turbine mode, simulation models have been used to predict the efficiency curves of the three analyzed pumps. Then, all the simulations have been performed to obtain the data necessary to evaluate the inverse characteristics.

The specific head ψ can be evaluated as:

$$\psi = \frac{gH}{n^2D^2} \tag{17}$$

The specific capacity φ depends by the flow Q and the impeller diameter D:

$$\varphi = \frac{Q}{nD^3} \tag{18}$$

The specific power is defined as:

$$\pi = \frac{P}{\rho n^3 D^5} \tag{19}$$

While the efficiency can be evaluated as:

$$\eta = \frac{P}{\rho QH} \tag{20}$$

where H (m), Q (m^3/s), and P (W) are the head, flow rate and power, respectively. The rotational speed is n (RPS) and D (m) is the impeller diameter. In the reverse mode simulation, the boundary conditions in pump and PAT mode were the same (declared data in pump mode). The boundary conditions in reverse mode are summarized in Table 2.

Table 2. Boundary conditions.

Boundary Conditions	Pump 1	Pump 2	Pump 3
Outlet pressure	1.9 bar	1.9 bar	1.9 bar
Inlet Volumetric Flow	90/210 m^3/h	30/85 m^3/h	48/90 m^3/h
T_{in}	293.15 K	293.15 K	293.15 K
P_{sat}	2886 Pa	2886 Pa	2886 Pa

The specific head, the specific power and the efficiency have been evaluated for both pump and turbine mode and for all the studied pumps. These are plotted versus specific capacity in Figure 9.

It is clear that at high capacity the specific head in reverse mode is always higher than in direct mode. In reverse mode, pumps have a larger power range than in direct mode. The trends are quite different and the curves arising at low flow rate and at higher capacities have a higher power value than in direct mode. In pump mode, the efficiency has a typical "bell shape" while in PAT mode its profile resembles that of a Francis turbine: it increases at low flow rates and reaches a maximum value at high flow rates. Moreover, the pump with the low specific speed works with low flow rates but high heads in direct mode. In reverse mode at high flow rates, the head is higher than the direct-mode value. The pump with the high specific speed works at high flow rates but low heads. In reverse mode, working with the same flow rates, the maximum head is also higher than the direct-mode value. For all pumps, the maximum value is lower than the direct-mode value. For low specific speed pumps, this maximum value is approximately equal to the direct-mode value, while for high specific speed pumps, it is lower.

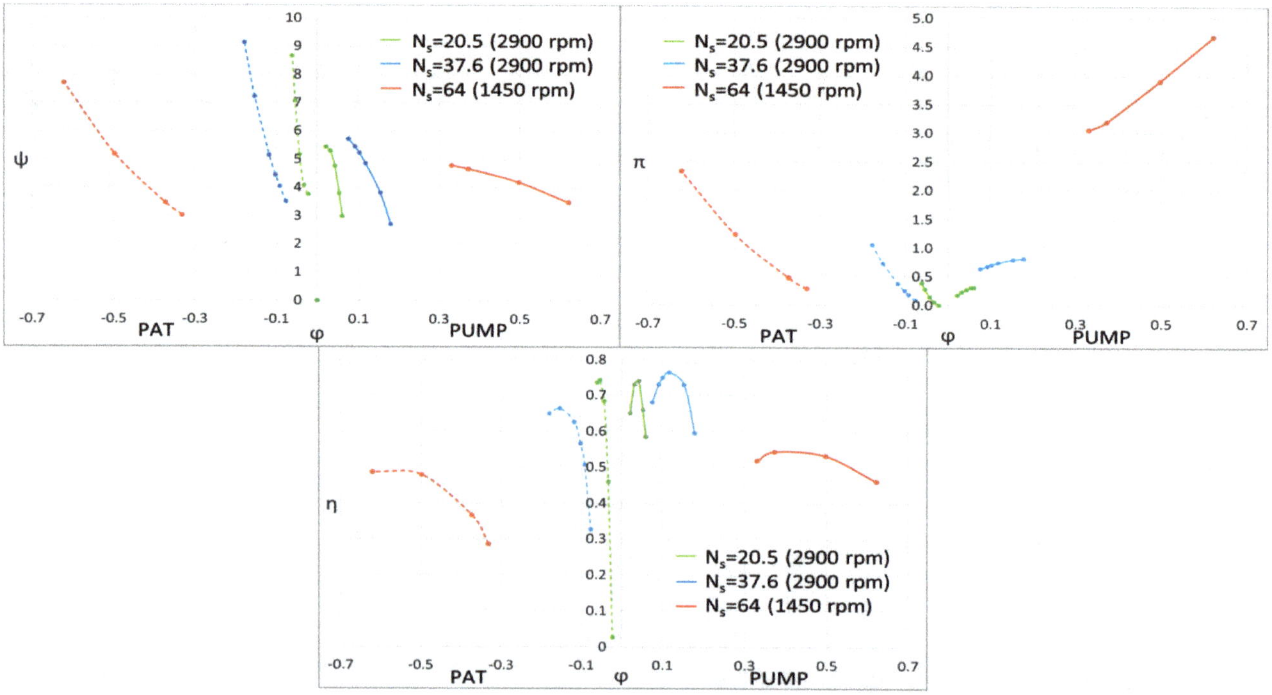

Figure 9. Specific head, specific power and efficiency.

In conclusion, in Table 3, the BEP values are summarized:

Table 3. Boundary conditions.

Boundary Conditions	Pump 1		Pump 2		Pump 3	
	Direct Mode	Reverse Mode	Direct Mode	Reverse Mode	Direct Mode	Reverse Mode
Head (m)	39	61	45.4	67	3.9	4.6
Capacity (m^3/s)	0.041	0.05	0.017	0.021	0.015	0.022
Power (kW)	20.5	19.98	10.01	10.24	1.05	0.75
Efficiency	0.787	0.663	0.743	0.741	0.543	0.487

6. Comparison of Prediction Methods

After the evaluation of the inverse characteristics, the results of the proposed methodology have been compared to the prediction methods available in literature. To this end, all the previously discussed methods have been applied to the three analyzed pumps. Some methods predict only the head and flow rate, while others also predict power and efficiency. In Table 4, all the results are shown for Pump 1.

The flow rates calculated using the methods of Sharma and of Hergt and Schmiedl are very close to those of the proposed CFD methodology. Grover's and Alatorre-Frenk's methods overestimate this value while the other methods underestimated the value by a margin of 10%–15%. All methods underestimate the head value, except for Grover's. These values diverge with errors of up to 30%.

Evaluating the percentage deviation as:

$$\text{percentage deviation} = \frac{\text{predicted value} - cfd\ \text{value}}{cfd\ \text{value}} \times 100 \tag{21}$$

In Figure 10 the deviations between the predictions of these methods and the simulated data are plotted vs. the pump specific speeds for all three pumps.

Table 4. Comparison for pump 1.

Methods	*H* (m)	*Q* (m^3/s)	*P* (kW)	η
Model results	61.42	0.05	19.98	0.663
Stepanoff	49.55	0.0463	22.5	0.787
Alatorre-Frenk	42.79	0.091	28.92	0.757
Sharma	51.99	0.0498	20.5	0.807
Schmiedl	43.94	0.0514	16.82	0.759
Grover	78.59	0.0657	36.92	0.729
Hergt	41.90	0.0508	-	-
Childs	49.55	0.0522	19.96	0.787
D&N	58.56	0.0411	18.17	0.769

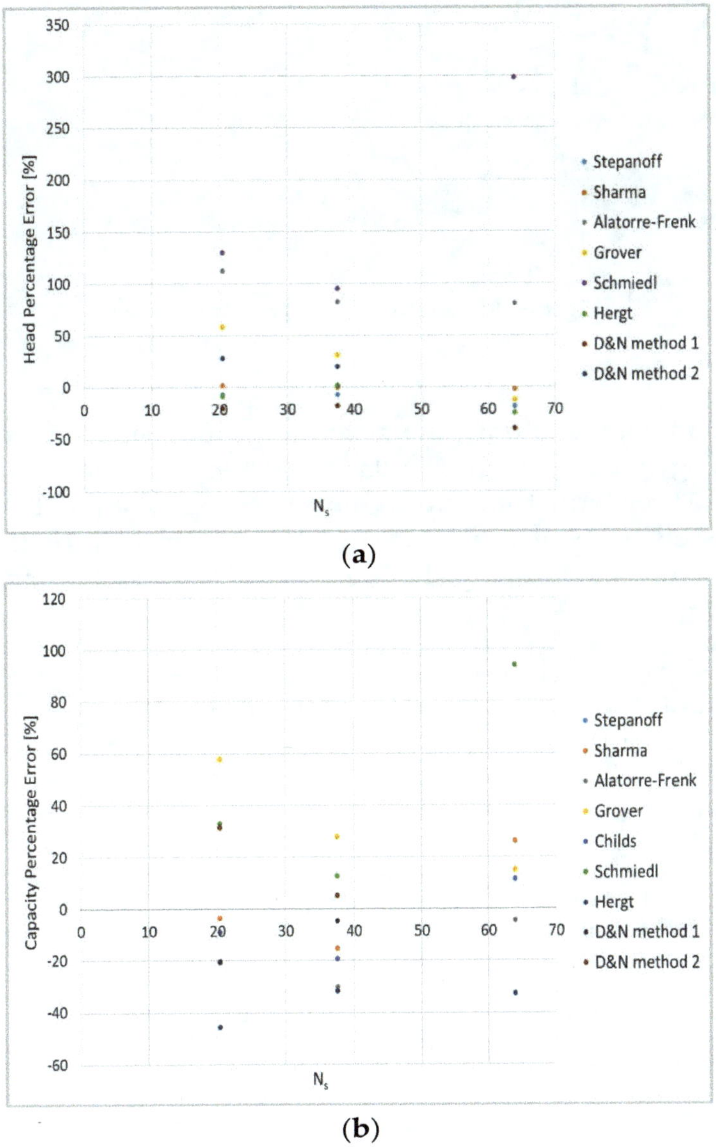

Figure 10. Prediction methods comparison. (**a**) Head Percentage Error; (**b**) Capacity Percentage Error.

It easy to observe that in some cases the deviation is very high. To evaluate the Derakhshan and Nourbakhsh efficiency [24], the CFD model results were used as shown in Figure 11. In this figure, the relative velocity magnitude (m/s) is shown and the angle between the relative and absolute velocity is highlighted in red, at BEP conditions and in pump mode.

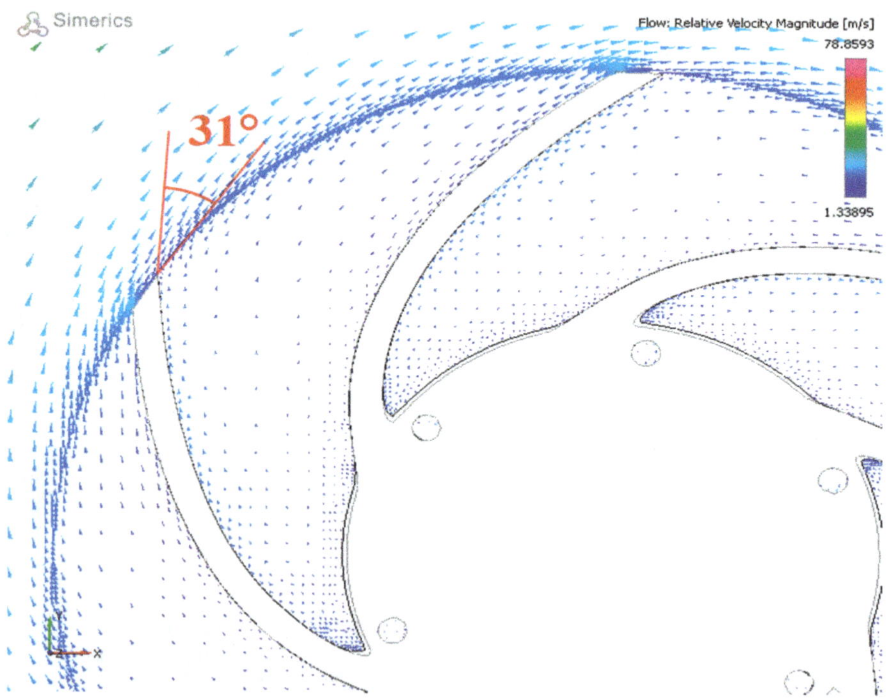

Figure 11. Relative velocity magnitude.

7. Conclusions

In this paper a methodology to predict the inverse characteristic of a centrifugal pump has been presented. This methodology is based on the results of a three-dimensional simulation model built with a commercial CFD code. Three industrial pumps have been analyzed, with different specific speeds. First, the simulation models have been validated with data supplied by the pump manufacturers. Then the results of an experimental campaign have been used to validate a model simulating the pump working in reverse conditions.

Starting from the CFD model results, the specific head, capacity, power and efficiency have been evaluated and the best efficient point of all the analyzed pumps was found. Furthermore, several prediction methods have been applied to the tested pumps and their predicted values were compared with those of the proposed methodology. Some methods (e.g., Childs' method) are not in accord while others (e.g., Stepanoff's method) show small relative differences.

Acknowledgments: This research has been supported by the Department of Industrial Engineering of the University of Naples "Federico II". We appreciate the technical contribution from Gennaro Stingo and Giuseppe Iovino of the Department of Industrial Engineering and from colleagues of the Department of Civil Construction and Environmental Engineering of the University of Naples "Federico II".

Author Contributions: Emma Frosina, Dario Buono and Adolfo Senatore conceived and designed the experiments; Dario Buono performed the experiments; Emma Frosina and Dario Buono analyzed the data; Adolfo Senatore contributed reagents/materials/analysis tools; Emma Frosina wrote the paper.

Nomenclature

PRV	Pressure Reducing Valve
PAT	Pump as Turbine
CFD	Computational Fluid Dynamic
BEP	Best Efficiency Point
H_p	Pump head
Q_p	Pump flow rate
P_p	Pump power

η_p	Pump overall efficiency
H_t	Turbine head
Q_t	Turbine flow rate
P_t	Turbine power
η_t	Turbine overall efficiency
ψ	Specific head
φ	Specific capacity
π	Specific power
η	Efficiency
N_s	Pump specific speed
N_{st}	Turbine specific speed
z	Impeller's blade number
P_{it}	Power losses due to leakage
P_{vt}	Volute power losses
P_{et}	Kinetic energy losses
P_{nt}	Turbine net power
P_{it}	Hydraulic losses of the impeller in turbine mode
n	Surface normal
k	Turbulence kinetic energy
p	Pressure (Pa)
Q	Flow rate (m^3/h)
rpm	Revolution per minute
U	Initial velocity
u	Velocity component (m/s)
u'	Component of v'
v	Velocity vector
v'	Turbulent fluctuation velocity
μ	Fluid viscosity (Pa-s)
ρ	Fluid density (kg/m^3)
$\tilde{\tau}$	Shear stress tensor
c_1, c_2	Constant
σ_k	Turbulent kinetic energy
σ_ε	Turbulent kinetic energy dissipation
S'_{ij}	Strain tensor
μ_t	Turbulent viscosity
G_t	Turbulent generation
τ_{ij}	Turbulent Reynolds stress
ε	Turbulence dissipation
Ω	Control volume

References

1. Williams, A. *Pumps as Turbines: A User's Guide*; Intermediate Technology Publications: London, UK, 1995.
2. Arriaga, M. Pump as turbine—A pico-hydro alternative in Lao People's Democratic Republic. *Renew. Energy* **2010**, *35*, 1109–1115. [CrossRef]
3. Derakhshan, S.; Nourbakhsh, A. Experimental study of characteristic curves of centrifugal pumps working as turbines in different specific speeds. *Exp. Therm. Fluid Sci.* **2008**, *32*, 800–807. [CrossRef]
4. Williams, A.A.; Rodrigues, A.; Singh, P.; Nestmann, F.; Lai, E. Hydraulic analysis of a pump as a turbine with CFD and experimental data. In Proceedings of the IMechE Seminar, Computational Fluid Dynamics for Fluid Machinery, London, UK, 18 November 2003.
5. Nautiyal, H.; Varun, A.K. Reverse running pumps analytical, experimental and computational study: A review. *Renew. Sustain. Energy Rev.* **2010**, *14*, 2059–2067. [CrossRef]
6. Nautiyal, H.; Varun, K.A.; Yadav, S. 2010 CFD analysis on pumps working as Turbines. *Hydro Nepal* **2010**, *6*, 35–37.

7. Carravetta, A.; Del Giudice, G.; Fecarotta, O.; Ramos, H. PAT design strategy for energy recovery in water distribution networks by electrical regulation. *Energies* **2013**, *6*, 411–424. [CrossRef]
8. Carravetta, A.; Del Giudice, G.; Fecarotta, O.; Ramos, H. Pump as Turbine (PAT) design in water distribution network by system effectiveness. *Water* **2013**, *5*, 1211–1225. [CrossRef]
9. Derakhshan, S.; Nourbakhsh, A.; Mohammadi, B. Efficiency improvement of centrifugal reverse pumps. *ASME J. Fluids Eng.* **2009**, *131*. [CrossRef]
10. Natanasabapathi, S.R.; Kshirsagar, J.T. *Pump as Turbine—An Experience with CFX 5.6*; Kirloskar Bros. Ltd., Corporate Research and Engineering Division: Pune, India, 2004.
11. Barrio, R.; Fernández, J.; Parrondo, J.; Blanco, E. Performance prediction of a centrifugal pump working in direct and reverse mode using Computational Fluid Dynamics. In Proceedings of the International Conference on Renewable Energies and Power Quality (ICREPQ'10), Granada, Spain, 23–25 March 2010.
12. Chapallaz, J.M.; Eichenberger, P.; Fischer, G. Manual on pumps used as turbines. In *Friedr Vieweg Sohn Verlagsgesellschaft, Braunschweig*; Informatica International, Inc.: Braunschweig, Germany, 1992.
13. Childs, S.M. Convert pumps to turbines and recover HP. *Hydro Carbon Process. Pet. Refin.* **1962**, *41*, 173–174.
14. McClaskey, B.M.; Lundquist, J.A. Can You Justify Hydraulic Turbines? *Hydrocarb. Process.* **1976**, *55*, 163–166.
15. Lobanoff, V.S.; Ross, R.R. Hydraulic power recovery turbines. In *Centrifugal Pumps: Design and Applications*; Gulf Publishing Company: Houston, TX, USA, 1985; pp. 246–282.
16. Sanjay, V.J.; Rajesh, N.P. Investigations on pump running in turbine mode: A review of the state-of-the-art. *Renew. Sustain. Energy Rev.* **2014**, *30*, 841–868.
17. Grover, K.M. *Conversion of Pumps to Turbines*; GSA Inter Corp.: Katonah, NY, USA, 1980.
18. Williams, A. The turbine performance of centrifugal pumps: A comparison of prediction methods. *J. Power Energy* **1994**, *208*, 59–66. [CrossRef]
19. Stepanoff, A.J. *Centrifugal and Axial Flow Pumps*; John Wiley: New York, NY, USA, 1957.
20. Alatorre-Frenk, C. Cost Minimization in Micro Hydro Systems Using Pumps-Asturbines. Ph.D. Thesis, University of Warwick, Coventry, UK, 1994.
21. Sharma, K. *Small Hydroelectric Project-Use of Centrifugal Pumps as Turbines*; Technical Report; Kirloskar Electric Co.: Bangalore, India, 1985.
22. Schmiedl, E. *Serien-Kreiselpumpen im Turbinenbetrieb*; Pumpentagung: Karlsruhe, Germany, 1988.
23. Lewinsky, K.; Heinz, P. Pumpen als Turbinen fur Kleinkraftwerke. *Wasserwirtschaft* **1987**, *77*, 531–537.
24. Derakhshan, S.; Nourbakhsh, A. Theoretical, numerical and experimental investigation of centrifugal pumps in reverse operation. *Exp. Therm. Fluid Sci.* **2008**, *32*, 1620–1627. [CrossRef]
25. Frosina, E.; Senatore, A.; Buono, D.; Santato, L. Analysis and Simulation of an Oil Lubrication Pump for the Internal Combustion Engine. In Proceedings of the Fluids Engineering Systems and Technologies, San Diego, CA, USA, 15–21 November 2013.
26. Frosina, E.; Senatore, A.; Buono, D.; Olivetti, M. A Tridimensional CFD Analysis of the Oil Pump of a High Performance Engine. In Proceedings of the SAE 2014 World Congress and Exhibition, Detroit, MI, USA, 8–10 April 2014.
27. Frosina, E.; Buono, D.; Senatore, A. Modeling methodology to study the internal fluid-dynamic of a gas filter. *Int. Rev. Model. Simul.* **2015**, *8*, 533. [CrossRef]
28. Frosina, E.; Senatore, A.; Buono, D.; Stelson, K.A.; Wang, F.; Mohanty, B.; Gust, M.J. Vane pump power split transmission: Three dimensional computational fluid dynamic modeling. In Proceedings of the ASME/BATH 2015 Symposium on Fluid Power and Motion Control, Chicago, IL, USA, 12–14 October 2015.
29. Frosina, E.; Senatore, A.; Buono, D.; Stelson, K.A. A Mathematical Model to Analyze the Torque Caused by Fluid-Solid Interaction on a Hydraulic Valve. *ASME J. Fluids Eng.* **2016**, *138*, 061103. [CrossRef]
30. Automated Meshing with Binary Tree. Available online: https://www.simerics.com/download/automated-meshing-with-binary-tree/ (accessed on 11 January 2017).
31. Frosina, E.; Buono, D.; Senatore, A.; Costin, I.J. A Simulation Methodology Applied on Hydraulic Valves for High Fluxes. *Int. Rev. Model. Simul.* **2016**, *9*, 217. [CrossRef]
32. Frosina, E.; Buono, D.; Senatore, A.; Stelson, K.A. A modeling approach to study the fluid dynamic forces acting on the spool of a flow control valve. *J. Fluids Eng.* **2016**, *139*, 11103–11115. [CrossRef]
33. Pellegri, M.; Vacca, A.; Frosina, E.; Buono, D.; Senatore, A. Numerical analysis and experimental validation of Gerotor pumps: A comparison between a lumped parameter and a computational fluid dynamics-based approach. *Inst. Mech. Eng. Part C J. Mech. Eng. Sci.* **2016**, *1989–1996*, 203–210. [CrossRef]

34. Pugliese, F.; De Paola, F.; Fontana, N.; Giugni, M.; Marini, G. Experimental characterization of two pumps as turbines for hydropower generation. *Renew. Energy* **2016**, *99*, 180–187. [CrossRef]
35. Su, X.; Huang, S.; Zhang, X.; Yang, S. Numerical research on unsteady flow rate characteristics of pump as turbine. *Renew. Energy* **2016**, *94*, 488–495. [CrossRef]
36. De Marchis, M.; Milici, B.; Volpe, R.; Messineo, A. Energy saving in water distribution network through pump as turbine generators: Economic and environmental analysis. *Energies* **2016**, *9*, 877. [CrossRef]
37. Wang, T.; Kong, F.; Chen, K.; Duan, X.; Gou, Q. Experiment and analysis of effects of rotational speed on performance of pump as turbine. *Trans. Chin. Soc. Agric. Eng.* **2016**, *32*, 67–74.
38. Frosina, E.; Senatore, A.; Buono, D.; Monterosso, F.; Olivetti, M.; Arnone, L.; Santato, L. A tridimensional CFD analysis of the lubrication circuit of a non-road application diesel engine. In Proceedings of the 11th International Conference on Engines and Vehicles, ICE 2013, Naples, Italy, 15–19 September 2013.
39. Venturini, M.; Alvisi, S.; Simani, S. Energy potential of pumps as turbines (PATs) in water distribution networks. In Proceedings of the 28th International Conference on Efficiency, Cost, Optimization, Simulation (ECOS) and Environmental Impact, Pau, France, 29 June–3 July 2015.
40. Barbarelli, S.; Amelio, M.; Florio, G. Predictive model estimating the performances of centrifugal pumps used as turbines. *Energy* **2016**, *107*, 103–121. [CrossRef]

14

Numerical Study of the Axial Gap and Hot Streak Effects on Thermal and Flow Characteristics in Two-Stage High Pressure Gas Turbine

Myung Gon Choi and Jaiyoung Ryu *

School of Mechanical Engineering, Chung-Ang University, 84, Heukseok-ro, Dongjak-gu, Seoul 06974, Korea; mgon1122@cau.ac.kr

* Correspondence: Jairyu@cau.ac.kr.

Abstract: Combined cycle power plants (CCPPs) are becoming more important as the global demand for electrical power increases. The power and efficiency of CCPPs are directly affected by the performance and thermal efficiency of the gas turbines. This study is the first unsteady numerical study that comprehensively considers axial gap (AG) in the first-stage stator and first-stage rotor (R1) and hot streaks in the combustor outlet throughout an entire two-stage turbine, as these factors affect the aerodynamic performance of the turbine. To resolve the three-dimensional unsteady-state compressible flow, an unsteady Reynolds-averaged Navier–Stokes (RANS) equation was used to calculate a $k-\omega$ SST γ turbulence model. The AG distance d was set as 80% (case 1) and 120% (case 3) for the design value case 2 (13 mm or d/Cs_1 = 0.307) in a GE-E^3 gas turbine model. Changes in the AG affect the overall flow field characteristics and efficiency. If AG decreases, the time-averaged maximum temperature and pressure of R1 exhibit differences of approximately 3 K and 400 Pa, respectively. In addition, the low-temperature zone around the hub and tip regions of R1 and second-stage rotor (R2) on the suction side becomes smaller owing to a secondary flow and the area-averaged surface temperature increases. The area-averaged heat flux of the blade surface increases by a maximum of 10.6% at the second-stage stator and 2.8% at R2 as the AG decreases. The total-to-total efficiencies of the overall turbine increase by 0.306% and 0.295% when the AG decreases.

Keywords: gas turbine; axial gap; hot streak; heat transfer

---

1. Introduction

The global electrical power demand is expected to increase by 46% from 2015 to 2040, and this will lead to a large increase in electricity generation. Natural gas will be used for electricity generation instead of coal to minimize environmental problems caused by CO_2 emissions. Electricity generation using natural gas is expected to increase from 5.22 trillion kWh in 2015 to 9.6 trillion kWh in 2040, an increase of 83.9%. Therefore, the importance of combined cycle power plants (CCPPs) that use natural gas is expected to considerably increase in the future. CCPPs consist of gas turbines that use natural gas and steam turbines that use steam, which is emitted from heat recovery steam generators. Increase in gas turbine power cause a similar increase of power in steam turbine and improves the overall efficiency of CCPP. The gas turbines' aerodynamic performance and thermal efficiency have a direct effect on the cost of generated power of CCPPs.

One method of increasing gas turbine efficiency is to increase the turbine inlet temperature (TIT). However, a TIT distribution that is higher than the melting point of the material of the turbine blades results in a high thermal load on the turbine blades. Without a suitable cooling system, this leads to high-temperature corrosion and becomes a major factor in reducing the life of a turbine. To increase the TIT, it is necessary to more closely analyze the heat transfer characteristics of the surface of the turbine blade through an analysis of the transformed temperature distribution of the turbine inlet. The TIT distribution is directly affected by the fluid flow characteristics and temperature distribution of the combustor outlets. The temperature distribution of the turbine inlet is called a hot streak (HS), and it creates a complex heat transfer environment in the fluid flow passage and blade surface of the turbine [1]. HS has a different effect on the turbine blade surface compared with uniform temperature distribution; therefore, it is important to consider HS to analyze the overall performance and efficiency. As such, various numerical analyses and experimental studies have already been performed to describe the effect of HS on the heat transfer phenomena of a turbine blade. Butler et al. observed that high-temperature gas is concentrated on the pressure side of the rotor owing to an increased incidence angle and discovered points where the heat transfer effect is weakened [2]. Povey and Qureshi performed an experimental study on the temperature distribution in a combustor outlet and developed enhanced OTDF (EOTDF), which has a temperature distribution ratio of 1.65 [3]. EOTDF was used to create HS and they studied the influence of uniform temperature distributions at the turbine inlet as well as changes in the position of HS in the radial direction on the stator and rotor [1,4]. Bai-Tao An et al. studied the effect of uniform temperature distributions and HS inlet conditions on aerodynamic parameters such as total temperature, static pressure, and velocity [5]. Feng et al. found that the second-stage stator exhibits a high efficiency and low thermal load when its clocking position and HS were aligned [6]. Smith found that the time-averaged heat load when the HS is aligned with the stator had a large effect on the stator and a small effect on the rotor compared with the case when the HS is aligned with the stator passage [7]. However, these studies did not consider the effect of axial gap (AG) on the thermal and flow characteristics of the blade surface.

To increase gas turbine efficiency, it is important to analyze not only HS but also factors that affect aerodynamic performance. One factor that affects aerodynamic performance is the AG, i.e., a length of the straight line from the stator to the rotor. AG is a factor that directly affects the design and operation of turbomachines. It not only determines the overall size, length, and weight of a turbomachine but it also affects the unsteady flow in the rotor, noise, and aerodynamic performance of the turbine blade [8,9]. Furthermore, if the AG is too short, problems such as reduced fatigue life due to high inlet temperature occur. Therefore, experimental and numerical studies have been performed on the AG as an element that affects the overall turbine performance and efficiency. The AG affects the heat transfer coefficient (HTC) of the blade midspan in large-scale axial-flow turbines as well as the flow at the hub [10,11]. Funazaki performed an experimental and numerical analysis on changes in the flow angle in the stator outlet according to the AG in the first stage of the turbine [12]. Syed performed a numerical study on the composite effect of tip clearance and the AG of a stator blade in a multistage compressor [9]. It was found that changes in the first-stage stator and rotor AG affected turbine performance, but changes in the distance between the first-stage rotor and second-stage stator did not have a considerable effect on performance improvement. Previous studies found that as the AG becomes shorter, it affects the rotor torque and improves aerodynamic performance. However, they did not consider the thermal and flow characteristics in which the turbine blade surface is affected by the nonuniform temperature distribution of the turbine inlet according to the AG.

Accurate predictions of thermal and flow characteristics in a high-pressure gas turbine at the turbine blade and passage in more than one stage can have a considerable effect on turbine design. However, numerical investigations on multistage gas turbines are still expensive, and not many studies have been conducted thus far [13]. Adel performed a numerical analysis on a two-stage gas turbine with steady and unsteady states. The first stage was not affected by the second stage, but the second stage was strongly affected by the first stage. It was found that the upstream flow caused distortion in the downstream flow along the circumferential direction, and the flow interacts with the secondary flow and tip leakage flow of the blade [14]. Therefore, it is necessary to accurately understand the effect of the first stage flow on the second-stage and predict the thermal and flow characteristics within the passage and the heat transfer distribution of the blade surface.

Previous numerical studies for the AG effects applied uniform inlet temperature distributions to examine the aerodynamic performance of a turbine. Nonuniform inlet temperatures have been applied to predict the heat transfer distribution of the blade surface and the thermal and flow characteristics, but the AG was not considered. Furthermore, numerical studies that considered AG or HS have been performed on turbines with 1.5 or fewer stages. The gas turbine efficiency is affected by various design elements such as the AG and temperature distribution and varies in each stage. Therefore, this study performed a numerical analysis to investigate the effect of the AG on the thermal and flow characteristics of the blade surface and passage when an HS is applied to a two-stage turbine.

2. Numerical Method

2.1. Numerical Model and Grid

The gas turbine configuration used in this study is a GE-E^3 gas turbine model. The actual gas turbine consists of 46 stators and 76 rotors in the first stage and 48 stators and 70 rotors in the second stage [15]. If one stator and two rotors have the same pitch angle, calculation errors during the numerical analysis can be minimized. As such, the number of blades was adjusted using a domain scaling method to create two rotors that correspond to the pitch of one stator in each stage [16]. In a state where the number of first stage rotors and the solidity of each blade are fixed, each chord length of the first stage stator (S1), second-stage stator (S2), and the rotor (R2) were magnified by 46/38, 48/38, and 70/76, respectively. Therefore, the number of blades used in the analysis was 38 for S1, 76 for the first stage rotor (R1), 38 for S2, and 76 for R2. Figure 1a shows the adjusted blade configuration for each stage and the computational domain used in this study, and Table 1 lists the information on each blade. The tip clearance of the rotor was 1% of the rotor height in R1 and 0.6% of the rotor height in R2.

In the numerical analysis, Ansys Turbogrid was used to create a hexahedral grid as shown in Figure 1b. To accurately predict the thermal and flow characteristics within the boundary layers, y^+ was set to less than 1 at all walls and less than 0.5 at the blade surface. A grid independence test was performed to determine the appropriate number of grids to be used in the study. Table 2 lists the number of meshes (to achieve the equal domain pitch angle, the R1, R2 meshes in the list is doubled) used in the test and the area-averaged heat flux of the blade. In the grid independence test, grids were created for each blade by multiplying 1.34 times based on Mesh-1. In Mesh-1 and 2 and Mesh-2 and 3, the relative errors of the area-averaged heat fluxes were both under 0.01% for S1, and they were 0.75% and 0.41% for R1, respectively; thus, the Mesh-2 grid was used for S1 and R1. In Mesh-2 and 3 and Mesh-3 and 4, the relative errors of area-averaged heat flux where 1.83% and 0.41% for S2, respectively, and 1.79% and 0.8% for R2, respectively, so the Mesh-3 grid was used for S2 and R2.

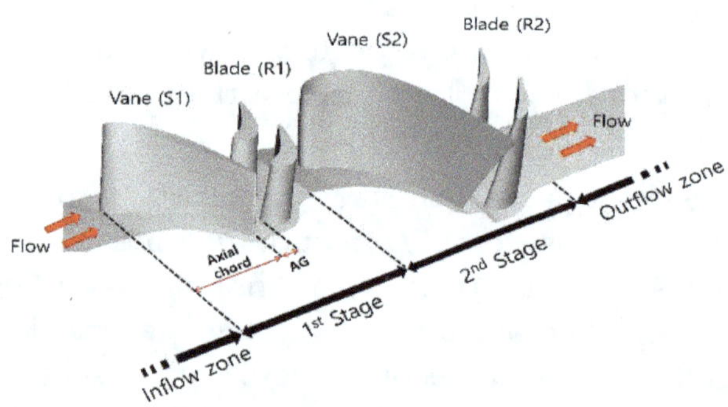

(a)

(b)

Figure 1. Two-stage stator and rotor geometry details (based on GE-E^3 (General Electric-energy efficient engine geometry [15]) turbine) (**a**) computational domain with domain scaling, and (**b**) computational grid of domain.

Table 1. Information on GE-E^3 turbine and domain.

	GE-E^3				Domain Scaling			
	S1	**R1**	**S2**	**R2**	**S1**	**R1**	**S2**	**R2**
Airfoil number	46	76	48	70	38	76	38	76
Chord length [mm]	70.412	41.821	78.370	40.734	85.211	41.821	98.991	37.504
TE pitch [mm]	47.042	28.452	45.337	31.066	56.646	28.452	57.268	28.613
Pitch chord ratio	0.668	0.680	0.579	0.668	0.668	0.680	0.579	0.668
Tip clearance [mm]	-	0.426	-	0.4191	-	0.426	-	0.4191

Table 2. Area-averaged heat flux at the surface of the four-row airfoil for applying the appropriate domain mesh (Mesh-2 is applied for 1st stage stator (S1) and rotor (R1), and Mesh-3 is applied for 2nd stage stator (S2) and rotor (R2)).

	Domain Node Number ($\times 10^6$)				Area Averaged Heat Flux [W/m^{2}K]				
	S1	**R1**	**S2**	**R2**	**S1**	**Relative Error**	**R1 (Relative Error)**	**S2 (Relative Error)**	**R2 (Relative Error)**
Mesh-1	0.99	2.48	0.97	2.34	277,226	Relative errors are less than 0.01%	382,851 (0.75%)	236,005 (1.83%)	185,058 (1.79%)
Mesh-2	**1.32**	**3.38**	1.33	3.12	277,022		385,727 **(0.41%)**	240,406 (0.96%)	188,448 (1.32%)
Mesh-3	1.82	4.40	**1.85**	**4.18**	276,993		387,291	242,747 **(0.52%)**	190,926 **(0.8%)**
Mesh-4	-	-	2.49	5.50	-		-	244,028	192,473

2.2. Numerical Details and Boundary Conditions

A continuity equation, momentum equation, and energy equation were used to analyze the compressible fluid flow of a three-dimensional unsteady state. These equations can be expressed as Equations (1)–(3), respectively.

$$\frac{\partial \rho}{\partial t} + \frac{\partial}{\partial x_i}(\rho u_i) = 0 \tag{1}$$

$$\frac{\partial}{\partial t}(\rho u_i) + \frac{\partial}{\partial x_i}(\rho u_i u_j) = -\frac{\partial P}{\partial x_i} + \frac{\partial}{\partial x_j}\left[\mu\left(\frac{\partial u_i}{\partial x_j} + \frac{\partial u_j}{\partial x_i} - \frac{2}{3}\delta_{ij}\frac{\partial u_k}{\partial x_k}\right)\right] + \frac{\partial}{\partial x_j}\left(-\rho\left\{u_i' u_j'\right\}\right) \tag{2}$$

$$\frac{\partial}{\partial t}(\rho E) + \frac{\partial}{\partial x_j}(u_j(\rho E + P)) = \frac{\partial}{\partial x_j}\left[\left(k_{eff}\right)\frac{\partial T}{\partial x_j}\right] + \frac{\partial}{\partial x_j}\left[u_i \mu_{eff}\left(\frac{\partial u_i}{\partial x_j} + \frac{\partial u_j}{\partial x_i} - \frac{2}{3}\delta_{ij}\frac{\partial u_k}{\partial x_k}\right)\right] \tag{3}$$

Here, ρ is the fluid's density. u is the fluid's velocity. P is the fluid's pressure. μ is the fluid's viscosity. In Equation (3), E is the specific internal energy. k_{eff} is the effective thermal conductivity. C_P is the specific heat capacity. μ_{eff} is the effective dynamic viscosity. To accurately predict the flow separation phenomena, a $\mathrm{k} - \omega$ SST γ turbulence model was used. To solve the governing equation, a commercial computational fluid dynamics software ANSYS CFX was used.

The boundary conditions used in this simulation are indicated following the GE-E^3 turbine test performance report [15], air was used as the working fluid, and a total pressure of 344,740 Pa was used at the inlet and a static pressure of 50,000 Pa was used at the outlet. The rotor's rotation speed was 3600 RPM. At the inlet, the Mach number was 0.11 (45.1 m/s). The Reynolds number was 210,000 based on S1's axial chord length (Cs_1 = 42.4 mm). The inlet turbulence intensity was set as 5%. A no-slip condition was used on the wall surfaces. To calculate the HTC, simulations for both isothermal and adiabatic wall conditions were conducted. Under the isothermal wall condition, the turbine blade temperature was 389.95 K.

To analyze the internal thermal and flow characteristics according to the inlet temperature field, a nonuniform HS inlet temperature condition with a maximum temperature of 838 K at the center was used, and a uniform inlet temperature condition with a temperature of 728 K was used, as shown in Figure 2. In addition, to examine the effects according to the AG when an HS was applied, three cases were analyzed in which the distance of AG, d, of S1 and R1 was set as the design value (case 2: 13 mm or $d/Cs_1 = 0.307$), 80% of the design value (case 1: 10.4 mm or $d/Cs_1 = 0.245$), and 120% of the design value (case 3: 15.6 mm or $d/Cs_1 = 0.368$).

For numerical simulations, a 96-core workstation (4 Intel(R) Xeon(R) CPU E7-8890 v4 @ 2.20 GHz, RAM 512 GB) and a 44-core workstation (2 Intel(R) Xeon(R) CPU E5-2699 v4 @ 2.20 GHz, RAM 256 GB) were used, and the computation time required for a case was about 60 hours when using the 36-core.

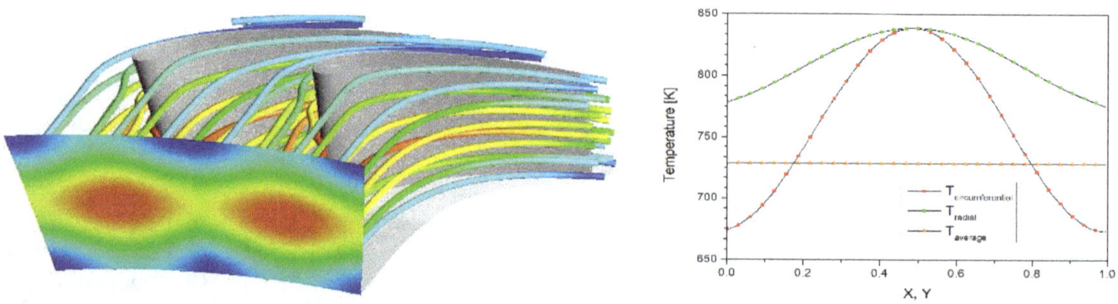

(a) (b)

Figure 2. Temperature profile at the turbine inlet: (**a**) hot streak (HS) temperature contour with streamlines, and (**b**) circumferential, radial, and average temperature distributions.

2.3. Unsteady State

A steady state analysis was performed by setting the rotor–stator interface as a frozen rotor, and the results were used as the initial conditions for the unsteady state analysis. The transient rotor–stator model was used for the rotor–stator interface for the unsteady simulations. The pressure values of R1 at the midspan were compared among 16, 32, and 50 time step unsteady simulations to determine the step of each cycle in which one R1 completely passes through a pitch of S1. Figure 3 shows the pressure distribution plot at the 4.7° and 9.45° positions when the 1 pitch angle of S1 is 9.45°. The relative error rate was calculated by substituting the pressure value of each step at the R1 midspan in Equations (4) and (5) below.

$$Relative\ error = |step32 - step16|/step32 \quad (4)$$

$$Relative\ error = |step50 - step32|/step50 \quad (5)$$

The maximum relative error rates at the 16 and 32 time-step simulations were 3.22% at R1 position 4.7° and 6.81% at 9.45°. The mean relative error rates were 1.27% at 4.7° and 0.43% at 9.45°. The maximum relative error rates at 32 and 50 time-step simulations were 1.27% at 4.7° and 4.28% at 9.45°. The mean relative error rates were 0.25% at 4.7° and 0.25% at 9.45°. The maximum and mean relative errors at 32 and 50 time-step simulations were significantly reduced compared with those at 16 and 32 time-step simulations. Thus, 32 time steps for one pitch were used in each cycle to perform the unsteady state flow analysis.

The temperature and pressure at one point near each blade pressure side wall were monitored to confirm periodically constant convergence state. The initial 20 pitches were excluded as the initial transient from the total 30 pitches' unsteady state simulation, and the remaining 10 pitches were used in the analysis of the results. Figure 4 shows the temperature and pressure measured for eight pitches, excluding the initial transient. The pressure and temperature for each pitch cycle (=32 time steps) in the unsteady state analysis appear to be periodic.

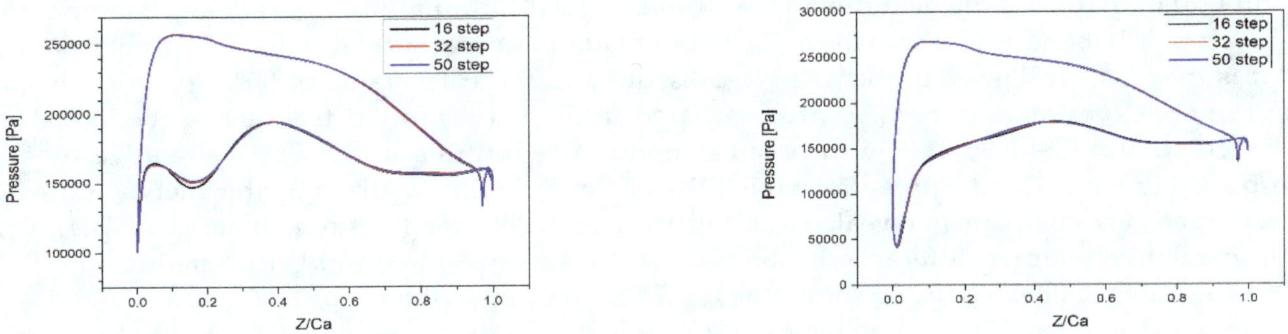

(a) (b)

Figure 3. Comparing pressure differences between 16, 32, and 50 time step simulations for each angle on the R1 midspan to find the appropriate step in the unsteady state flow analysis: (**a**) $\theta = 4.7°$, and (**b**) $\theta = 9.45°$.

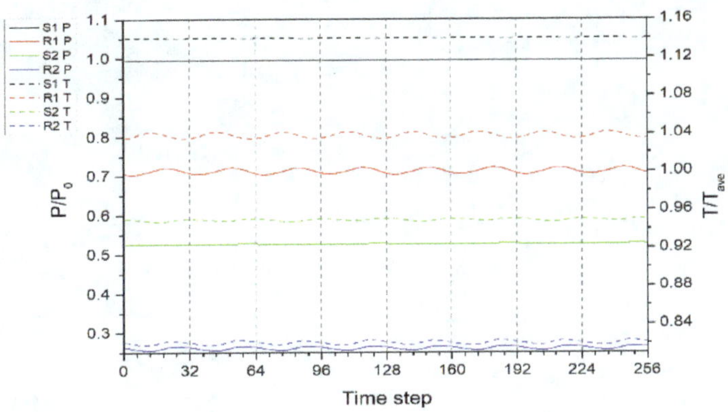

Figure 4. Periodic variation in temperature and pressure at four monitoring points (32 steps = 1 pitch, P_0 = 344,740 Pa, T_{ave} = 728 K).

2.4. Validation of the Turbulence Model

Deciding the appropriate turbulence model is important for more precise numerical analysis. Direct numerical simulations (DNS) and large eddy simulations (LES) provide a flow database for detailed turbulence statistics, but they require a high computational cost [17–21]. To replace such high-cost models, a Reynolds-average Navier-Stokes (RANS) model ($k - \varepsilon$, $k - \omega$, SST, SST γ etc. . . .) is used in turbomachinery simulation and especially, SST γ and SST $\gamma - \theta$ transition models can predict more accurate transitional flows [22,23]. To determine the turbulence model for this study, a steady state analysis was performed. The validation for the model was performed using the relative pressure and HTC in comparison with Hylton's experiment values [24]. The configuration and boundary conditions can be seen in the experiment section of Hylton et al. [24]. The 50% span experiment values of the C3X cascade No. 4311 experimental stator were used for comparison. As for the computation domain used in validation, the inflow zone, which is between the inlet and the stator leading edge (LE), was set to be same as the C3X vane axial chord (AC), and the outflow zone, which is between the trailing edge (TE) and the outlet, was set at twice the AC. A grid independence test was performed and 4,015,728 hexahedron meshes were applied. The boundary conditions used in the simulation were set such that the uniform inlet condition's total pressure was 244,763 Pa, the Mach number was 0.17, the Reynolds number based on the C3X vane's AC was 3.9×10^5, the temperature was 802 K, and the turbulence intensity was 6.5%. The outlet conditions were set such that the static pressure was 131,800 Pa, and the outlet Mach number was 0.91. To calculate the HTC, simulations with adiabatic and isothermal wall (temperature 537 K) conditions were conducted. The HTC in this study was calculated using Equation (6), as given below:

$$h = q/(T_w - T_{aw}) \tag{6}$$

In Equation (6), q is the heat flux in the isothermal wall simulation, T_{aw} is the wall temperature value in the adiabatic wall simulation, T_w is the blade temperature value for the isothermal case, and h is the HTC. In Figure 5, the relative pressure and the HTC value of 50% span of the stator found in the experiment paper [24] are compared to the values found from using the $k-\omega$, SST, $k-\omega$ SST γ, and $k-\omega$ SST $\gamma-\theta$ turbulence models used in this study. For the relative pressure distribution in Figure 5a, the pressure at the C3X vane midspan normalized by the inlet total pressure P_0 were used. The comparison revealed a strong agreement between the simulations and experimental data. A relatively larger difference in the area of the suction side (SS) is attributed to the strong unsteadiness of the flow. Figure 5b shows the HTC distribution. In the $k-\omega$, SST, and $k-\omega$ SST $\gamma-\theta$ turbulence models, the SS transition region was different from the experiment values. The $k-\omega$ SST γ turbulence model shows a similar tendency with regard to the experiment values overall, including the SS transition regions. Therefore the $k-\omega$ SST γ turbulence model was used in this study with the onset Reynolds number of 150 [23].

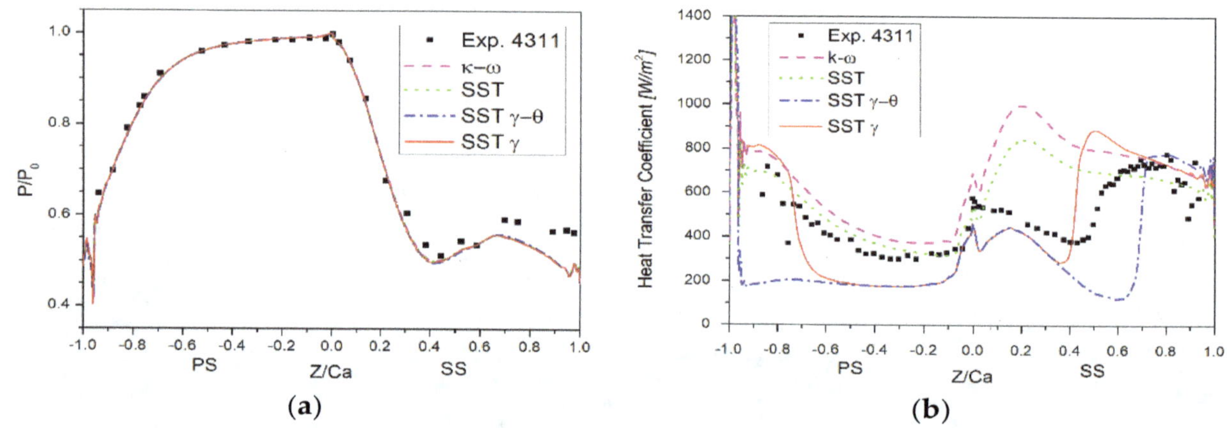

Figure 5. Comparison of experimental data with different turbulence models at 50% span of C3X vane (experimental geometry [24]): (**a**) relative pressure distribution (P_0 = 244,763 Pa), and (**b**) heat transfer coefficient (HTC) distribution.

3. Results and Discussions

3.1. *Effect of Inlet Temperature Field*

To examine the effect that the turbine inlet temperature conditions have on the blade, two steady state analyses were performed under uniform and nonuniform temperature conditions. In the uniform conditions, the turbine inlet temperature was maintained at 728 K. In the nonuniform conditions, the temperature distribution was nonuniformed with an average inlet temperature of 728 K, as shown in Figure 2. Figure 6 shows the surface temperature distribution of the S1 surface's according to each inlet condition on adiabatic walls. In the uniform inlet temperature conditions shown in Figure 6a, the maximum temperature was approximately 729 K, and the temperature distribution showed a trend of decreasing at the PS in the streamwise direction. In Figure 6b, it can be seen that the high temperature gas was centered on the midspan because of the temperature distribution caused by the HS. It is also clear that the high temperature gas formed in the radially inward direction due to the secondary flow transport effect that occurs at the tip and hub near the trailing edge of the SS. This led to fluid-mixing at the stator endwall and the temperature became lower around the hub and tip of the S1 TE. When an HS was applied, the S1 maximum surface temperature was 839 K, which shows a difference of over 110 K compared to the uniform inlet conditions.

Figure 7a shows a comparison of the temperature distributions along the span direction at AC 50% in the PS of S1 and R1. When a uniform temperature distribution was applied, there were few temperature changes along the span direction in S1 and R1. When an HS was applied, higher temperatures are observed at the midspan than around the endwall in S1 and R1. Furthermore, the temperature between 20% and 70% of the span of R1 was higher than the temperature of S1 when a uniform temperature was applied. The maximum temperature differences according to the inlet conditions were 110 K and 75 K for S1 and R1, respectively. Both Figure 7b,c shows the temperature contours at the first midspan according to each inlet temperature condition. Overall higher temperatures were observed in Figure 7c as compared to Figure 7b. Compared to the uniform inlet temperature distribution, the HS inlet temperature gradient, which formed differently along the radial and circumferential directions, had a direct effect on the overall temperature distribution of the blade. Figures 6 and 7 show that it is important to consider the HS inlet conditions in the numerical analysis to understand the thermal and flow characteristics.

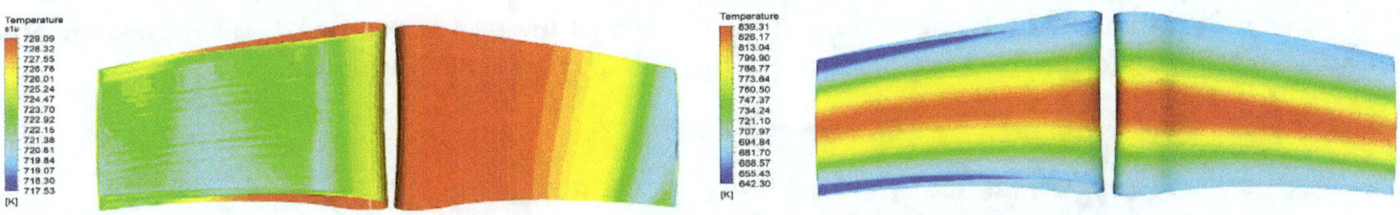

(a) (b)

Figure 6. S1 surface temperature contour caused by inlet conditions (Right is pressure side (PS)): (**a**) uniform inlet temperature of 728 K, and (**b**) HS with an average temperature of 728 K.

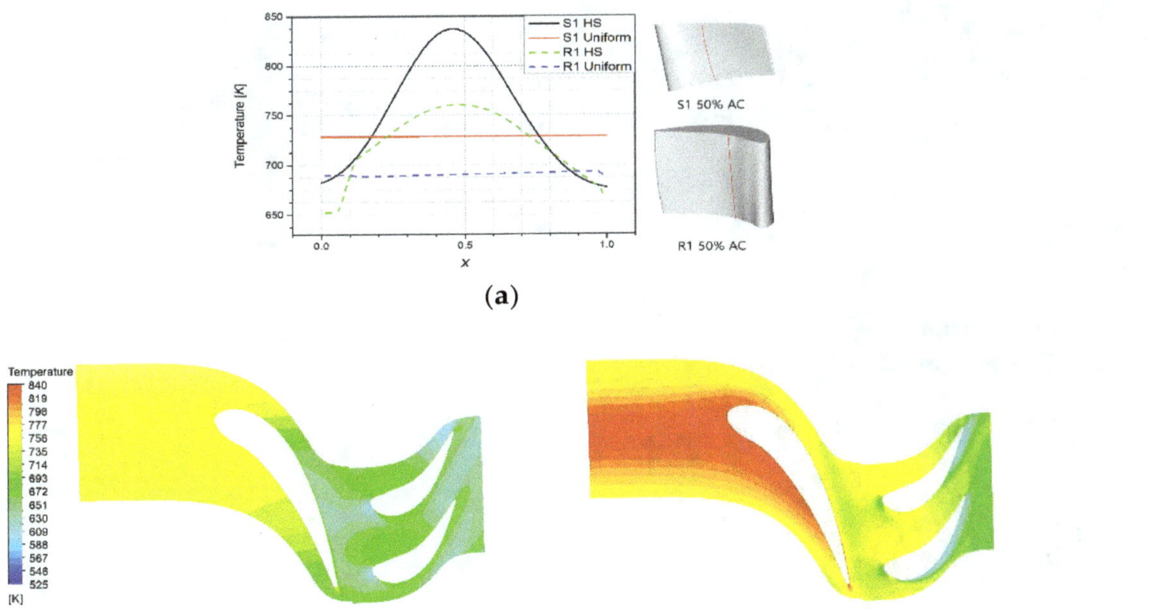

(a)

(b) (c)

Figure 7. Temperature distribution and contour on the first stage with different inlet condition: (**a**) temperature distribution on S1, R1 PS axial chord (AC) 50%, (**b**) uniform inlet temperature of 728 K at the midspan, and (**c**) HS with an average temperature of 728 K at the midspan.

3.2. *Flow and Thermal Characteristics at R1, R2*

Flow in a turbine passage affects the surrounding blade surfaces. Figure 8a–c shows the time-averaged streamline of the SS for the surface of R1. Figure 8d–f shows the time-averaged streamline of the SS for the surface of R2. Compared to the PS where the S1 downstream flow acts directly, the flow characteristics of the surface according to the AG were greater at SS. In the figure, the part indicated with the solid line is the recirculation zone. If Figure 8a,c is compared, it can be seen that the AG was reduced (from c to a), and recirculation zone became larger and farther from the LE. This is because in case 3 the mass flow at the tip was 6.32 g/s and in case 1 it was 6.42 g/s. Thus, the mass flow that passes through the tip clearance increased by 1.61%. Owing to the decreased AG, the total pressure at the front of the R1 LE was 219,828 Pa for case 3 and 222,395 for case 1, which is an increase of 1.16%. In the R2 surface's time-averaged streamline, the flow effect at R2 was distributed such that the effect of the secondary flow and recirculation zone was not observed, unlike R1, which was strongly affected by the inlet. As the AG reduced, some streamlines formed downward from the tip (Figure 8d), as opposed to the most streamline formed around the LE in Figure 8f. The change in the AG of S1 and R1 is a factor that affects the rotor surface flow and the creation of secondary flow at the turbine passage after R1 owing to changes in the mass flow at the tip and R1 LE's total pressure.

These phenomena are also caused by a difference in velocity distribution in the main flow of the S1 downstream as the HS passes S1 until it reaches R1. Figure 9 shows the time-averaged velocity distribution at the R1 inlet for each case. The part indicated by the dotted line is where the S1 downstream flow exhibits high-speed flow in the vicinity of the R1 LE. As the AG decreases, the area with higher velocity forms from the midspan to the hub and exerts its influence.

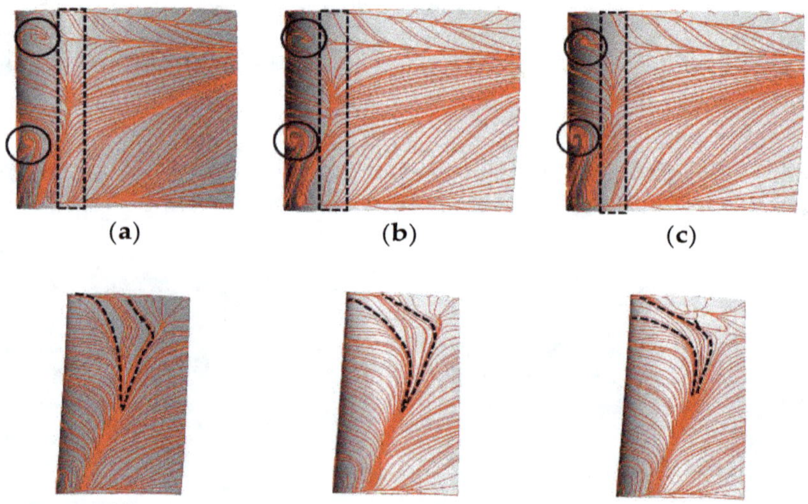

Figure 8. Surface streamlines on R1, R2 suction side (SS): (**a**) R1 case 1, (**b**) R1 case 2, (**c**) R1 case 3, (**d**) R2 case 1, (**e**) R2 case 2, and (**f**) R2 case 3.

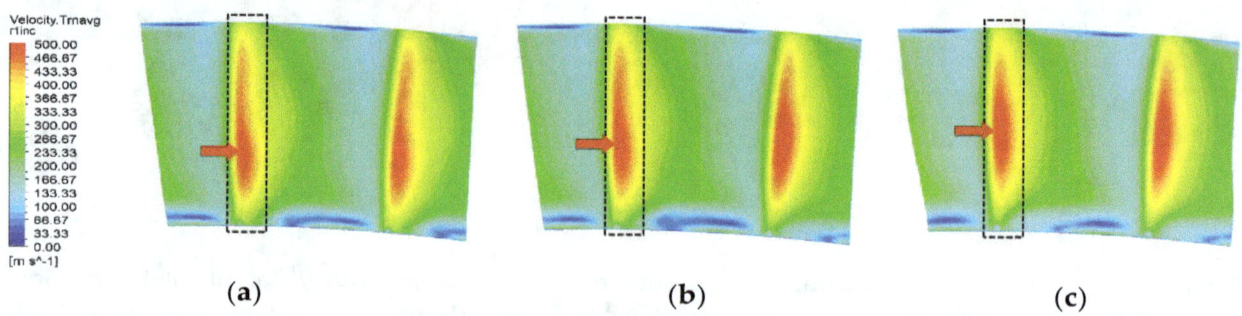

Figure 9. Time-averaged velocity profile at R1 inlet in three cases: (**a**) case 1, (**b**) case 2, and (**c**) case 3.

Figure 10 shows the time-averaged temperature contours of the PS and SS of the surface of R1. At the PS, where the HS turbine inlet condition is primarily concentrated, a high temperature area was observed; however, the effect of the axial gap was not large. At the SS, the high temperature area became larger at the TE in Figure 10a compared to Figure 10c owing to the secondary flow and tip leakage flow around the tip and hub as shown in Figure 8. Figure 11 shows the temperature distribution along the span at AC of 50% and 80% of the SS. Observing the effect that the AG had on the SS surface temperature, it can be seen that the effect of the tip leakage vortex at the tip was larger than the effect of the secondary flow at the hub. Near the tip, case 3 (in which the AG was large) was lower than case 1 by a maximum of 27 K at AC of 50% and a maximum of 43 K at AC of 80%.

The time-averaged temperature and pressure distribution results for each R1 were compared to analyze the thermal and flow characteristics according to the AG in a two-stage gas turbine under HS inlet conditions. Figure 12 shows the area-averaged and maximum temperature and pressure of the surface of R1. The time-averaged values in the area-averaged temperature distribution shown in Figure 12a were 689.15 K for case 1 and 687.69 K for case 3, which shows a difference of approximately 1.5 K. The maximum temperature distributions of case 1 and case 3 differed by over 3 K. The area-averaged pressure distributions in Figure 12b were 181,601 Pa for case 1, which has a short AG, and 179,779 Pa for case 3. This yields a difference of approximately 2000 Pa. In the maximum pressure distributions, there was a difference of approximately 400 Pa between case 1 and case 3, and a relatively small difference compared to the area-averaged pressure distribution was observed. Overall, a trend can be seen in which the AG decreased and the area-averaged and maximum pressure and temperature increased.

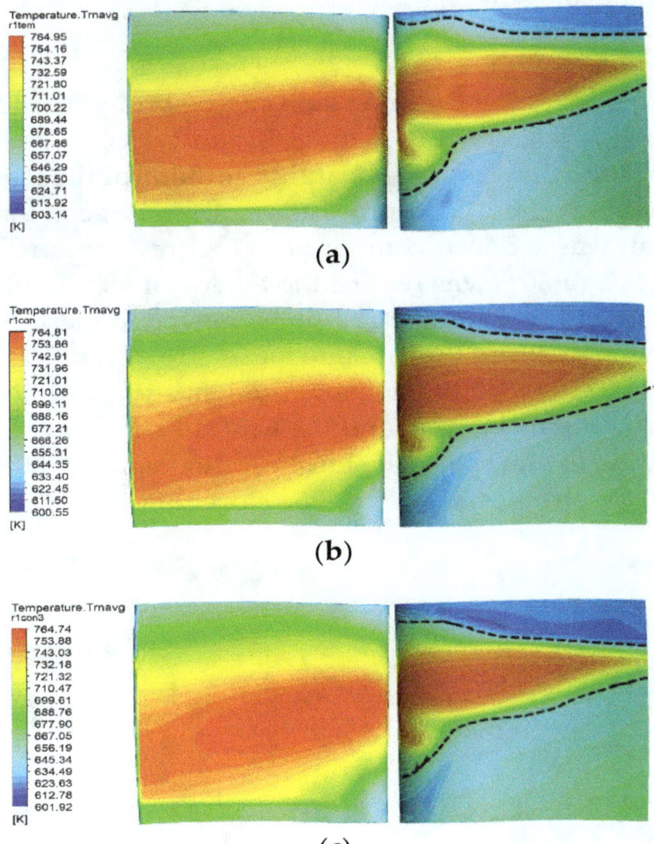

Figure 10. Time-averaged temperature contours on R1, PS, and SS (Left is PS) (**a**) case 1, (**b**) case 2, and (**c**) case 3.

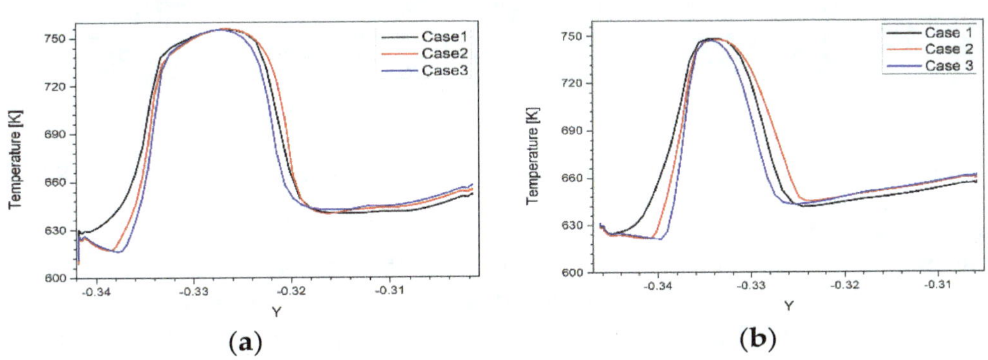

Figure 11. R1 SS span direction temperature distribution: (**a**) AC 50%, and (**b**) AC 80%.

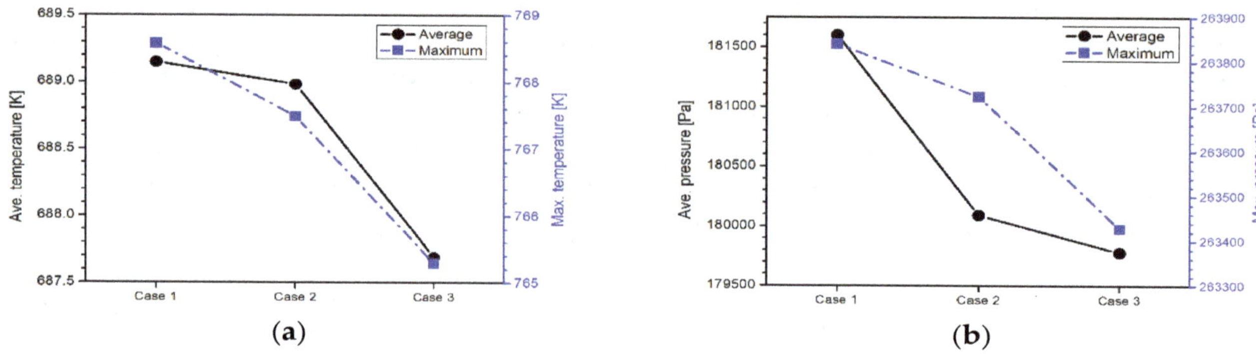

Figure 12. Time-averaged temperature and pressure distributions on R1 surface in three cases: (**a**) the distribution of area-averaged and maximum temperature on R1 surface, and (**b**) the distribution of area-averaged and maximum pressure on R1 surface.

3.3. Effect of Heat Flux on the Vane and the Blade Surface

The heat flux characteristics of the turbine blade can be understood by accurately quantifying the heat flux regions of each blade according to the AG, and a cooling technology can be developed accordingly. Figure 13 shows the time-average heat flux contour of the R1 and R2 surfaces. In the R1 PS shown in Figure 13a, in which the AG becomes close, it can be seen that the low heat flux region from the tip to the 40% span was reduced compared to Figure 13c. Furthermore, the low heat flux region that extended to an AC of 80% was reduced to an AC of 50% in Figure 13a. On the SS, there was a high heat flux around the LE in Figure 13a. When it moved to the TE side, and a contour formed in the radially inward direction and heat flux became lower. However, in Figure 13c, which has a long AG, a lower heat flux was formed at the LE than in Figure 13a, and a low heat flux region was centered on the TE. At R2's PS, the AG became closer, and the heat flux increased on the LE near the turbine hub. It was centered on the hub to mid span. The low heat flux region at 30% span formed at 40% span and the heat flux increased. On the SS, the AG became closer and heat flux increased at 20% span.

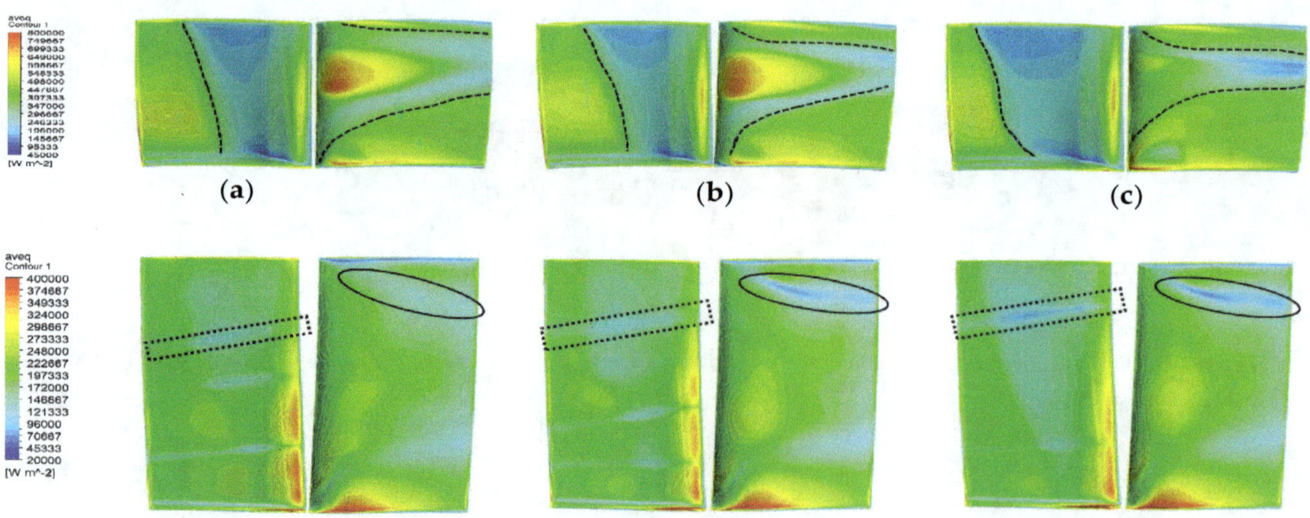

Figure 13. Time-averaged R1, R2 surface heat flux contour (Left is PS): (**a**) R1 case 1, (**b**) R1 case 2, (**c**) R1 case 3, (**d**) R2 case 1, (**e**) R2 case 2, and (**f**) R2 case 3.

Figure 14 shows the time-averaged distribution of the area-averaged heat flux for each blade (including the tip). At S1, the heat flux based on the AG showed a difference of less than 0.1%, thereby confirming that the S1-R1 AG did not have a significant effect on S1. At S2 and R2, the heat flux increased by 10.6% and 2.8%, respectively, as the AG decreased from case 3 to case 1. At R1, the heat flux was slightly lower in case 1 than in case 2. This is because the area-averaged heat flux at case 2's tip was higher than that of case 1. At the second-stage of S2 and R2, as the AG became closer, the heat flux became larger and the effect on the surface became larger.

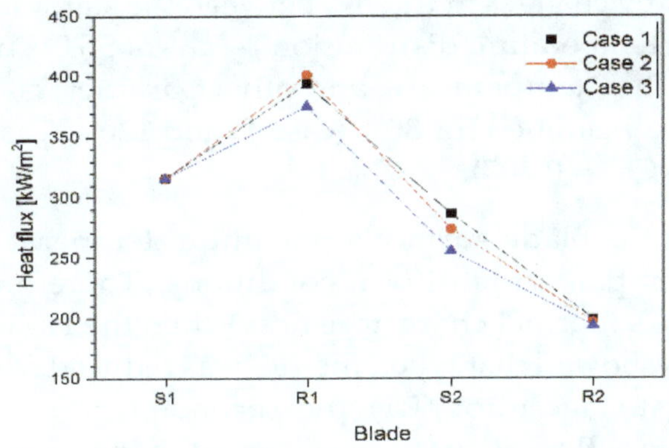

Figure 14. Area-averaged heat flux on the surface of S1, R1, S2, and R2.

3.4. Total-to-Total Efficiency

To quantitatively evaluate the aerodynamic performance of the two-stage turbine with regard to the AG, the total-to-total efficiency was calculated. The efficiency was calculated using Equation (7):

$$\eta = \frac{T\omega}{\dot{m}c_pT_1\left\{1 - (P_{02}/P_{01})^{\frac{\kappa-1}{\kappa}}\right\}} \quad (7)$$

Here, η is the total-to-total efficiency, T is the torque, ω is the angular velocity, $\dot{m}$ is the mass flow rate, c_p is the specific heat capacity of the turbine inlet, T_1 is the mean temperature of the turbine inlet, P_{02} is the mass-averaged total pressure of the turbine outlet, P_{01} is the total pressure of the turbine inlet, and the κ is the ratio of specific heat. Both T and P_{02} directly affect the efficiency whereas other parameter values remain constant throughout the whole cases. The X axis in Figure 15 indicates that the AG distance was divided by the axial chord length of S1, and the Y axis indicates the efficiency, torque, and pressure outlet (P_{02}) normalized by the value of case 2. The normalized values of efficiency, torque, and pressure outlet in cases 1 and 3 are 1.00295 and 0.99694, 1.00218 and 0.99735, and 1.00171 and 0.99911, respectively. The efficiency according to the decrease in the AG was increased. As the AG decreases, the turbine efficiencies increased by 0.306% and 0.295% for case 2 and case 1, respectively. In case 1, which had the shortest AG, the efficiency was 0.601% higher than case 3; however, R1's surface maximum temperature was the highest, as shown in Figure 12. Hence, it leads to an increase in thermal load, which has an effect on the blade's fatigue life.

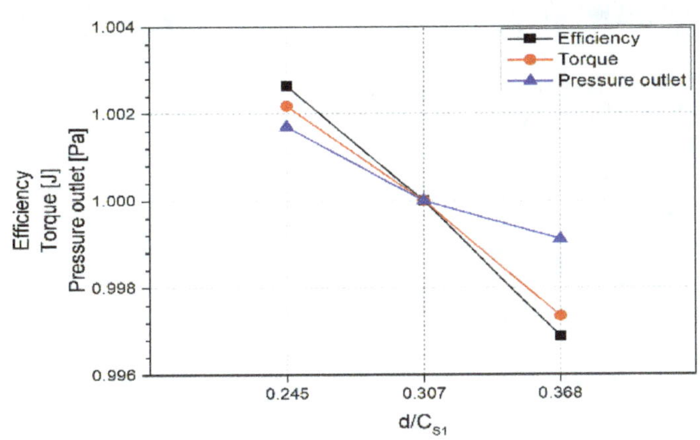

Figure 15. Variation in the total-to-total efficiency, torque, and pressure outlet with changes in the axial gap for the three cases.

4. Conclusions

Herein, we presented a numerical study on the unsteady state to understand how turbine blades and their passages are affected by changes in the AG between the stator and rotor of a GE-E^3 two-stage gas turbine that has a HS inlet temperature distribution. A $k - \omega$ SST γ turbulence model was used to examine the flow fields, heat flux, temperature, and other parameters, at the walls and the passage. In this process, the results were examined for 80% (case 1) and 120% (case 3) of the designed value for the AG of case 2 (13 mm or $d/Cs_1 = 0.307$).

The results showed that the blade surface's maximum temperature increased by over 110 K for the HS inlet conditions compared to uniform conditions. Therefore, an HS must be applied to clearly understand the turbine's thermal characteristics. When the HS inlet conditions were applied, the results according to the AG showed that when the AG was reduced from case 3 to case 1, the surface suction side streamline of first stage rotor (R1) and second-stage rotor (R2) become closer to the endwall because of the secondary flow. Thus, it was formed in the radially outward direction at R1's trailing edge. The area-averaged and maximum temperatures in R1 surface increased by 1.5 K and 3 K, respectively, and the area-averaged and maximum pressure increased by 2000 Pa and 400 Pa, respectively. The low temperature region near the tip and hub decreased. In the second-stage stator and R2, the area-averaged heat fluxes increased by 10.6% and 2.8%, respectively. As the AG decreased, the turbine overall efficiencies increased by 0.306% and 0.295%, respectively; however, this increases the blade surface's thermal load and reduces the turbine blade fatigue life. If an appropriate cooling technology is developed, it will lead to the development of a higher efficiency design using a shorter AG and reduce long-term operating costs.

The unsteady RANS model used in this study has a low computational cost; however, it has the drawback of being unable to precisely predict vortices and flow separation phenomena that occur around the tips and walls. DES is more accurate; however, it has a high computational cost and previous studies using DES have mostly focused on the local flow phenomena rather than full blade simulations [25–27]. With significant advances in computational resources, it is expected that DES or LES will soon be used to accurately predict the complex unsteady flow physics near the tips and walls within full blade simulations.

Author Contributions: Conceptualization, methodology, investigation: M.G.C. and J.R.; validation, formal analysis, writing—original draft preparation, visualization: M.G.C.; writing—review and editing, supervision, project administration, funding acquisition: J.R.

Nomenclature

C_{S1}	Axial chord length of S1 [mm]
d	Axial chord length [mm]
q	Heat flux [W/m^2]
T_w	Wall temperature [K]
T_{aw}	Adiabatic wall temperature [K]
P	Pressure [Pa]
P_0	Inlet total pressure [Pa]
T	Temperature [K]
T_0	Inlet total temperature [K]
X, Y, Z	Cartesian coordinates

Abbreviations

HP	High pressure
HS	Hot streak
S1, R1, S2, R2	first stage stator, rotor, second-stage stator, rotor
HTC, h	Heat transfer coefficient
AG	Axial gap

LE	Leading edge
TE	Trailing edge
PS	Pressure side
SS	Suction side
AC, Ca	Axial chord
Subscripts	
ave	Averaged
max	Maximum value
min	Minimum value

References

1. Wang, Z.; Liu, Z.; Feng, Z. Influence of mainstream turbulence intensity on heat transfer characteristics of a high pressure turbine stage with inlet hot streak. *J. Turbomach.* **2016**, *138*, 041005. [CrossRef]
2. Butler, T.L.; Sharma, O.P.; Dring, R.P. Redistribution of an inlet temperature distortion in an axial flow turbine stage. *J. Propuls. Power* **1989**, *5*, 64–71. [CrossRef]
3. Povey, T.; Qureshi, I. A hot-streak (combustor) simulator suited to aerodynamic performance measurements. *Proc. Inst. Mech. Eng. Part G J. Aerosp. Eng.* **2008**, *222*, 705–720. [CrossRef]
4. Gundy-Burlet, K.L.; Dorney, D.J. Effects of radial location on the migration of hot streaks in a turbine. *J. Propuls. Power* **2000**, *16*, 377–387. [CrossRef]
5. An, B.T.; Liu, J.J.; Jiang, H.D. Numerical investigation on unsteady effects of hot streak on flow and heat transfer in a turbine stage. In Proceedings of the ASME Turbo Expo 2008: Power for Land, Sea, and Air, Berlin, Germany, 9–13 June 2008.
6. Feng, Z.; Liu, Z.; Shi, Y.; Wang, Z. Effects of hot streak and airfoil clocking on heat transfer and aerodynamic characteristics in gas turbine. *J. Turbomach.* **2016**, *138*, 021002. [CrossRef]
7. Smith, C.I.; Chang, D.; Tavoularis, S. Effect of inlet temperature nonuniformity on high-pressure turbine performance. In Proceedings of the ASME Turbo Expo 2010: Power for Land, Sea, and Air, Glasgow, UK, 14–18 June 2010.
8. Yamada, K.; Funazaki, K.; Kikuchi, M.; Sato, H. Influences of axial gap between blade rows on secondary flow and aerodynamic performance in a turbine stage. In Proceedings of the ASME Turbo Expo 2009: Power for Land, Sea, and Air, Orlando, FL, USA, 8–12 June 2009.
9. Danish, S.N.; Qureshi, S.R.; Imran, M.M.; Khan, S.U.; Sarfraz, M.M.; El-Leathy, A.; Al-Ansary, H.; Wei, M. Effect of tip clearance and rotor–stator axial gap on the efficiency of a multistage compressor. *Appl. Therm. Eng.* **2016**, *99*, 988–995. [CrossRef]
10. Dring, R.P.; Joslyn, H.D.; Hardin, L.W.; Wagner, J.H. Turbine rotor-stator interaction. *J. Eng. Power* **1982**, *104*, 729–742. [CrossRef]
11. Gaetani, P.; Persico, G.; Dossena, V.; Osnaghi, C. Investigation of the flow field in a HP turbine stage for two stator-rotor axial gaps: Part I—3D time averaged flow field. In Proceedings of the ASME Turbo Expo 2006: Power for Land, Sea and Air, Barcelona, Spain, 8–11 May 2006.
12. Funazaki, K.I.; Yamada, K.; Kikuchi, M.; Sato, H. Experimental studies on aerodynamic performance and unsteady flow behaviors of a single turbine stage with variable rotor-stator axial gap: Comparisons with time-accurate numerical simulation. In Proceedings of the ASME Turbo Expo 2007: Power for Land, Sea, and Air, Montreal, QC, Canada, 14–17 May 2007.
13. Sasao, Y.; Kato, H.; Yamamoto, S.; Satsuki, H.; Ooyama, H.; Ishizaka, K. Numerical and experimental investigations of unsteady 3-D flow through two-stage cascades in steam turbine model. In Proceedings of the International Conference on Power Engineering, Kobe, Japan, 16–20 November 2009.
14. Ghenaiet, A.; Touil, K. Characterization of component interactions in two-stage axial turbine. *Chin. J. Aeronaut.* **2016**, *29*, 893–913. [CrossRef]
15. Timko, L.P. *Energy Efficient Engine High Pressure Turbine Component Test Performance Report*; Technical Report; National Aeronautics and Space Administration (NASA): Cincinnati, OH, USA, 1 January 1984.
16. Arnone, A.; Benvenuti, E. Three-dimensional Navier-Stokes analysis of a two-stage gas turbine. In Proceedings of the ASME 1994 International Gas Turbine and Aeroengine Congress and Exposition, The Hague, The Neterlands, 13–16 June 1994.
17. Ryu, J.; Livescu, D. Turbulence structure behind the shock in canonical shock–vortical turbulence interaction. *J. Fluid Mech.* **2014**, *756*, R1. [CrossRef]

18. Ryu, J.; Lele, S.K.; Viswanathan, K. Study of supersonic wave components in high-speed turbulent jets using an LES database. *J. Sound Vib.* **2014**, *333*, 6900–6923. [CrossRef]
19. Celik, I.; Yavuz, I.; Smirnov, A. Large eddy simulations of in-cylinder turbulence for internal combustion engines: A review. *Int. J. Engine Res.* **2001**, *2*, 119–148. [CrossRef]
20. Meloni, R.; Naso, V. An insight into the effect of advanced injection strategies on pollutant emissions of a heavy-duty diesel engine. *Energies* **2013**, *6*, 4331–4351. [CrossRef]
21. Haworth, D.C.; Jansen, K. Large-eddy simulation on unstructured deforming meshes: Towards reciprocating IC engines. *Comput. Fluids* **2000**, *29*, 493–524. [CrossRef]
22. Hao, Z.R.; Gu, C.W.; Ren, X.D. The application of discontinuous Galerkin methods in conjugate heat transfer simulations of gas turbines. *Energies* **2014**, *7*, 7857–7877. [CrossRef]
23. Kim, J.; Park, J.G.; Kang, Y.S.; Cho, L.; Cho, J. A study on the numerical analysis methodology for thermal and flow characteristics of high pressure turbine in aircraft gas turbine engine. *Korean Soc. Fluid Mach.* **2013**, *17*, 46–51. [CrossRef]
24. Hylton, L.; Mihelc, M.; Turner, E.; Nealy, D.; York, R. *Analytical and Experimental Evaluation of the Heat Transfer Distribution over the Surfaces of Turbine Vanes*; Technical Report; National Aeronautics and Space Administration (NASA): Indianapolis, IN, USA, 1 May 1983.
25. Martini, P.; Schulz, A.; Bauer, H.J.; Whitney, C.F. Detached eddy simulation of film cooling performance on the trailing edge cutback of gas turbine airfoils. In Proceedings of the ASME Turbo Expo 2005: Power for Land, Sea, and Air, Reno, NV, USA, 6–9 June 2005.
26. Effendy, M.; Yao, Y.F.; Yao, J.; Marchant, D.R. DES study of blade trailing edge cutback cooling performance with various lip thicknesses. *Appl. Therm. Eng.* **2016**, *99*, 434–445. [CrossRef]
27. Mao, X.; Dal Monte, A.; Benini, E.; Zheng, Y. Numerical study on the internal flow field of a reversible turbine during continuous guide vane closing. *Energies* **2017**, *10*, 988. [CrossRef]

Permissions

All chapters in this book were first published in MDPI; hereby published with permission under the Creative Commons Attribution License or equivalent. Every chapter published in this book has been scrutinized by our experts. Their significance has been extensively debated. The topics covered herein carry significant findings which will fuel the growth of the discipline. They may even be implemented as practical applications or may be referred to as a beginning point for another development.

The contributors of this book come from diverse backgrounds, making this book a truly international effort. This book will bring forth new frontiers with its revolutionizing research information and detailed analysis of the nascent developments around the world.

We would like to thank all the contributing authors for lending their expertise to make the book truly unique. They have played a crucial role in the development of this book. Without their invaluable contributions this book wouldn't have been possible. They have made vital efforts to compile up to date information on the varied aspects of this subject to make this book a valuable addition to the collection of many professionals and students.

This book was conceptualized with the vision of imparting up-to-date information and advanced data in this field. To ensure the same, a matchless editorial board was set up. Every individual on the board went through rigorous rounds of assessment to prove their worth. After which they invested a large part of their time researching and compiling the most relevant data for our readers.

The editorial board has been involved in producing this book since its inception. They have spent rigorous hours researching and exploring the diverse topics which have resulted in the successful publishing of this book. They have passed on their knowledge of decades through this book. To expedite this challenging task, the publisher supported the team at every step. A small team of assistant editors was also appointed to further simplify the editing procedure and attain best results for the readers.

Apart from the editorial board, the designing team has also invested a significant amount of their time in understanding the subject and creating the most relevant covers. They scrutinized every image to scout for the most suitable representation of the subject and create an appropriate cover for the book.

The publishing team has been an ardent support to the editorial, designing and production team. Their endless efforts to recruit the best for this project, has resulted in the accomplishment of this book. They are a veteran in the field of academics and their pool of knowledge is as vast as their experience in printing. Their expertise and guidance has proved useful at every step. Their uncompromising quality standards have made this book an exceptional effort. Their encouragement from time to time has been an inspiration for everyone.

The publisher and the editorial board hope that this book will prove to be a valuable piece of knowledge for researchers, students, practitioners and scholars across the globe.

List of Contributors

Carlo Cravero and Davide De Domenico
Department of Mechanical Engineering, DIME, Università di Genova, 16145 Genova, Italy

Jesús Rodríguez and Ernesto Amores
Centro Nacional del Hidrógeno (CNH2), Prolongación Fernando El Santo s/n, 13500 Puertollano, Ciudad Real, Spain

Weimin Song, Yufei Zhang and Haixin Chen
School of Aerospace Engineering, Tsinghua University, Beijing 100084, China

Xiangdong Han and Deng Li
Key Laboratory of Hydraulic Machinery Transients, Ministry of Education, Wuhan University, Wuhan 430072, China
Hubei Key Laboratory of Waterjet Theory and New Technology, Wuhan University, Wuhan 430072, China
School of Power and Mechanical Engineering, Wuhan University, Wuhan 430072, China

Yong Kang
Key Laboratory of Hydraulic Machinery Transients, Ministry of Education, Wuhan University, Wuhan 430072, China
Hubei Key Laboratory of Waterjet Theory and New Technology, Wuhan University, Wuhan 430072, China
School of Power and Mechanical Engineering, Wuhan University, Wuhan 430072, China
Collaborative Innovation Center of Geospatial Technology, Wuhan 430079, China

Weiguo Zhao
School of Energy and Power Engineering, Lanzhou University of Technology, Lanzhou 730050, China

Ming Gao
School of Energy and Power Engineering, Shandong University, Jinan 250061, China

Peixin Dong
School of Energy and Power Engineering, Shandong University, Jinan 250061, China
School of Mechanical and Mining Engineering, University of Queensland, Brisbane 4067, Australia

Shenghui Lei
Thermal Management Research Group, Efficient Energy Transfer (ηET) Department, Bell Labs Ireland, Nokia, Dublin D15 Y6NT, Ireland

Ali Turan
School of Mechanical, Aerospace and Civil Engineering, University of Manchester, Manchester M60 1QD, UK

Iñigo Errasti, Unai Fernández-Gamiz, Pablo Martínez-Filgueira and Jesús María Blanco
Nuclear Engineering and Fluid Mechanics Department, University of the Basque Country, E-01009 Vitoria-Gasteiz, Spain

Wonseok Oh
Graduate School of Engineering, Department of Architecture, The University of Tokyo, 4-6-1 Komaba, Meguro-ku, Tokyo 153-8505, Japan

Shinsuke Kato
Institute of Industrial Science, The University of Tokyo, 4-6-1 Komaba, Meguro-ku, Tokyo 153-8505, Japan

Ji Tae Kim, Ki-Bae Hong and Hong Sun Ryou
School of Mechanical Engineering, Chung-Ang University, Seoul 06974, Korea

Manuel Rigosi
Casappa S.p.A., Via Balestrieri 1, Lemignano di Collecchio, 43044 Parma, Italy

Lei Zhang, Liang Zhang, Qian Zhang, Kuan Jiang, Yuan Tie and Songling Wang
School of Energy, Power and Mechanical Engineering, North China Electric Power University, Baoding 071003, China

Sun-Min Kim, Jun-Ho Jo, Ye-Eun Lee and Yeong-Seok Yoo
Division of Environmental and Plant Engineering, Korea Institute of Civil Engineering and Building Technology, 283, Goyang-daero, Ilsanseo-gu, Goyang-si, Gyeonggi-do 10223, Korea

Chang Guo and Kun Wang
School of Energy and Power Engineering, Shandong University, Jinan 250061, China

Dongyue Lu
State Nuclear Power Engineering Company, Shanghai 200233, China

Emma Frosina, Dario Buono and Adolfo Senatore
Department of Industrial Engineering, University of Naples Federico II, Via Claudio, 21-80125 Naples, Italy

Myung Gon Choi and Jaiyoung Ryu
School of Mechanical Engineering, Chung-Ang University, 84, Heukseok-ro, Dongjak-gu, Seoul 06974, Korea

Index

www.ingramcontent.com/pod-product-compliance
Lightning Source LLC
LaVergne TN
LVHW081017210926
848485LV00004B/884